# The Chemistry of PCB's

**O. Hutzinger**
National Research Council
Halifax, Nova Scotia

**S. Safe**
Department of Chemistry
University of Guelph
Guelph, Ontario

**V. Zitko**
Fisheries and Marine Service
St. Andrews, New Brunswick

**published by:**

CRC PRESS

18901 Cranwood Parkway, Cleveland, Ohio 44128

International Standard Book Number 0-87819-049-X
Library of Congress Card Number 73-88621

Printed in the United States

# PREFACE

In the last few years polychlorinated biphenyls (PCB's) have attained the dubious honor of surpassing the chlorinated insecticides as the most talked-about organochlorine pollutants.

Research on various aspects of the PCB problem has grown and continues to grow with remarkable intensity. Opinions on the importance of these ubiquitously distributed industrial compounds, however, still vary greatly. The authors do not wish to enter this discussion. The main aim of this book is to present a comprehensive summary of the chemistry of chlorobiphenyls. Included are the nomenclature, composition of technical mixtures, syntheses of individual chlorobiphenyls, chemical reactions, photochemical and metabolic alterations, and spectroscopic properties. Different methods of PCB analysis are presented and evaluated in a separate chapter. Toxicological and general biological effects are frequently referred to but are not a proper topic for this book.

Certain areas of PCB research are developed further than others. For example, almost half of the 209 possible chlorobiphenyls have been synthesized and properly described using a variety of methods. Much is to be learned, on the other hand, on the metabolism of chlorobiphenyls and other routes of degradation, including photochemical, and the complex PCB mixtures will continue to challenge analytical chemists. In yet other areas, for example toxicological significance and exact environmental impact, our knowledge is only fragmentary at the moment.

A considerable amount of information included in this book had not been published when the manuscript was written. Although some of this information comes from our own laboratories, we have to acknowledge the help of many scientists who have generously supplied us with their unpublished data. It is a great pleasure to thank the following colleagues for this courtesy: Drs. M. Bolgar, E. J. Bonelli, G. W. Bowes, R. H. DeVos, R. Edwards, W. Ernst, V. H. Freed, A. M. Gardner, A. V. Holden, W. D. Jamieson, S. Jensen, W. Klein, J. B. Knight, F. Korte, R. A. Lidgett, L. L. Miller, D. R. Osborne, I. Pomerantz, L. O. Ruzo, D. L. Stalling, J. D. Stewart, G. Sundström, A. C. Tas, G. M. Telling, P. R. Wallnöfer, and R. G. Webb.

We would like to express our appreciation to Dr. K. L. Loening of Chemical Abstracts Service for advice with nomenclature and the Monsanto Company for supplying samples and information. Thanks are also due to CRC Press for their cooperation; particularly for allowing us to submit the last chapter (Recent Developments) months after the original manuscript.

Part of the contribution by one of us (O. Hutzinger) was written while on sabbatical leave at the Institute of Ecological Chemistry. This author would like to express his thanks for the hospitality extended to him and also gratefully acknowledge financial support from the Alexander von Humboldt-Stiftung.

The literature for this book was covered, as it was available to the authors, until the end of 1973.

## THE AUTHORS

**Otto Hutzinger** is a Scientist at the National Research Council (Atlantic Regional Laboratory), Halifax, Nova Scotia and Adjunct Professor. He holds an Ing. Chem. degree from the Institute of Chemical Technology in Vienna, Austria, and M.Sc. and Ph.D. degrees from the University of Saskatchewan, Canada. He is presently a Senior Fellow of the Alexander von Humboldt Foundation at the Institute of Ecological Chemistry, Technical University of Munich, Germany. His research interests are the chemistry of pollutant compounds, particularly pesticides and industrial chemicals, and the application of mass spectrometry to these studies. He is author or co-author of over 100 scientific publications.

**Stephen Safe** is Associate Professor in the Department of Chemistry at the University of Guelph, Guelph, Ontario. He holds B.Sc. and M.Sc. degrees from Queen's University, Canada, and a D.Phil. from Oxford University. His work includes research in the chemistry and metabolism of pesticides and pollutants, the application of mass spectrometry in environmental and other fields, and the chemistry and microbiology of primary metabolites in fungi and bacteria. He is author or co-author of over 80 scientific papers.

**Vladimir Zitko** is the Leader of the Toxicology Section of the Fisheries and Marine Service, Biological Station, St. Andrews, New Brunswick. He holds Ing. Chem. and C.Sc. (Ph.D.) degrees from the Slovak Academy of Sciences, Bratislava, Czechoslovakia. His experience includes work in physical, carbohydrate, and lipid chemistries. He is author or co-author of approximately 80 scientific publications.

TABLE OF CONTENTS

Chapter 1

# INTRODUCTION

## Historical and Literature

Chlorinated biphenyls were known before the turn of the century and the useful industrial properties of mixtures obtained by chlorination of biphenyl were recognized early (*cf.* 77). From a purely chemical point of view, the chlorobiphenyls are a rather unattractive class of compounds and limited attention was paid to them by the scientific community. In 1966, a note in the *New Scientist* about the discovery of the widespread occurrence of chlorinated biphenyls in the Swedish environment (Figure 1) and the following events were responsible for the truly dramatic upsurge of interest in these compounds.[3,46] In 1967, Widmark reported mass spectroscopic data as unambiguous proof of the chemical nature of these new contaminants[106] and at about the same time PCB's were found in various parts of the world.[42,43,54,56,89]

Even before the reports mentioned above, analysts were aware of "spurious peaks" in electron capture gas chromatography which did not correspond to any of the known chlorinated pesticides.[37,41,92,93,105] These peaks corresponding to components of PCB are not separated well under GC conditions usually used in pesticide analysis (Figures 2 and 3) and it is quite likely that many reports gave falsely high results for some pesticide residues (*cf.* 55).

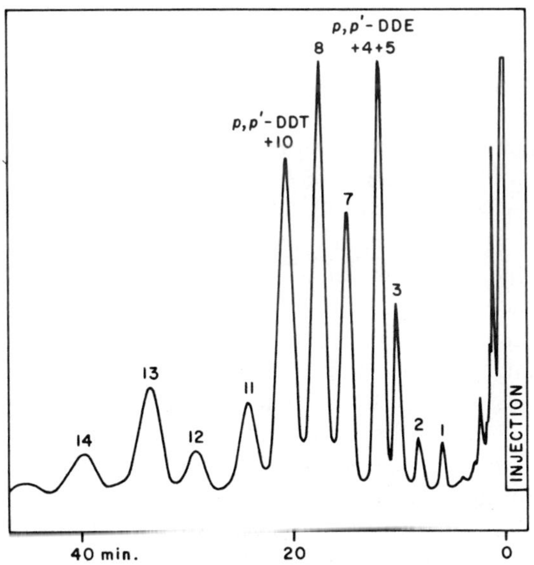


FIGURE 1. Gas chromatogram from purified extract of a white-tailed eagle found dead in the archipelago of Stockholm. All peaks except DDT and DDE were unknown. (Courtesy of Dr. Jensen.)

It is now established that because of chemical properties (lipid solubility and resistance to degradation) PCB's accumulate in food chains and are distributed worldwide, similar to the chlorinated pesticides such as DDT.

The literature on chlorinated biphenyl is growing rapidly and a number of reviews on various aspects of the problem are available. Generally, chlorinated biphenyls have been discussed in the context of chlorinated pesticides[32,54,66,91] and specifically.[9,22-24,34,44,75,76,90,100,108,109] A bibliography (to 1971) of articles dealing with chlorobiphenyls has been published[83] and useful information can be obtained from conference reports,[4,26,33] a review of the Panel on Hazardous Trace Substances,[73] a recent Monsanto report,[68] and a mimeographed publication *PCB Newsletter*.[96] Analytical methods have been discussed.[67,86-88,107]

## Toxicology of PCB's

Long ago, toxicity of PCB's to humans was recognized as causing yellow atrophy of the liver in fatal cases, dermatitis, and fatty degeneration of the liver in chronic exposure.[14] Threshold limit values of 0.5 and 1 mg/m$^3$ for PCB's containing 42 and 54% chlorine, respectively, were established for industrial environments.[97]

In 1966, the discovery of PCB's in environmental samples spurred renewed interest in the toxicity of PCB's and a mass of information on the acute and chronic toxicity of PCB's to a variety of animals has been obtained since then. Most of the studies were carried out using commercial PCB

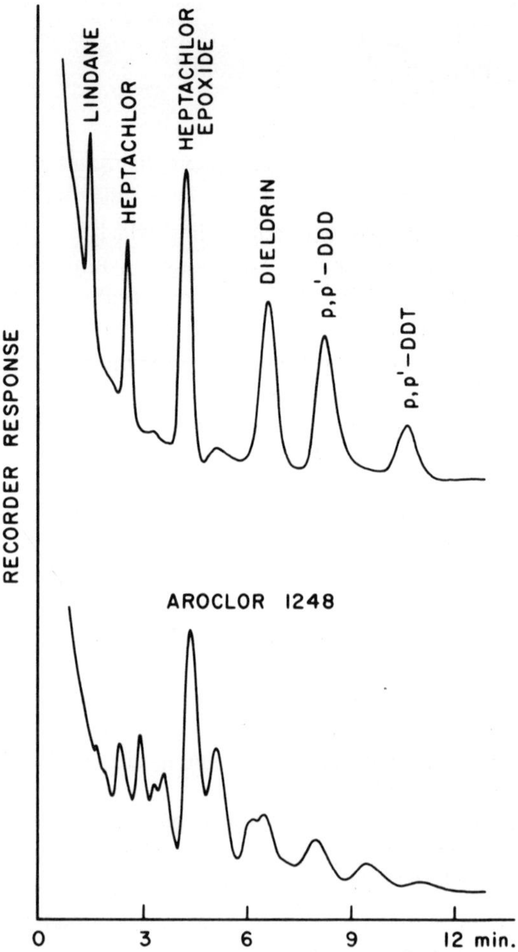


FIGURE 2. Gas chromatograms of Aroclor® 1248 (bottom) and some common pesticides (top). Conditions: 40% SE-30 or Chromosorb W® 60/80; 200°; carrier gas, nitrogen 60 ml/min.

preparations and very little is known about the toxicity of individual chlorobiphenyls.

PCB's accumulate in living matter, particularly in tissues and organs rich in lipids. The accumulation appears to be higher in the case of penta- and more highly chlorinated biphenyls. Tetra and less chlorinated biphenyls are hydroxylated and excreted (see Chapter 7). The toxicity of PCB's varies a great deal from species to species, probably as a result of differences in metabolic rates and capabilities, physiological differences, etc. Thus, PCB's are more toxic to some aquatic invertebrates than to fish and more toxic to salmon eggs than to juvenile Atlantic salmon. Chicken eggs may fail to hatch if they contain a high concentration of PCB's and mink fail to

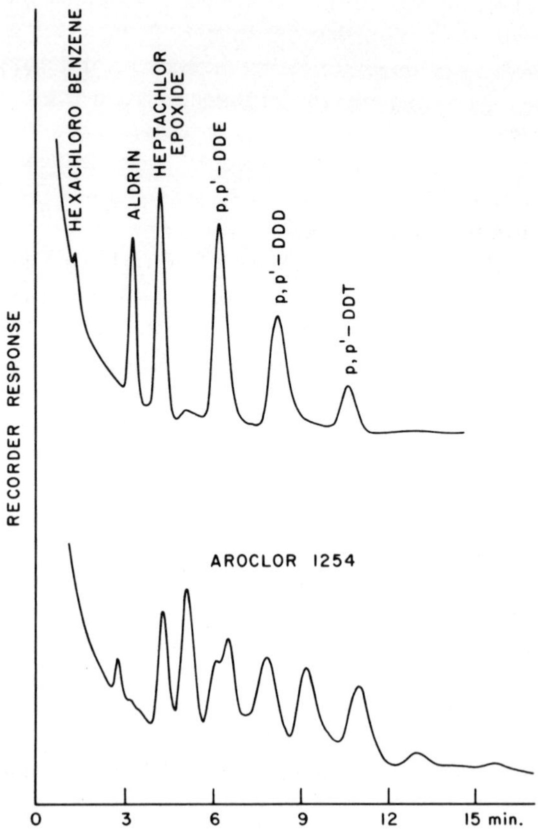


FIGURE 3. Gas chromatograms of Aroclor 1254 (bottom) and some common pesticides (top). Conditions identical to those of Figure 2.

reproduce when fed a PCB-contaminated diet. In humans, the appearance of symptoms of PCB poisoning was observed after the consumption of approximately 0.5 g of PCB's in rice oil.

The mechanism of the toxic action of PCB's is not known. It is very likely that the mechanism is different for various chlorobiphenyls. Some highly chlorinated biphenyls may be toxic simply by their presence in the tissues and less chlorinated biphenyls may be toxic by taking part in metabolic processes. It is possible that chlorobiphenylols, the metabolic products of chlorobiphenyls, are more toxic than the parent compounds. (Chlorophenols are usually more toxic than chlorobenzenes.)

Review articles on PCB's cited in this book usually contain a section on the toxicity of PCB's. Readers who are particularly interested in this subject should consult the review *Polychlorinated Biphenyls – Environmental Impact*, compiled by the Panel on Hazardous Trace Substances,[73] the report of the Interdepartmental Task Force on PCB's,[5] the published Proceedings of the Conference on PCB's, organized by the National Institute of Environmental Health Sciences,[4] and papers presented at the Symposium on PCB's, sponsored by the American Chemical Society, available in preprint form,[35] soon to be published as a book,[33] and a review by Isono and Fujiwara.[45] An excellent review of the toxicity of chlorinated hydrocarbons including PCB's was recently published by Kimbrough.[50] The effects of PCB's on the aquatic fauna and on the wildlife in general were summarized by Walsh[104] and Dustman et al.[22] Additional papers on the toxicity, biological, and biochemical effects of PCB's, published in 1972 and 1973 are summarized below.

Relatively few papers deal with the effects of PCB's on microorganisms,[1,18] algae and phytoplankton,[25,49,69,70,99] insects,[29,79] and aquatic invertebrates.[64,94]

The uptake and excretion of PCB's by fish is the subject of several papers.[38,39,65] The effects of PCB's on the metabolism[47] and activity of ATP-ase in fish[21,53] have been studied. The whole body residue of PCB's (Aroclor® 1254) in Cayuga Lake, Ithaca, New York, was an exponential function of age PCB = 1.031 $e^{0.259a}$ (a = age in years) with a correlation coefficient of 0.86. PCB-length and PCB-weight correlation coefficients were 0.85 and 0.80, respectively.[6]

The effects of PCB's on birds, particularly chicks, pheasants, and cockerels continued to attract attention,[7,8,16,19,20,36,61,80,82,84,85] with the main emphasis on tissue distribution, pathology, persistence, and sublethal effects.

Among mammals, rats and mice were used most frequently to study the morphological changes, enzyme induction and other sublethal effects caused by the administration of PCB's.[10–13,15,17,30,36,40,48,51,57,59,62,63,71,74,78,98,101] Hepatocarcinogenicity[72] and neoplastic changes[52] in rat liver induced by PCB's have been reported. The effect of PCB's on rabbit and boar has been studied.[58,81,102] Individual chlorobiphenyls (3,3′,4,4′-tetra and 2,2′,4,4′,5,5′-hexachlorobiphenyl) were also used in rabbit toxicity studies.[60,103] With the former, the effects of skin application were similar to those of a commercial PCB preparation. The latter caused a more severe liver injury than Aroclor 1260. The excretion of PCB's from cows was the subject of three papers.[27,28,95]

PCB's were reported to induce gastric mucosal hyperplasia in primates[2] and a paper on PCB-induced fetopathy was published.[31]

**Nomenclature**

The biphenyl (diphenyl, phenylbenzene) ring system is numbered as shown in Figure 4. The order of citation of locants is as follows:

1. An unprimed locant is of lower order (i.e., preferred) than its corresponding primed locant. For example, 2 is lower than 2′, but 2′ is lower than 3.
2. Locants as low as possible should be given to all substituents, ignoring primes at this stage.
3. As few primed locants as possible should be used.
4. When the number of chlorine substituents in each ring is the same, the ring with the lower numbered substituents receives the unprimed numbers. If everything else is equal, the locant cited first is unprimed.

An example of a correctly named chlorobiphenyl (2,2′,3,4′-tetrachlorobiphenyl) is given in Figure 5. Chloro-substituents are always cited as prefixes as are the other halogens, nitro, nitroso and alkoxy groups.

There is an important difference between the nomenclature of chlorobiphenyls and that of two important groups of substitution products, chloro-

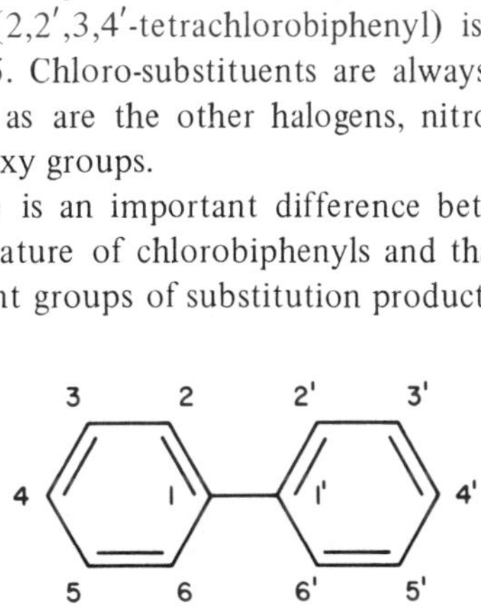


FIGURE 4. Numbering in the biphenyl ring system.

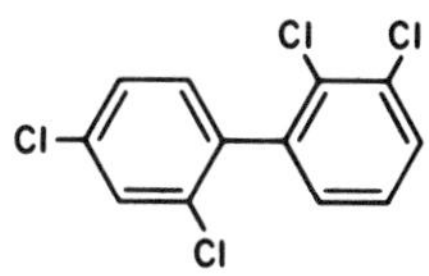


2,2',3,4'– tetrachlorobiphenyl

FIGURE 5. Structure and correct nomenclature for an arbitrarily chosen tetrachlorobiphenyl.

biphenylamines (aminochlorobiphenyls) and chlorobiphenylols (chlorohydroxybiphenyls, chlorophenylphenols). The amino and hydroxy groups represent the principal function in these compounds and they are to be cited, therefore, as suffixes rather than prefixes in the names. The rules for assigning locants described above are still valid, but they must be applied in turn, *first* to the substituents cited as suffixes and only then to those cited as prefixes. Examples for amino (Figure 6) and hydroxy (Figure 7) derivatives will illustrate these rules.

When a compound contains more than one kind of principal function, a certain order applies which, for the purpose of this book (the list is not complete), can be given as: carboxylic acids (-COOH), sulfonic acids (-$SO_3H$), phenols (-OH), and amines (-$NH_2$). The principal function of the higher order in a compound is then cited as a suffix, all others as prefixes. Two examples (Figure 8) will again illustrate these rules.

Throughout the book the abbreviation "PCB" will only be used for the technical chlorobiphenyl mixtures or samples (extracts from environmental sources, etc.) having similar GC characteristics. Individual compounds will be referred to by their

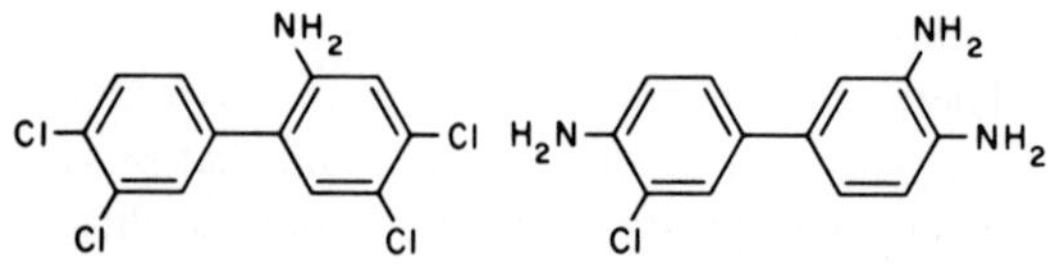


3',4,4',5 - tetrachloro - -2-biphenylamine

3'-chloro-3,4,4'- -biphenyltriamine

FIGURE 6. Structure and correct nomenclature for arbitrarily chosen chlorobiphenylamines (aminochlorobiphenyls).

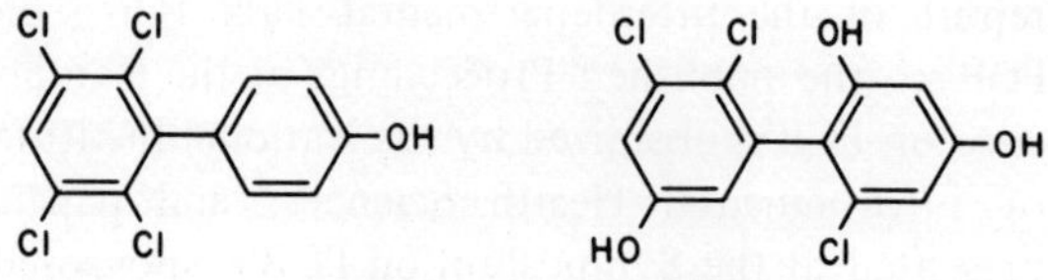


2',3',5',6' - tetrachloro - -4-biphenylol

5',6,6'-trichloro-2,3',4- -biphenyltriol

FIGURE 7. Structure and correct nomenclature for arbitrarily chosen chlorobiphenylols (chlorohydroxybiphenyls).

proper names or generally as "chlorobiphenyls" or "chlorinated biphenyls."

Attention should be drawn to two terms which have frequently been used incorrectly in the literature dealing with chlorinated biphenyls. First, the term isomer applies only to certain compounds of *identical molecular composition*. It is ambiguous to refer to technical PCB as "mixture of isomers." Second, the term *homolog* (or homologous series) is reserved for a group of chemical compounds differing only by a methylene ($CH_2$) group as for instance in a series of compounds having alkyl substituents methyl-, ethyl-, propyl-, etc.

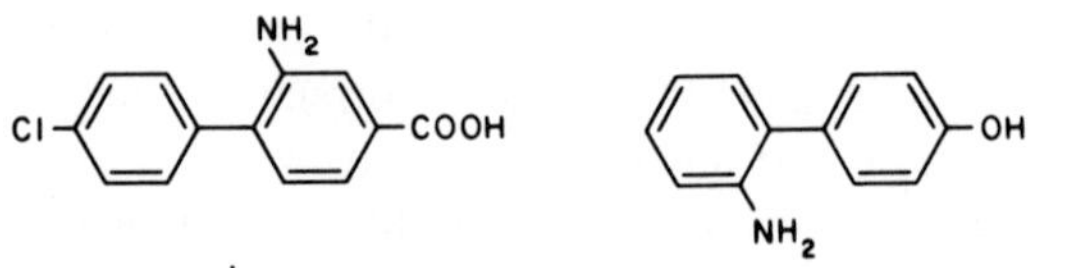


2-amino-4'-chloro- 4-biphenylcarboxylic acid

2'-amino -4-biphenylol

FIGURE 8. Structure and correct nomenclature for two biphenyl derivatives having more than one principal function.

# REFERENCES


1. **Ahmed, M. and Focht, D. D.,** *Can. J. Microbiol.,* 19, 47, 1973.
2. **Allen, J. R. and Norback, D. H.,** *Science,* 179, 498, 1973.
3. **Anonymous,** *New Scientist,* 612, 1966.
4. **Anonymous,** *Environ. Health Perspect.,* Issue No. 1, 1972.
5. **Anonymous,** *Polychlorinated Biphenyls and the Environment,* Interdepartmental Task Force on PCB's, Washington, D.C. 1972, National Technical Information Service, COM-72-10419, Department of Commerce, Springfield, Virginia 22151.
6. **Bache, C. A., Serum, J. W., Youngs, W. D., and Lisk, D. J.,** *Science,* 177, 1191, 1972.
7. **Bagley, G. E. and Cromartie, E.,** *J. Chromatogr.,* 75, 219, 1973.
8. **Bailey, S. and Bunyan, P. J.,** *Nature,* 236, 34, 1972.
9. **Bauer, U.,** *Gwf-wasser/abwasser,* 113, 58, 1972.
10. **Benthe, H. F., Knop, J., and Schmoldt, A.,** *Arch. Toxikol.,* 29, 85, 1972.
11. **Benthe, H. F., Schmoldt, A., and Schmidt, H.,** *Arch. Toxikol.,* 29, 97, 1972.
12. **Benthe, H. F. and Schmoldt, A.,** *Arch. Toxikol.,* 30, 207, 1973.
13. **Bickers, D. R., Harber, L. C., Kappas, A. and Alvares, A. P.,** *Res. Commun. Chem. Pathol. Pharmacol.,* 3, 505, 1972.
14. **Brooks, V. J. and Jacobs, M. B.,** *Poisons – Properties, Chemical Identification, Symptoms, and Emergency Treatment,* 2nd ed., Van Nostrand Reinhold Co., New York, 1958.
15. **Bruckner, J. V., Khanna, K. L., and Cornish, H. H.,** *Toxicol. Appl. Pharmacol.,* 24, 434, 1973.
16. **Carlson, R. W. and Duby, R. T.,** *Bull. Environ. Contam. Toxicol.,* 9, 261, 1973.
17. **Cecil, H. C., Harris, S. J., Bitman, J., and Fries, G. F.,** *Bull. Environ. Contam. Toxicol.,* 9, 179, 1973.
18. **Cooley, N. R., Keltner, J. M., and Forester, J.,** *J. Protozool.,* 19, 636, 1972.
19. **Dahlgren, R. B., Bury, R. J., Linder, R. L., and Reidinger, R. F., Jr.,** *J. Wildl. Manage.,* 36, 524, 1972.
20. **Dahlgren, R. B., Linder, R. L., and Tucker, W. L.,** *J. Wildl. Manage.,* 36, 974, 1972.
21. **Desaiah, D., Cutkomp, L. K., Yap, H. H., and Koch, R. B.,** *Biochem. Pharmacol.,* 21, 857, 1972.
22. **Dustman, E. H., Stickel, L. F., Blus, L. J., Reichel, W. L., and Wiemeyer, S. N.,** *Trans. N. Am. Wildl. Nat. Resour. Conf.,* Washington, D.C., 118, March 7–10, 1971.
23. **Edwards, R.,** *Chem. Ind. (Lond.),* 1340, 1971.
24. **Fishbein, L.,** *J. Chromatogr.,* 68, 345, 1972.
25. **Fisher, N. S., Graham, L. B., and Carpenter, E. J.,** *Nature,* 241, 548, 1973.
26. **Friberg, L.,** PCB Conference II, Stockholm, 1972.
27. **Fries, G. F., Marrow, G. S., Jr., and Gordon, C. H.,** *Bull. Environ. Contam. Toxicol.,* 7, 252, 1972.
28. **Fries, G. F., Marrow, G. S., Jr., and Gordon, C. H.,** *J. Agric. Food Chem.,* 21, 117, 1973.
29. **Fuhremann, T. W. and Lichtenstein, E. P.,** *Toxicol. Appl. Pharmacol.,* 22, 628, 1972.
30. **Fukuda, T.,** *Fukuoka Igaku Zasshi,* 63, 378, 1972.
31. **Funatsu, I., Yamashita, F., Ito, Y., Tsugawa, S., Funatsu, T., Yoshikane, T., and Hayashi, M.,** *Kurume Med. J.,* 19, 43, 1972.
32. **Goldberg, E. D., Butler, P., Meier, P., Menzel, D., Paulik, G., Risebrough, R., and Stickel, L. F.,** *Chlorinated Hydrocarbons in the Marine Environment,* Natl. Acad. Sci., Washington, 1971.
33. **Gustafson, C. G., Ed.,** *PCB in the Environment,* Marcel Dekker, New York, in press.
34. **Gustafson, C. G.,** *Environ. Sci. Technol.,* 4, 814, 1970.
35. **Gustafson, C. G.,** Symposium on PCB's, Am. Chem. Soc. 164th National Meeting, Division of Water, Air, and Waste Chemistry, 12, 1972.
36. **Harris, J. R. and Rose, L.,** *J. Am. Vet. Med. Assoc.,* 161, 1584, 1972.
37. **Harrison, R. B.,** *J. Sci. Food Agric.,* 17, 10, 1966.
38. **Hattula, M. L. and Karlog, O.,** *Acta Pharmacol. Toxicol.,* 31, 238, 1972.
39. **Hattula, M. L. and Karlog, O.,** *Acta Pharmacol. Toxicol.,* 32, 237, 1973.
40. **Hayashi, K. and Ooma, H.,** *Fukuoka Igaku Zasshi,* 63, 416, 1972.
41. **Holden, A. V.,** *J. Appl. Ecol.,* 3, 45, 1966.
42. **Holden, A. V. and Marsden, K.,** *Nature,* 216, 1274, 1967.
43. **Holmes, D. C., Simmons, J. H., and Tatton, J. O. G.,** *Nature,* 216, 227, 1967.
44. **Hutzinger, O., Safe, S., and Zitko, V.,** *Analabs Res. Notes,* 12, 1, 1972.
45. **Isono, N. and Fujiwara, K.,** *Kagaku,* 42, 396, 1972.
46. **Jensen, S.,** *Ambio,* 123, 1972.
47. **Johansson, N., Larsson, A., and Lewander, K.,** *Comp. Gen. Pharmacol.,* 3, 310, 1972.
48. **Kato, T., Yakushiji, M., Tuda, H., Arima, A., Miyahara, M., Takahashi, K., Shimomura, M., Adachi, M., Tashiro, Y., et al.,** *Kurume Med. J.,* 19, 53, 1972.
49. **Keil, J. E., Sandifer, S. H., Graber, C. D., and Priester, L. E.,** *Water Res.,* 6, 837, 1972.
50. **Kimbrough, R. D.,** *Arch. Environ. Health.,* 25, 125, 1972.
51. **Kimbrough, R. D., Linder, R. E., and Gaines, T. B.,** *Arch. Environ. Health.,* 25, 354, 1972.
52. **Kimura, N. T. and Baba, T.,** *Gann,* 64, 105, 1973.

53. Koch, R. B., Desaiah, D., Yap, H. H., and Cutkomp, L. K., *Bull. Environ. Contam. Toxicol.*, 7, 87, 1972.
54. Koeman, J. H., Ed., *TNO-nieuws,* 27, 527, 1972.
55. Koeman, J. H. and van Genderen, H., *J. Appl. Ecol.*, 3, 99, 1966.
56. Koeman, J. H., Ten Noever de Brauw, M. C., and De Vos, R. H., *Nature,* 221, 1126, 1969.
57. Koga, T., *Fukuoka Igaku Zasshi,* 63, 382, 1972.
58. Koller, L. D. and Zinkl, J. G., *Am. J. Pathol.*, 70, 363, 1973.
59. Komatsu, F., *Fukuoka Igaku Zasshi,* 63, 374, 1972.
60. Komatsu, F., and Kikuchi, M., *Fukuoka Igaku Zasshi,* 63, 384, 1972.
61. Lincer, J. L., and Peakall, D. B., *Environ. Pollut.*, 4, 59, 1973.
62. Litterst, C. L., Farber, T. M., Baker, A. M., and van Loon, E. J., *Toxicol. Appl. Pharmacol.*, 23, 112, 1972.
63. Litterst, C. L. and van Loon, E. J., *Proc. Soc. Exp. Biol. Med.*, 141, 765, 1972.
64. Lowe, J. I., Parrish, P. R., Patrick, J. M., Jr., and Forester, J., *Mar. Biol.*, 17, 209, 1973.
65. Mestres, R. and Illes, S., *Trav. Soc. Pharm. Montp.*, 31, 237, 1971.
66. Miller, M. W. and Berg, G. G., Eds., *Chemical Fallout,* Charles C Thomas, Springfield, 1969.
67. Monsanto Chemical Co., *Methodology for Aroclor,* 1971.
68. Monsanto Chemical Co., *Presentation to the Interdepartmental Task Force on PCB's,* Washington, 1972.
69. Mosser, J. L., Fisher, N. S., Teng, T-C., and Wurster, C. F., *Science,* 175, 191, 1972.
70. Mosser, J. L., Fisher, N. S., and Wurster, C. F., *Science,* 176, 533, 1972.
71. Nagai, J., Higuchi, K., and Yae, Y., *Fukuoka Igaku Zasshi,* 63, 367, 1972.
72. Nagasaki, H., Tomii, S., Mega, T., Marugami, M., and Ito, N., *Gann,* 63, 805, 1972.
73. Nelson, N., (Chairman), *PCB's – Environmental Impact. Environ. Res.*, 5, 1972.
74. Nishihara, H. and Yamamoto, T., *Fukuoka Igaku Zasshi,* 63, 352, 1972.
75. Peakall, D. B. and Lincer, J. L., *Bioscience,* 20, 958, 1970.
76. Peakall, D. B., *Residue Rev.*, 44, 1, 1972.
77. Penning, C. H., *Ind. Eng. Chem.*, 22, 1180, 1930.
78. Phillips, W. E., Hatina, G., Villeneuve, D. C., and Grant, D. L., *Environ. Physiol. Biochem.*, 2, 165, 1972.
79. Plapp, F. W., Jr., *Environ. Entomol.*, 1, 58, 1972.
80. Platonow, N. S. and Funnell, H. S., *Can. J. Comp. Med.*, 36, 89, 1972.
81. Platonow, N. S., Liptrap, R. M., and Geissinger, H. D., *Bull. Environ. Contam. Toxicol.*, 7, 358, 1972.
82. Platonow, N. S., Karstad, L. H., and Saschenbrecker, P. W., *Can. J. Comp. Med.*, 37, 90, 1973.
83. Quinby, G. E., *Effects on health and environment. Part I. Bibliography 1881 – 1971,* Oak Ridge National Laboratory, Oak Ridge, Tenn. April 1972; Part II, Review, in preparation.
84. Rehfeld, B. M., Bradley, R. L., Jr., and Sunde, M. L., *Poult. Sci.*, 51, 435, 1972.
85. Rehfeld, B. M., Bradley, R. L., Jr., and Sunde, M. L., *Poult. Sci.*, 51, 488, 1972.
86. Reynolds, L. M., *Residue Rev.*, 34, 27, 1971.
87. Risebrough, R. W., *International Symposium on Identification and Measurement of Environmental Pollutants,* Westley, B., Ed., National Research Council of Canada, Ottawa, 1971, 147.
88. Risebrough, R. W., Rieche, P., and Olcott, H. S., *Bull. Environ. Contam. Toxicol.*, 4, 192, 1969.
89. Risebrough, R. W., Rieche, P., Peakall, D. B., Herman, S. G., and Kirven, M. N., *Nature,* 220, 1098, 1968.
90. Risebrough, R. W. and Brodine, V., *Environment,* 12, 16, 1970.
91. Robinson, J., *Chem. Br.*, 7, 472, 1971.
92. Robinson, J., Richardson, A., Crabtree, A. N., Coulson, J. C., and Potts, G. R., *Nature,* 214, 1307, 1967.
93. Roburn, J. *Analyst,* 90, 467, 1965.
94. Sanders, H. O. and Chandler, J. H., *Bull. Environ. Contam. Toxicol.*, 7, 257, 1972.
95. Saschenbrecker, P. W., Funnell, H. S., and Platonow, N. S., *The Veterinary Record,* 90, 100, 1972.
96. Snarski, V. M., *PCB Newsletter,* U.S. Environmental Protection Agency, Duluth, Minnesota.
97. Sunshine, I., Ed., *Handbook of Analytical Toxicology,* The Chemical Rubber Co., Cleveland, 1969.
98. Tanaka, K. and Komatsu, F., *Fukuoka Igaku Zasshi,* 63, 360, 1972.
99. Van Leeuwen, B. L. J., *Kon. Ned. Akad. Wetensch., Versl. Gewone Vergad. Afd. Natuurk.*, 81, 46, 1972.
100. Veith, G. D. and Lee, G. F., *Water Res.*, 4, 265, 1970.
101. Villeneuve, D. C., Grant, D. L., and Phillips, W. E. J., *Bull. Environ. Contam. Toxicol.*, 7, 264, 1972.
102. Vos, J. G. and de Roij, T., *Toxicol. Appl. Pharmacol.*, 21, 549, 1972.
103. Vos, J. G. and Notenboom-Ram, E., *Toxicol. Appl. Pharmacol.*, 23, 563, 1972.
104. Walsh, G. E., *J. Wash. Acad. Sci.*, 62, 122, 1972.
105. Walker, C. H., Hamilton, G. A., and Harrison, R. B., *J. Sci. Food Agric.*, 18, 123, 1967.
106. Widmark, G., *J. Assoc. Offic. Anal. Chem.*, 50, 1069, 1967.
107. Wrabetz, K., *Zum Stand der PCB-Analytik bei den Farbenfabriken Bayer,* December 1971.
108. Zitko, V., *Fish Res. Board Can. Tech. Rep.*, No. 1083, 1970.
109. Zitko, V., and Choi, P. M. K., *Fish Res. Board Can. Tech. Rep.*, No. 272, 1971.

Chapter 2

# COMMERCIAL PCB PREPARATIONS: PROPERTIES AND COMPOSITION

## Introduction

The polychlorinated biphenyls (PCB) are a class of chlorinated, aromatic compounds which have found widespread applications because of their general stability and inertness as well as their excellent dielectric properties.

Most information on the technical preparation, chemical and physical properties, and general characteristics of PCB comes from trade publications[3,25,26] and articles in technical encyclopedias.[18,42]

PCB's have been prepared industrially since 1929 and are now produced in many industrial countries (Table 1). Most information by far is available on Monsanto's PCB preparations (the Aroclors) and this particular brand will serve to discuss various aspects of PCB in general.

All Aroclor products are characterized by a four-digit number. The first two digits represent the type of molecule; 12 = chlorinated biphenyl, 54 = chlorinated terphenyl.* Aroclor 25-- and 44-- are blends of PCB and chlorinated terphenyls (75% and 60% PCB, respectively). The last two digits give the weight percent of chlorine (see Table 2).

Recently, a new PCB product of the Aroclor series "1016" was introduced[26] which contains 41% chlorine per weight but in which the penta-, hexa-, and heptachlorobiphenyl content has been significantly reduced.

TABLE 1

**The World's Major Producers of PCB**

| Producer | Country | Tradename of PCB |
|---|---|---|
| Monsanto | U.S.A. and Great Britain | Aroclor® |
| Bayer | Germany | Clophen® |
| Prodelec | France | Phenoclor and Pyralene® |
| Kanegafuchi | Japan | Kanechlor® |
| Mitsubishi-Monsanto | Japan | Santotherm® |
| Caffaro | Italy | Fenclor® |
| Sovol | U.S.S.R. | |
| Chemko | Czechoslovakia | |

## General Information and Properties of PCB

### *Use of PCB*

The outstanding physical and chemical characteristics of PCB are their thermal stability, resistance to oxidation, acid, bases, and other chemical agents as well as their excellent dielectric (electrically insulating) properties.

These and other desirable properties have led to numerous uses of PCB[7,18,25] such as dielectric fluids (capacitors, transformers), industrial fluids (use in hydraulic systems, gas turbines, and vacuum pumps), fire retardants, heat transfer applications, plasticizers (adhesives, textiles, surface coatings, sealants, printing, copy paper). Some uses of PCB classified according to grade of Aroclor are shown in Table 3.

Some chlorobiphenyls were shown to have insecticidal[9] and fungistatic[5] activity; however, they were apparently never used as pesticides although recommended for incorporation into pesticide formulations.[17,32,40]

PCB are also reported to increase the insecticidal properties of DDT,[23] lindane,[33] organophosphorus compounds,[13] and carbaryl.[31]

### *Production Figures for PCB*

Little information is available on the worldwide production and use of PCB. The Monsanto Company has released figures[27] for domestic sales for their products (Aroclor) for the period 1957–1972.

The data are shown in graph form in Figure 1 (by use) and Figure 2 (by PCB grade). The drop in output after 1970 is due to the voluntary restriction of sales by Monsanto essentially to uses in closed systems (capacitor and transformer applica-

*Gas chromatographic and mass spectrometric analysis showed Aroclor 5460 to be a mixture of chlorinated terphenyls. Aroclor 5442, however, was found to contain chlorinated biphenyls as well.[19]

TABLE 2

**Chlorine Content of Aroclor Preparations***

| Aroclor | % Cl | Average number of Cl per molecule | Average molecular weight |
|---|---|---|---|
| 1221 | 20.5–21.5 | 1.15 | 192 |
| 1232 | 31.5–32.5 | 2.04 | 221 |
| 1242 | 42 | 3.10 | 261 |
| 1248 | 48 | 3.90 | 288 |
| 1254 | 54 | 4.96 | 327 |
| 1260 | 60 | 6.30 | 372 |
| 1262 | 61.5–62.5 | 6.80 | 389 |
| 1268 | 68 | 8.70 | 453 |

*Manufacturers specifications.

TABLE 3

**Use of PCB Classified to Grade of Aroclor**

| Current use of PCB | Grade of Aroclor used |
|---|---|
| Electrical capacitors | 1016 (1221, 1254) |
| Electrical transformers | 1242, 1254, 1260 |
| Vacuum pumps | 1248, 1254 |
| Gas-transmission turbines | 1221, 1242 |
| Former use of PCB | |
| Hydraulic fluids | 1232, 1242, 1248, 1254, 1260 |
| Plasticizer in synthetic resins | 1248, 1254, 1260, 1262, 1268 |
| Adhesives | 1221, 1232, 1242, 1248, 1254 |
| Plasticizer in rubbers | 1221, 1232, 1242, 1248, 1254, 1268 |
| Heat transfer systems | 1242 |
| Wax extenders | 1242, 1254, 1268 |
| Dedusting agents | 1254, 1260 |
| Pesticide extenders, inks, lubicants, cutting oils | 1254 |
| Carbonless reproducing paper | 1242 |

tions). Estimations on the Japanese production[20] give values of approximately 26 million pounds per year production. Of this, 40 to 50% is used for capacitors, 15% for transformer oils, 10 to 15% for heat transfer fluids, 5% for plasticizers, 15% for carbonless copying paper, and 5 to 10% for export. The annual consumption of PCB in Finland is about 250 metric tons.[15]

*Preparation and Properties of PCB*

PCB's are prepared industrially* by the chlorination of biphenyl with anhydrous chlorine and iron filings or ferric chloride as catalyst.[18,21] The crude product is generally purified to remove color, traces of hydrogen chloride, and catalyst which is usually achieved by treatment with alkali and distillation. The resulting product is a compli-

*The preparation of chlorobiphenyl mixtures from hexachlorocyclohexane is described in patents.[22,29,41] However, this process is not being used for any large-scale preparation of PCB.

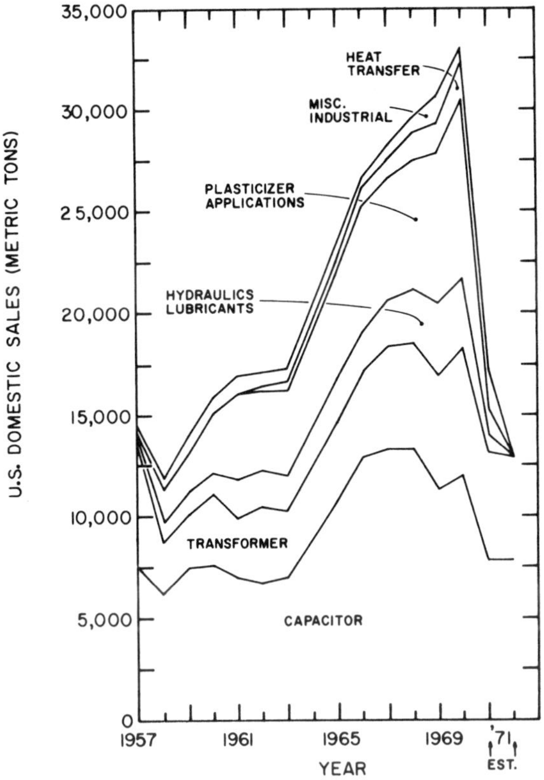


FIGURE 1. Domestic sales of Monsanto's Polychlorinated Biphenyls in the U.S. (by use). (Courtesy of Monsanto Company.)

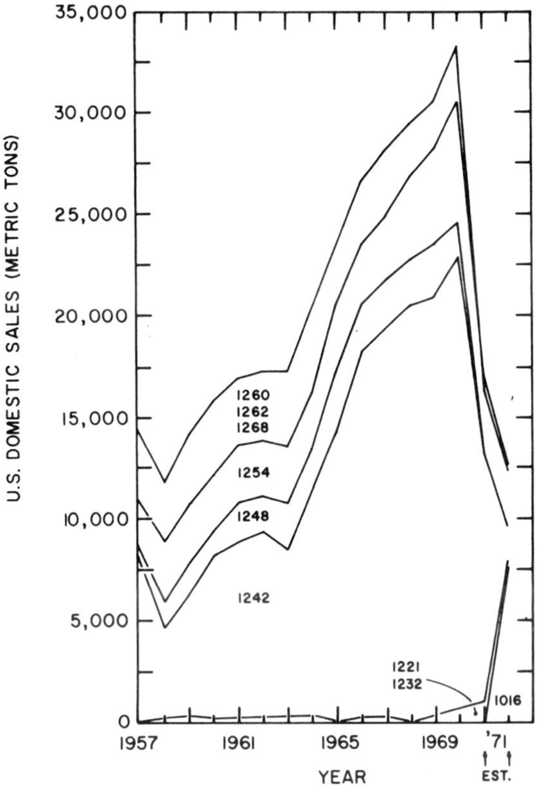


FIGURE 2. Domestic sales of Monsanto's Polychlorinated Biphenyls in the U.S. (specific Aroclors). (Courtesy of Monsanto Company.)

cated mixture of chlorobiphenyls with different numbers of chlorine atoms per molecule and their isomers.[18,24] This fact is responsible for the physical state of PCB preparations; most individual chlorobiphenyls are solids at room temperature whereas commercial mixtures are mobile oils (e.g., Aroclor 1221, 1232, 1242, and 1248); viscous liquids (e.g., Aroclor 1254) or sticky resins (e.g., Aroclor 1260 and 1262) due to the mutual depression of melting points of their components.

With the exception of Aroclor 1221 and 1268, PCB's do not crystallize but show a "pour point" below which the material changes into a resinous state. Some chemical and physical properties of Aroclor products are summarized in Table 4.[25] Electrical properties of some Aroclors are given in Table 5.[18]

The most important physical properties of PCB's from an environmental point of view, are solubility and vapor pressure. Solubilities of various Aroclors in a number of solvents are shown in Table 6.

The solubility of PCB's in water is low and decreases with increasing chlorine content. Values given by Monsanto[30] are 200 ppb for 1242, 100 ppb for 1248, 40 ppb for 1254 and about 25 ppb for 1260. Freed[12] found a solubility of 43 ppb for Aroclor 1248 at 26°. These results are biased in each case by selective solution of components of lower chlorine content. Evidence for this comes from solubilization experiments of Zitko[49,50] and Freed.[12] Table 7 shows the relative peak height of saturated aqueous solutions of Aroclor 1254. The first few peaks which generally correspond to PCB components of lower chlorine content are significantly higher than in standard Aroclor 1254.

This variation of solubility between individual PCB components is also evident from studies on the water solubility of individual chlorobiphenyls[16,44] (Table 8).

Studies on the solubility of PCB in water are complicated by the fact that these compounds are strongly adsorbed onto various surfaces. PCB was shown to adsorb relatively rapidly onto woodburn,[12] plastic,[49] glass,[14,44] and container and/or silt.[10]

TABLE 4

**Chemical and Physical Properties of Representative Aroclor Plasticizers and Resins**

| Property | Aroclor 1221 | Aroclor 1232 | Aroclor 1242 | Aroclor 1248 | Aroclor 1254 | Aroclor 1260 |
|---|---|---|---|---|---|---|
| Appearance | Clear, mobile oil | Clear, mobile oil | Clear, mobile oil | Clear, mobile oil | Light-yellow viscous liquid | Light-yellow soft, sticky resin |
| Color, maximum | 100 APHA | 100 APHA | 100 APHA | 100 APHA | 100 APHA | 150 APHA |
| Chlorine, percent | 20.5–21.5 | 31.4–32.5 | 42 | 48 | 54 | 60 |
| Acidity, mg KOH/g maximum | 0.014 | 0.014 | 0.015 | 0.010 | 0.010 | 0.014 |
| Moisture, ppm, maximum | – | – | 50 | 50 | 50 | 50 |
| Ave. coefficient of expansion, cc/cc/°C | 0.00071 (15°–40°C) | 0.00073 (25°–100°C) | 0.00068 (25°–65°C) | 0.00070 (25°–65°C) | 0.00066 (25°–65°C) | 0.00067 (20°–100°C) |
| Specific gravity | 1.182–1.192 (25°/15.5°C) | 1.270–1.280 (25°/15.5°C) | 1.381–1.392 (25°/15.5°C) | 1.405–1.415 (65°/15.5°C) | 1.495–1.505 (65°/15.5°C) | 1.555–1.566 (90°/15.5°C) |
| Density, lb/gal, 25°C | 9.85 | 10.55 | 11.50 | 12.04 | 12.82 | 13.50 |
| Distillation range, °C, corrected (ASTM D-20, modified) | 275–320 | 290–325 | 325–366 | 340–375 | 365–390 | 385–420 |
| Evaporation loss, %, 100°C, 6 hr | 1.0–1.5 | 1.0–1.5 | 0–0.4 | 0–0.3 | 0–0.2 | 0–0.1 |
| (ASTM D-6, mod.) 163°C, 5 hr | | | 3.0–3.6 | 3.0–4.0 | 1.1–1.3 | 0.5–0.8 |
| Flash point °C (Cleveland Open °F Cup) | 141–150<br>286–302 | 152–154<br>305–310 | 176–180<br>348–356 | 193–196<br>379–384 | None to boiling point | None to boiling point |
| Fire point °C (Cleveland Open °F Cup) | 176<br>349 | 238<br>460 | None to boiling point | None to boiling point | None to boiling point | None to boiling point |

| Property | Aroclor 1221 | Aroclor 1232 | Aroclor 1242 | Aroclor 1248 | Aroclor 1254 | Aroclor 1260 |
|---|---|---|---|---|---|---|
| Pour point °C<br>(ASTM E-97) °F | 1 (crystals)<br>34 (crystals) | −35.5<br>−32 | −19<br>2 | −7<br>19.4 | 10<br>50 | 31<br>88 |
| Softening point °C<br>(ASTM E-28) °F | – | – | – | – | – | – |
| Refractive index,<br>n D-20 20°C | 1.617–1.618 | 1.620–1.622 | 1.627–1.629 | 1.630–1.631 | 1.629–1.641 | 1.647–1.649 |
| Viscosity, sec<br>Saybolt Universal<br>(ASTM D-83) | | | | | | |
| 100°F (37.8°C) | 38–41 | 44–51 | 82–92 | 185–240 | 1800–2500 | |
| 130°F (54.4°C) | 35–37 | 39–41 | 49–56 | 73–80 | 260–340 | 3200–4500 |
| 210°F (98.9°C) | 30–31 | 31–32 | 34–35 | 36–37 | 44–48 | 72–78 |

| Property | Aroclor 1262 | Aroclor 1268 | Aroclor 2565 | Aroclor 4465 | Aroclor 5442 | Aroclor 5460 |
|---|---|---|---|---|---|---|
| Appearance | Light-yellow sticky, viscous resin | White to off-white powder | Black, opaque brittle resin | Clear, light-yellow resin | Clear yellow sticky resin | Clear, yellow-to-amber, brittle resin or flakes |
| Color, maximum | 150 APHA | 1.5 NPA (molten) | – | 2 NPA (molten) | 2 NPA (molten) | 2 NPA (molten) |
| Chlorine, percent | 61.5–62.5 | 68 | 65 | 65 | 42 | 58.5–60.6 |
| Acidity, mg KOH/g maximum | 0.014 | 0.05 | 1.4 | 0.05 | 0.05 | 0.05 |
| Moisture, ppm, maximum | – | – | – | – | – | – |
| Ave. coefficient of expansion, cc/cc/°C | 0.00064<br>(25°–65°C) | 0.00067<br>(20°–100°C) | 0.00066<br>(25°–65°C) | 0.00061<br>(25°–65°C) | 0.00123<br>(25°–99°C) | 0.00179<br>(25°–124°C) |
| Specific gravity | 1.572–1.583<br>(90°/15.5°C) | 1.804–1.811<br>(25°/25°C) | 1.734<br>(25°/25°C) | 1.670<br>(25°/25°C) | 1.470<br>(25°/25°C) | 1.670<br>(25°/25°C) |

TABLE 4 (continued)

| Property | Aroclor 1262 | Aroclor 1268 | Aroclor 2565 | Aroclor 4465 | Aroclor 5442 | Aroclor 5460 |
|---|---|---|---|---|---|---|
| Density, lb/gal, 25° | 13.72 | 15.09 | 14.44 | 13.91 | 12.24 | 13.91 |
| Distillation range, °C, corrected (ASTM D-20) modified) | 390–425 | 435–450 | – | 230–320 (4 mm Hg) | 215–300 (4 mm Hg) | 280–335 (5 mm Hg) |
| Evaporation loss, % 100°C, 6 hr | 0-0.1 | 0–0.06 | | 0–0.02 | 0.01 | |
| (ASTM D-6, mod.) 163°C, 5 hr | 0.5–0.6 | 0.1–0.2 | 0.2–0.3 | 0.2–0.3 | 0.2 | 0.03 |
| Flash point °C (Cleveland Open °F Cup) | None to boiling point | None to boiling point | – | None to boiling point | 247<br>477 | None to boiling point |
| Fire point (Cleveland Open °C Cup) °F | None to boiling point | None to boiling point | – | None to boiling point | > 350<br>> 662 | None to boiling point |
| Pour point °C (ASTM E-97) °F | 38–38<br>99 | – | – | – | 46<br>115 | – |
| Softening point °C ASTM E-28) °F | – | 150–170<br>302-338 (hold point) | 66–72<br>149–162 | 60–66<br>140–151 | 46–52<br>115–126 | 98–105<br>208–222 |
| Refractive index n D-20 20°C | 1.6501-1.6517 | – | – | 1.664–1.667 | – | 1.660–1.665 |
| Viscosity seconds Saybolt Universal (ASTM D-83) | | | | | | |
| 100°F (37.8°C) | 600–850 | | | | | |
| 130°F (54.4°C) | (160°F: 71°C) | – | – | 90–150 | | – |
| 210°F (98.9°C) | 86–100 | | | (265°F; 130°C) | 300–400 | |

TABLE 5

**Electrical Properties of Some Aroclors**

| Aroclor | Dielectric constant at 1000 cycles[a] | | Volume resistivity,[b] 8 cm at 100°C 500 V, de | Dielectric strength,[c] kV | Power factor,[a] 100°C, 1000 cycles, % |
|---|---|---|---|---|---|
| | 25°C | 100°C | | | |
| 1232 | 5.7 | 4.6 | | | |
| 1242 | 5.8 | 4.9 | above 500 x $10^9$ | >35 | <0.1 |
| 1248 | 5.6 | 4.6 | above 500 x $10^9$ | >35 | <0.1 |
| 1254 | 5.0 | 4.3 | above 500 x $10^9$ | >35 | <0.1 |
| 1260 | 4.3 | 3.7 | above 500 x $10^9$ | >35 | <0.1 |
| 1268 | 2.5 | | | | |
| 5442 | 3.0 | 4.9 | above 500 x $10^9$ | | |
| 5454 | 2.7 | 4.2 | | | |
| 5460 | 2.5 | 3.7 | | | |
| 4465 | 2.7 | 3.3 | | | |

[a]ASTM D-150-47T
[b]ASTM D-257-46
[c]ASTM D-149-44

Nonionic surfactants (e.g. ethylene oxide adduct of lautic acid, Corexit® 7664) solubilize PCB;[49,50] a similar effect has been observed with 4,4′-dichlorobiphenyl using Tween® 80 (see reference 44 and Table 8).

Vapor pressures (Figure 3) and vaporization rates (Table 9) of several Aroclors are shown below. Similar to the solubility data, the vapor pressures of the Aroclors may be biased by the components of lower chlorine content which are generally more volatile. Freed's data[12] (Table 10) show that the earlier GC-peaks from Aroclor 1254 decrease more rapidly when Aroclor 1254 is heated with water on a steam bath (1 mg Aroclor 1254, 300 ml water; conditions of codistillation). A much smaller difference was observed however when Aroclor 1254 (1 mg) was heated on a petri-dish (steam bath) without added water.

Since gas chromatography is the preferred analytical method for the determination of PCB,[11] GC-properties of six Aroclor preparations on two different phases, which have recently been reported,[1] will be given here.

Gas chromatograms for the Aroclors are shown in Figures 4, 5, and 6, employing the following conditions: support material, Chromosorb® W HP 80/100 mesh; stationary phase, 10% DC-200 (6 ft x 4 mm I.D. column); carrier gas, nitrogen, 120 ml/min; temperature, 200°; detector, electron capture (tritium). Accurate retention time data for the Aroclors and the GC conditions described are given in Table 11.

Figures 7, 8, and 9 show gas chromatograms of the same six Aroclors using a different stationary phase (1:1 15% QF-1/10% DC-200) under conditions which were otherwise identical. Accurate retention time data for the Aroclors using this stationary phase are given in Table 12. A gas chromatogram of the new Aroclor 1016 is shown in Figure 10.

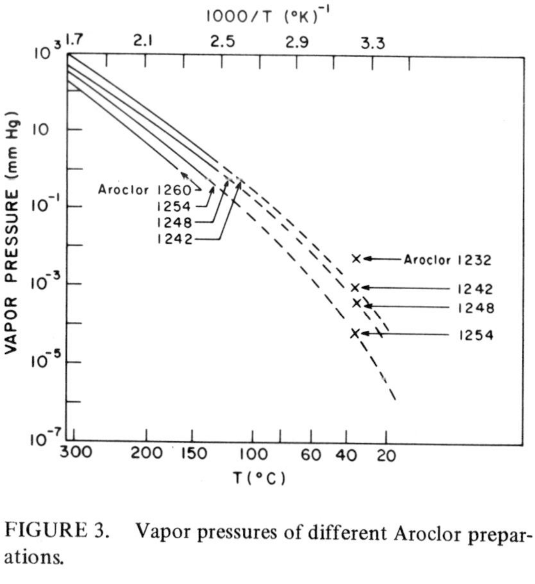


FIGURE 3. Vapor pressures of different Aroclor preparations.

TABLE 6
**Solubilities of Aroclor Plasticizers in Various Solvents***

| Solvent | Aroclor 1242 | | Aroclor 1248 | | Aroclor 1254 | | Aroclor 1268 | | Aroclor 4465 | | Aroclor 5450 |
|---|---|---|---|---|---|---|---|---|---|---|---|
| | 25°C | Hot | 25°C | Hot | 25°C | Hot | Cold | Hot | Cold | Hot | 25°C |
| Acid | | | | | | | | | | | |
| Acetic Acid | S | S | – | – | S | S | – | – | SS | S | – |
| Oleic Acid | S | S | – | – | S | S | – | – | S | VS | – |
| Benzoic Acid | $10.0^{31}$°C | – | $10.0^{32}$°C | – | – | – | – | – | – | – | – |
| Aldehyde | | | | | | | | | | | |
| 40% Formaldehyde | 1 | 1 | 1 | 1 | 1 | 1 | 1 | 1 | 1 | 1 | – |
| Furfural | VS | VS | VS | VS | VS | VS | SS | SS | VS | VS | – |
| Amine | | | | | | | | | | | |
| Aniline | S | S | – | – | S | S | – | – | VS | VS | – |
| Pyridine | $132.5^{30}$°C | $440^{99}$°C | – | – | $114^{31}$°C | $425^{100}$°C | – | – | VS | VS | – |
| Chloro derivatives | | | | | | | | | | | |
| Amyl chlorides – mixed | S | S | S | S | S | S | – | – | VS | VS | – |
| Carbon Tetrachloride | S | S | S | S | S | S | – | – | VS | VS | 156 |
| Chloroform | S | S | S | S | S | S | S | S | VS | VS | – |
| Dichloroethylene | – | – | – | – | – | – | – | – | VS | VS | – |
| Ethylene Dichloride | S | S | S | S | S | S | S | S | VS | VS | – |
| Monochlorobenzene | S | S | S | S | S | S | – | – | VS | VS | – |
| Orthodichlorobenzene | – | – | – | – | – | – | – | – | VS | VS | – |
| Tetrachlorethane | S | S | S | S | S | S | – | – | VS | VS | – |
| Trichlorethane | S | S | S | S | S | S | – | – | VS | VS | – |
| Drying Oil | | | | | | | | | | | |
| Tung Oil | S | S | S | S | S | S | – | – | VS | VS | – |
| Linseed Oil | S | S | S | S | S | S | 1 | S | VS | VS | – |
| Ester | | | | | | | | | | | |
| Amyl Acetate | S | S | S | S | S | S | S | S | VS | VS | – |
| Butyl Acetate | S | S | S | S | S | S | S | S | VS | VS | – |
| "Cellosolve" Acetate | S | S | S | S | S | S | – | – | VS | VS | – |
| Cottonseed Oil | S | S | S | S | S | S | – | – | S | VS | – |
| Dibutyl Phthalate | S | S | S | S | S | S | S | S | S | VS | – |
| Diethyl Phthalate | S | S | S | S | S | S | – | – | S | VS | – |
| Ethyl Acetate | S | S | S | S | S | S | SS | SS | S | VS | – |
| Ethyl lactate | S | S | S | S | S | S | 1 | S | VS | VS | – |
| Ethylene Glycol Diacetate | S | S | S | S | S | S | – | – | VS | VS | – |

| Solvent | Aroclor 1242 25°C | Aroclor 1242 Hot | Aroclor 1248 25°C | Aroclor 1248 Hot | Aroclor 1254 25°C | Aroclor 1254 Hot | Aroclor 1268 Cold | Aroclor 1268 Hot | Aroclor 4465 Cold | Aroclor 4465 Hot | Aroclor 5450 25°C |
|---|---|---|---|---|---|---|---|---|---|---|---|
| Methyl Acetate | S | S | S | S | S | S | SS | SS | S | S | – |
| Tricresyl Phosphate | S | S | S | S | S | S | – | – | SS | S | – |
| Ether: Ethyl Ether | S | S | S | S | S | S | S | S | S | – | – |
| Ether Alcohol | | | | | | | | | | | |
| Carbitol | $224^{31\,°C}$ | $307^{99\,°C}$ | VS | VS | $173^{26\,°C}$ | $259^{98\,°C}$ | 1 | S | SS | – | – |
| Cellosolve | S | S | S | S | S | S | 1 | S | S | – | – |
| Diethylene Glycol | – | – | – | – | – | – | – | – | S | – | – |
| *p,p′*-Dihydroxy Ethyl Ether | $16.9^{23\,°C}$ | $19^{99\,°C}$ | SS | SS | $8^{30\,°C}$ | $10^{100\,°C}$ | – | – | SS | – | – |
| Hydrocarbon | | | | | | | | | | | |
| Benzene | VS | VS | VS | VS | VS | VS | S | S | VS | VS | 143 |
| Gasoline | VS | VS | VS | VS | VS | VS | – | – | VS | VS | – |
| Kerosene | VS | VS | VS | VS | VS | VS | SS | S | VS | VS | – |
| Mineral Spirits | VS | VS | VS | VS | VS | VS | S | S | VS | VS | – |
| Paraffin | $2.0^{27.5\,°C}$ | S | $2.0^{28\,°C}$ | S | – | S | – | – | 5.0 | S | – |
| Pine Oil | S | S | VS | VS | S | S | S | S | S | S | – |
| Toluene | VS | VS | VS | VS | VS | VS | S | S | VS | VS | 142 |
| Turpentine | VS | VS | VS | VS | VS | VS | S | S | VS | VS | – |
| Xylene | VS | VS | VS | VS | VS | VS | S | S | VS | VS | 178 |
| Hydroxy derivatives | | | | | | | | | | | |
| Amyl Alcohol | S | S | – | – | S | S | SS | S | S | S | – |
| *n*-Butyl Alcohol | S | S | – | – | S | S | 1 | SS | SS | S | – |
| Ethyl Alcohol (3-A) | $23.3^{29\,°C}$ | $80.0^{70\,°C}$ | – | – | $10^{27\,°C}$ | $28^{75\,°C}$ | 1 | 1 | SS | – | – |
| Glycerine | 1 | 1 | 1 | 1 | 1 | 1 | 1 | 1 | 1 | 1 | – |
| Methyl Alcohol | $42.5^{29\,°C}$ | $88.5^{60\,°C}$ | – | – | $15^{26\,°C}$ | $22.2^{85\,°C}$ | – | – | SS | – | – |
| Phenol – 90% | $194^{30\,°C}$ | S | – | – | SS | S | SS | S | S | S | – |
| Ketone: Acetone | S | S | – | – | S | S | 1 | 1 | S | S | 260 |
| Miscellaneous | | | | | | | | | | | |
| Carbon Disulfide | S | S | – | – | S | S | S | S | VS | VS | – |
| Nitrobenzene | S | S | – | – | S | S | – | – | VS | – | – |
| Water | 1 | 1 | 1 | 1 | 1 | 1 | 1 | 1 | 1 | 1 | – |

1 = Insoluble; SS = Slightly Soluble; S = Soluble; VS = Very Soluble.

*Figures show g of Aroclor/100 ml of solvent at 25°C unless otherwise indicated.

TABLE 7

**Relative Peak Heights (Peak 5 = 100) in Saturated Aqueous Solutions of Aroclor 1254**

| Peak No. | Saturated aqueous solution (26°C) | Saturated aqueous solution (4°C) | Aroclor 1254 standard |
|---|---|---|---|
| 1 | 172 | 144 | 35 |
| 2 | 91 | 72 | 16 |
| 3 | 47 | 41 | 30 |
| 4 | 14 | 9 | 1 |
| 5 | 100 | 100 | 100 |
| 6 | 33 | 28 | 23 |
| 7 | 57 | 59 | 55 |
| 8 | 5 | 5 | 10 |
| 9 | 21 | 24 | 25 |
| 10 | 8 | 13 | 31 |
| 11 | 4 | 4 | 6 |
| 12 | 11 | 24 | 50 |
| 13 | 6 | 10 | 11 |

TABLE 8

**Solubility of Chlorobiphenyls in Water**

| Compound | Solubility mg/ℓ (ppm) |
|---|---|
| Monochlorobiphenyls | |
| 2- | 5.9 |
| 3- | 3.5 |
| 4- | 1.19 |
| Dichlorobiphenyls | |
| 2,4- | 1.40 |
| 2,2′- | 1.50 |
| 2,4′- | 1.88 |
| 4,4′- | 0.08 |
| Trichlorobiphenyls | |
| 2,4,4′- | 0.085 |
| 2′,3,4- | 0.078 |
| Tetrachlorobiphenyls | |
| 2,2′,5,5′- | 0.046 |
| 2,2′,3,3′- | 0.034 |
| 2,2′,3,5′- | 0.170 |
| 2,2′,4,4′- | 0.068 |
| 2,3′,4,4′- | 0.058 |
| 2,3′,4′,5- | 0.041 |
| 3,3′,4,4′- | 0.175 |
| Pentachlorobiphenyls | |
| 2,2′,3,4,5′- | 0.022 |
| 2,2′,4,5,5′- | 0.031 |
| Hexachlorobiphenyl | |
| 2,2′,4,4′,5,5′- | 0.0088 |
| Octachlorobiphenyl | |
| 2,2′,3,3′,4,4′,5,5′- | 0.0070 |
| Decachlorobiphenyl | 0.015 |
| 4,4′-Dichlorobiphenyl | |
| + Tween 80 0.1% | 5.9 |
| + Tween 80 1% | ⩾10.0 |
| + Humic acid extract | 0.07 |

## TABLE 9

### Vaporization Rates of Aroclors

| Aroclor (Surface area: 12.28 cm²) | Wt. loss (g) | Exposure at 100°C (hr) | Vaporization rate (g/cm²/hr) |
|---|---|---|---|
| 1221 | 0.5125 | 24 | 0.00174 |
| 1232 | 0.2572 | 24 | 0.000874 |
| 1242 | 0.0995 | 24 | 0.000338 |
| 1248 | 0.0448 | 24 | 0.000152 |
| 1254 | 0.0156 | 24 | 0.000053 |
| 1262 | 0.0039 | 24 | 0.000013 |
| 1260 | 0.0026 | 24 | 0.000009 |
| 4465 | 0.0064 | 72 | 0.000007 |
| 5442 | 0.0039 | 72 | 0.000004 |
| 5460 | 0.0032 | 72 | 0.000004 |

## TABLE 10

### Percent Loss in Area of Seven Chromatogram Peaks of Aroclor After Heating

| | % Peak Remaining After Heating | | |
|---|---|---|---|
| | with water | | without water |
| Aroclor 1254 peak | 25 min | 60 min | 10 min |
| 1 | 34 | 17 | 13 |
| 2 | 59 | 26 | 15 |
| 3 | 78 | 27 | 20 |
| 4 | 60 | 46 | 20 |
| 5 | 86 | 49 | 27 |
| 6 | 100 | 85 | 28 |
| 7 | 100 | 67 | 16 |

TABLE 11

**GLC Retention Time Data for Six Aroclors***

| Aroclor 1221 | Aroclor 1242 | Aroclor 1248 | Aroclor 1254 | Aroclor 1260 | Aroclor 1262 |
|---|---|---|---|---|---|
| 0.21 | | | | | |
| 0.26 | | | | | |
| 0.31 | | | | | |
| 0.36 | | | | | |
| 0.39 | 0.39 | | | | |
| 0.52 | 0.51 | 0.51 | | | |
| 0.59 | 0.57 | 0.57 | | | |
| 0.64 | | | | | |
| 0.69 | 0.67 | 0.67 | | | |
| 0.75 | 0.72 | | | | |
| | | 0.80 | | | |
| 0.88 | 0.86 | 0.85 | 0.87 | | |
| 0.99 | 0.96 | 0.96 | 0.98 | | |
| | 1.03 | 1.03 | 1.05 | | |
| 1.27 | 1.22 | 1.23 | 1.27 | 1.28 | 1.26 |
| 1.42 | 1.39 | 1.39 | | | |
| 1.52 | 1.49 | 1.49 | 1.52 | 1.50 | 1.50 |
| 1.76 | 1.74 | 1.74 | 1.78 | | |
| 1.86 | 1.83 | 1.84 | 1.88 | 1.86 | 1.85 |
| 2.08 | | | | 2.07 | 2.07 |
| 2.22 | 2.20 | 2.20 | 2.20 | 2.21 | 2.22 |
| | 2.56 | 2.54 | | | |
| 2.65 | | | 2.63 | 2.63 | 2.61 |
| | | | | 2.84 | 2.82 |
| 3.10 | | 3.04 | 3.08 | 3.08 | 3.06 |
| | | | 3.60 | 3.50 | 3.50 |
| | | | 4.10 | 4.10 | 4.10 |
| | | | 4.30 | | |
| | | | 4.90 | 4.90 | 4.90 |
| | | | 5.80 | 5.80 | 5.80 |
| | | | | | 6.40 |
| | | | | 6.50 | 6.60 |
| | | | | 7.80 | 7.80 |
| | | | | 9.10 | 9.10 |

*Retention sequence relative to aldrin from solvent peak on 10% DC-200 (Figures 4, 5, and 6).

## TABLE 12

**GLC Retention Time Data for Six Aroclors***

| Aroclor 1221 | Aroclor 1242 | Aroclor 1248 | Aroclor 1254 | Aroclor 1260 | Aroclor 1262 |
|---|---|---|---|---|---|
| 0.20 | | | | | |
| 0.29 | | | | | |
| 0.36 | | | | | |
| 0.39 | 0.41 | | | | |
| 0.53 | 0.53 | 0.53 | | | |
| 0.57 | 0.61 | 0.61 | | | |
| 0.71 | 0.72 | 0.72 | | | |
| 0.78 | | | | | |
| | 0.81 | 0.81 | | | |
| 0.88 | 0.90 | 0.90 | 0.87 | | |
| 1.02 | 1.02 | 1.02 | 1.02 | | |
| 1.10 | 1.09 | 1.09 | | | |
| 1.31 | 1.34 | 1.34 | 1.32 | 1.31 | 1.31 |
| 1.53 | 1.52 | 1.52 | 1.52 | 1.53 | 1.53 |
| 1.82 | 1.82 | 1.82 | 1.82 | | 1.84 |
| | | | | 1.86 | |
| 1.96 | 1.97 | 1.97 | 1.96 | | |
| 2.08 | | | | | 2.08 |
| | | | | 2.14 | |
| 2.26 | | | | 2.26 | 2.24 |
| 2.34 | 2.36 | 2.36 | 2.34 | | |
| 2.68 | | | 2.68 | 2.66 | 2.66 |
| 2.84 | 2.80 | 2.80 | 2.80 | 2.88 | 2.88 |
| 3.22 | 3.24 | 3.24 | 3.20 | 3.22 | 3.22 |
| | | | 3.50 | 3.50 | 3.50 |
| | | | 3.90 | 3.90 | 3.90 |
| | | | 4.20 | 4.20 | 4.20 |
| | | | 5.00 | 5.00 | 5.00 |
| | | | 6.10 | 6.10 | 6.10 |
| | | | | 6.50 | 6.50 |
| | | | | | 7.40 |
| | | | | 8.00 | |
| | | | | 9.50 | 9.30 |
| | | | | | 11.9 |

*Retention sequence relative to aldrin from solvent peak on 1:1 15% QF-1/10% DC-200 (Figures 7, 8, and 9).

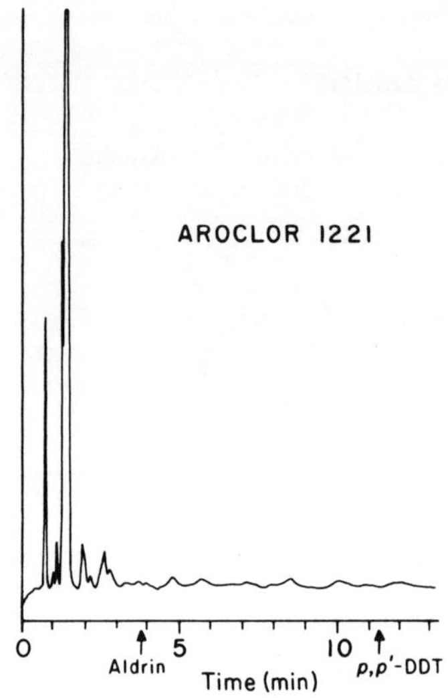


FIGURE 4. GLC separation on 10% DC-200 column of 120 ng Aroclor 1221. GLC conditions are given in the text. (From *J. Chromatogr.,* 72, 275, 1972. With permission.)

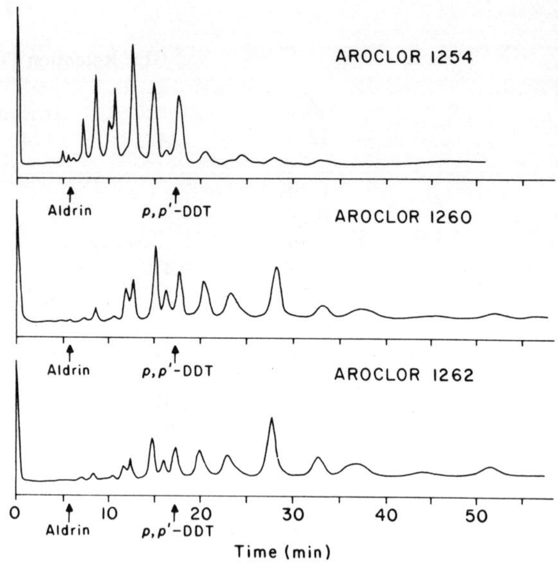


FIGURE 6. GLC separation on 10% DC-200 column of 32 ng Aroclor 1254, 20 ng Aroclor 1260, and 20 ng Aroclor 1262. GLC conditions are given in the text. (From *J. Chromatogr.,* 72, 275, 1972. With permission.)

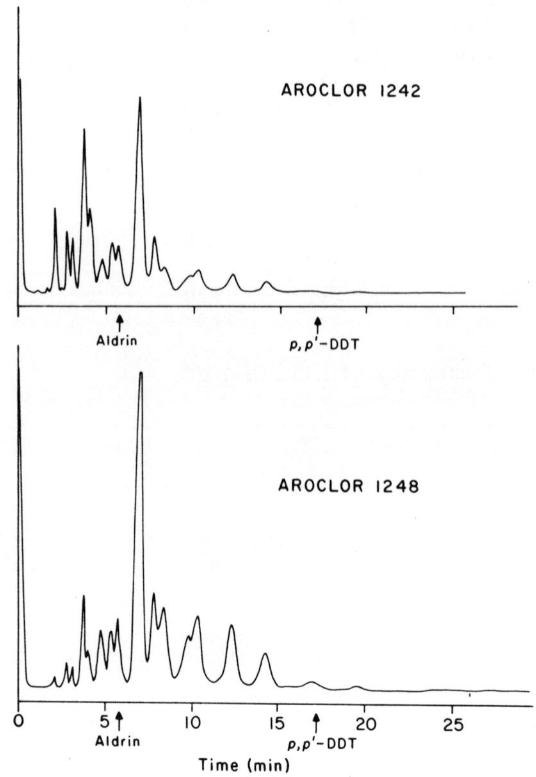


FIGURE 5. GLC separation on 10% DC-200 column of 50 ng Aroclor 1242 and 50 ng Aroclor 1248. GLC conditions are given in the text. (From *J. Chromatogr.,* 72, 275, 1972. With permission.)

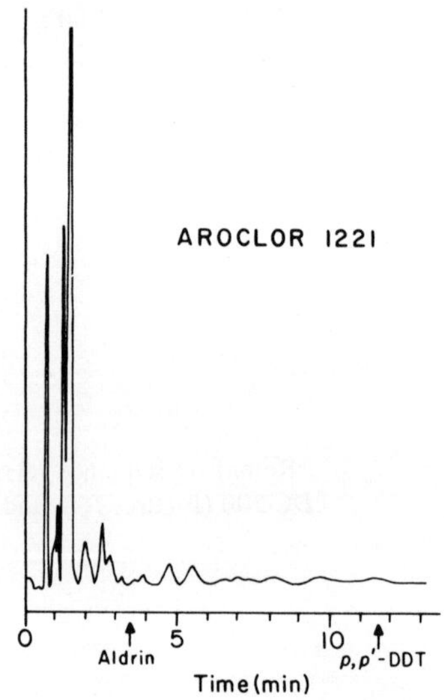


FIGURE 7. Gas chromatographic separation on 1:1 15% QF-1/10% DC-200 of 120 ng Aroclor 1221. GLC conditions are given in the text. (From *J. Chromatogr.,* 72, 275, 1972. With permission.)

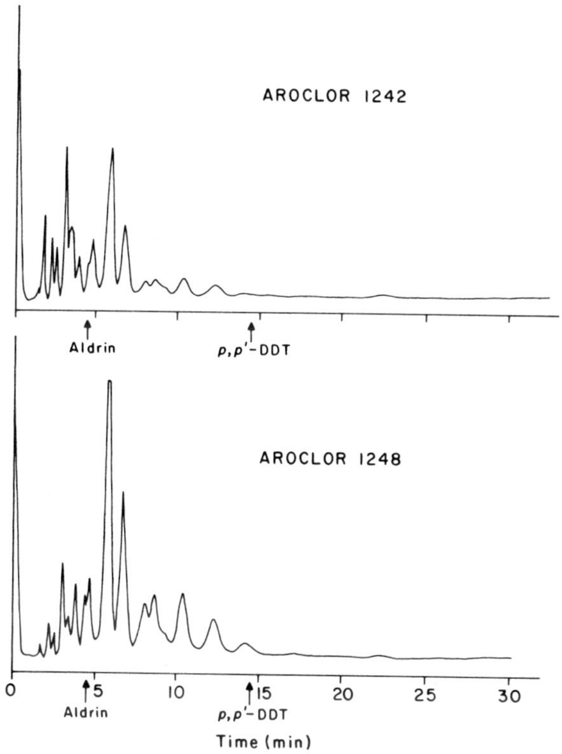


FIGURE 8. Gas chromatographic separation on 1:1 15% QF-1/10% DC-200 of 50 ng Aroclor 1242 and 50 ng Aroclor 1248. GLC conditions are given in the text. (From *J. Chromatogr.*, 72, 275, 1972. With permission.)

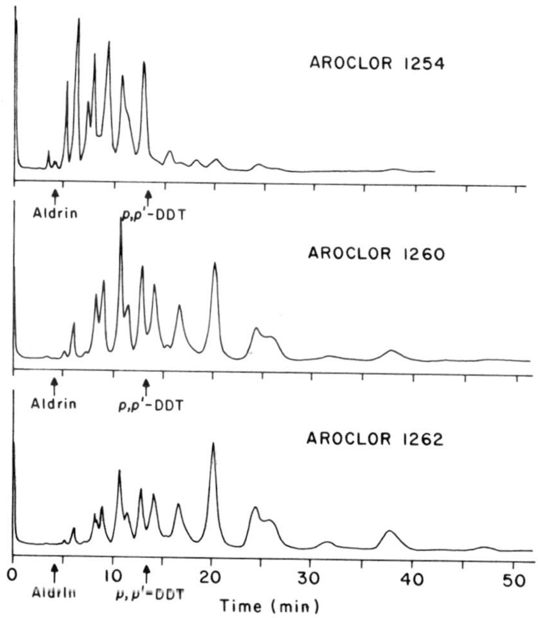


FIGURE 9. Gas chromatographic separation on 1:1 15% QF-1/10% DC-200 of 32 ng Aroclor 1254, 20 ng Aroclor 1260, and 20 ng Aroclor 1262. GLC conditions are given in the text. (From *J. Chromatogr.*, 72, 275, 1972. With permission.)

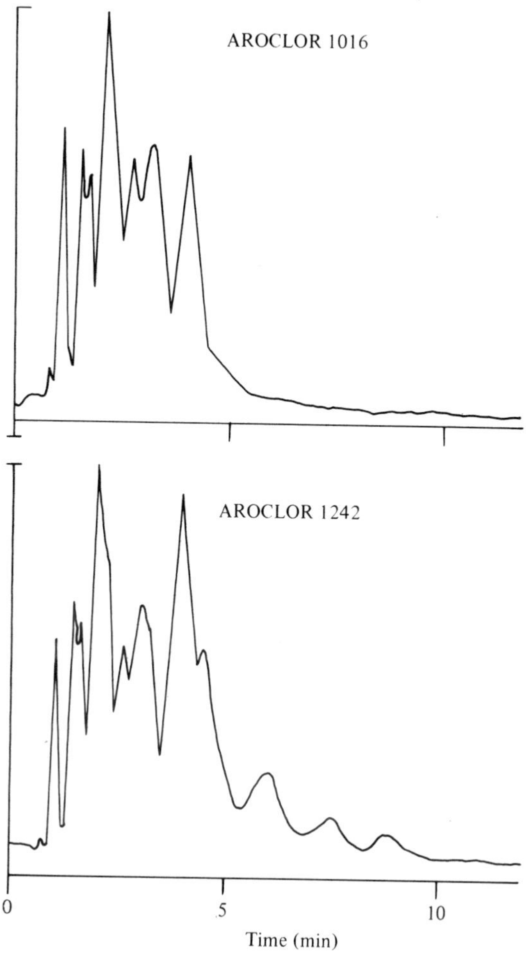


FIGURE 10. Gas chromatographic separation of Aroclor 1016 (40 mg) and Aroclor 1242. GLC conditions: 6' x 0.5" glass column packed with 4% SE-30 on chromosorb W, oven temperature: 200° electron capture detector.

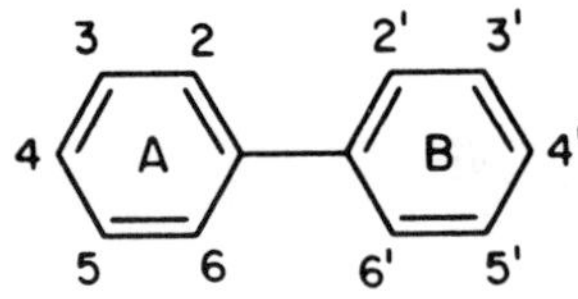


| CHLORINE ATOMS ON RING B | | CHLORINE ATOMS ON RING A: 0 | 1 | 2 | 3 | 4 | 5 |
|---|---|---|---|---|---|---|---|
| | 0 | 1 | 3 | 6 | 6 | 3 | 1 |
| | 1 | | 6 | 18 | 18 | 9 | 3 |
| | 2 | | | 21 | 36 | 18 | 6 |
| | 3 | | | | 21 | 18 | 6 |
| | 4 | | | | | 6 | 3 |
| | 5 | | | | | | 1 |

FIGURE 11. Possible distribution of chlorine atoms in the two rings of biphenyl.

**Chemical Composition of PCB Mixtures**

Two hundred and nine different chlorobiphenyls do theoretically exist[8,46] (Figure 11). However, it is unlikely on the basis of both mechanistic and statistical grounds that all of them be formed in the technical chlorination processes. A chlorobiphenyl fully chlorinated on one ring and not at all on the other (2,3,4,5,6-pentachlorobiphenyl) is an extreme example for a compound which is most likely absent. Some, but not all theoretical predictions[46] as to which compounds are the most significant in the PCB mixture were later verified by experimental data.

Empirical formulas for PCB components of different numbers of chlorine atoms, their molecular weight, and chlorine content (percent, based on Cl = 35.45) are given in Table 13.

Distribution of PCB components in Aroclor according to molecular composition[39] is shown in Table 14. The most recent results on the composition of several Aroclors[26] are shown in Table 15.

Other than chlorobiphenyls, no major components have been reported to be present in PCB preparations. Products formed by the addition (rather than substitution) of chlorine to the biphenyl molecule to give partially saturated structures have been detected (>10%) in laboratory chlorinations (e.g., 4). Apparently, these addition products are either not formed in the industrial chlorination process or destroyed in the purification step. However, trace quantities of polychloronaphthalenes and polychlorodibenzofurans were found in some samples of PCB,[43] and the presence of the latter type of compound may be of considerable toxicological significance.

The first attempt to identify individual components in the complicated mixtures obtained by the chlorination of biphenyl was by Russian workers[48] who oxidized two chlorobiphenyl mixtures containing an average of five and seven chlorine atoms, respectively, with nitric acid (d = 1.4). Up to 100 hr reaction time at reflux temperatures with frequent change of the nitric acid was required to convert 200 g of the chlorobiphenyl mixture to 40 g of chlorobenzoic acids. The major components in this acid mixture were identified as 3,4-dichlorobenzoic acid and 2,4,5-trichlorobenzoic acid from the PCB averaging five chlorine atoms and 2,4,5-trichlorobenzoic acid and 2,3,4,5-tetrachlorobenzoic acid from the heptachloro PCB (Figure 12). The acids were identified by comparison with authentic samples after repeated recrystallization and purification as silver salts and amides.

Although it is obvious that selective oxidation of certain isomers may have taken place, these early experiments have correctly identified some of the major components in PCB mixtures.

TABLE 13

**Empirical Formula, Molecular Weights, and Percent Chlorine of Chlorinated Biphenyls**

| Empirical formula | Molecular weight* | Percent chlorine* |
|---|---|---|
| $C_{12}H_{10}$ | 154.21 | 0 |
| $C_{12}H_9Cl$ | 188.65 | 18.79 |
| $C_{12}H_8Cl_2$ | 223.10 | 31.77 |
| $C_{12}H_7Cl_3$ | 257.54 | 41.30 |
| $C_{12}H_6Cl_4$ | 291.99 | 48.56 |
| $C_{12}H_5Cl_5$ | 326.43 | 54.30 |
| $C_{12}H_4Cl_6$ | 360.88 | 58.93 |
| $C_{12}H_3Cl_7$ | 395.32 | 62.77 |
| $C_{12}H_2Cl_8$ | 429.77 | 65.98 |
| $C_{12}HCl_9$ | 464.21 | 68.73 |
| $C_{12}Cl_{10}$ | 498.66 | 71.18 |

*Based on Cl = 35.45

TABLE 14

**Molecular Composition of Some Aroclors[39]**

| Chlorobiphenyl composition | Presence (%) in Aroclor | | | |
|---|---|---|---|---|
| | 1242 | 1248 | 1254 | 1260 |
| $C_{12}H_9Cl$ | 3 | | | |
| $C_{12}H_8Cl_2$ | 13 | 2 | | |
| $C_{12}H_7Cl_3$ | 28 | 18 | | |
| $C_{12}H_6Cl_4$ | 30 | 40 | 11 | |
| $C_{12}H_5Cl_5$ | 22 | 36 | 49 | 12 |
| $C_{12}H_4Cl_6$ | 4 | 4 | 34 | 38 |
| $C_{12}H_3Cl_7$ | | | 6 | 41 |
| $C_{12}H_2Cl_8$ | | | | 8 |
| $C_{12}HCl_9$ | | | | 1 |

TABLE 15

**Molecular Composition of Some Aroclors[26]**

| Chlorobiphenyl composition | Presence in Aroclor* | | | |
|---|---|---|---|---|
| | 1221 | 1016 | 1242 | 1254 |
| $C_{12}H_{10}$ | 11 | <0.1 | <0.1 | <0.1 |
| $C_{12}H_9Cl$ | 51 | 1 | 1 | <0.1 |
| $Cl_{12}H_8Cl_2$ | 32 | 20 | 16 | 0.5 |
| $Cl_{12}H_7Cl_3$ | 4 | 57 | 49 | 1 |
| $C_{12}H_6Cl_4$ | 2 | 21 | 25 | 21 |
| $C_{12}H_5Cl_5$ | <0.5 | 1 | 8 | 48 |
| $C_{12}H_4Cl_6$ | ND** | <0.1 | 1 | 23 |
| $C_{12}H_3Cl_7$ | ND | ND | <0.1 | 6 |
| $C_{12}H_2Cl_8$ | ND | ND | ND | ND |

*Per cent (W/W) by GC/mass using area correction factors by homolog response
**None detected, <0.01% = ND

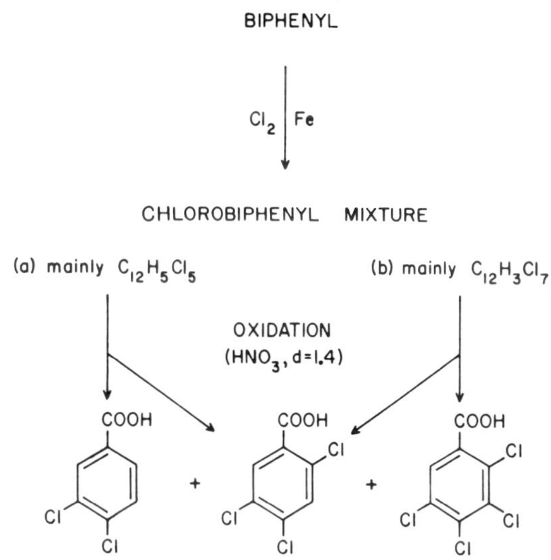


FIGURE 12. First attempt to identify structures of chlorobiphenyls in PCB mixtures: chlorobenzoic acids obtained by the oxidation of PCB with nitric acid.

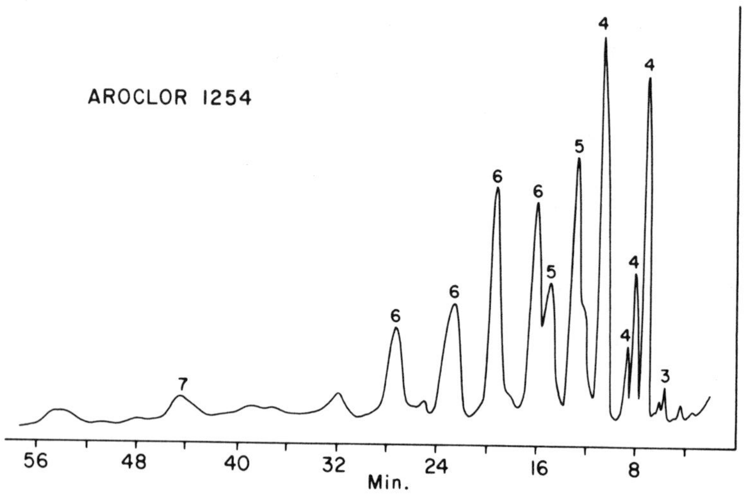


FIGURE 13. Total ion current chromatogram of Aroclor 1254 (1.5 μg). The numbers of chlorine atoms per biphenyl molecule are indicated by numbers above the peaks. GC-conditions: 9′ x 0.25″ SE-30 column, oven temperature 180°, helium flow 35 ml/min. (From *J. Assoc. Offic. Anal. Chem.*, 53, 251, 1970. With permission.)

In the last few years, GC-MS was used to investigate the molecular composition (i.e., number of chlorine atoms per molecule) of individual peaks in gas chromatograms of PCB preparations. Packed[2,28] as well as capillary[6,36] columns were used for this purpose and some of the results are shown in Figures 13 and 14.

A number of groups have investigated the detailed structure (i.e., position of chlorine atoms) of individual components. The commercial preparations of the Aroclor series (Monsanto Co.) have been most thoroughly explored.

The qualitative and quantitative composition of Aroclor 1221 has been elucidated by Willis and Addison by comparing the major peaks of this mixture to synthetic materials.[47] Results are shown in Table 16. The 7.3% which were unaccounted for may be due, in part, to experimental error and to the presence of minor components (an unidentified trichlorobiphenyl, for instance, was reported to be present in the mixture).

In contrast to the more highly chlorinated mixtures, none of the major components of Aroclor 1221 had chlorine substitution of position 3 of the biphenyl system.

Five major high-chlorine components of Kanechlor® 400 were separated from the technical mixture by distillation to give three fractions (b.p.$_{3mm}$ 155 to 165°, 165 to 170°, 170 to 175°) and subsequent preparative gas chromatography of the fraction 170 to 175°.[34] The structures were investigated by mass spectrometry, UV spectroscopy and GC retention times and finally proven (Figure 15) by comparison with synthetic materials (Chapter 3).

Seven major components of Phenoclor® DP6 (the Prodelec PCB preparation containing 60% chlorine) were identified by a Dutch group[37,38] (Figure 16). The components were separated by preparative GC and the samples so obtained compared to authentic synthetic materials.

Webb and McCall[45] have separated 27 components from Aroclors 1221, 1242 and 1254 by preparative GC and identified them by comparison (GC-retention times and infrared spectra) with synthetic materials.

TABLE 16

**The Major Components of Aroclor 1221**

| Compound | % in Aroclor 1221* |
|---|---|
| biphenyl | 12.7 |
| 2-chlorobiphenyl | 28.4 |
| 4-chlorobiphenyl | 18.7 |
| 2,2′-dichlorobiphenyl | 9.2 |
| 2,4 -dichlorobiphenyl | 3.5 |
| 2,4′-dichlorobiphenyl | 13.6 |
| 4,4′-dichlorobiphenyl | 6.2 |

*Total % Aroclor 1221 accounted for: 92.3

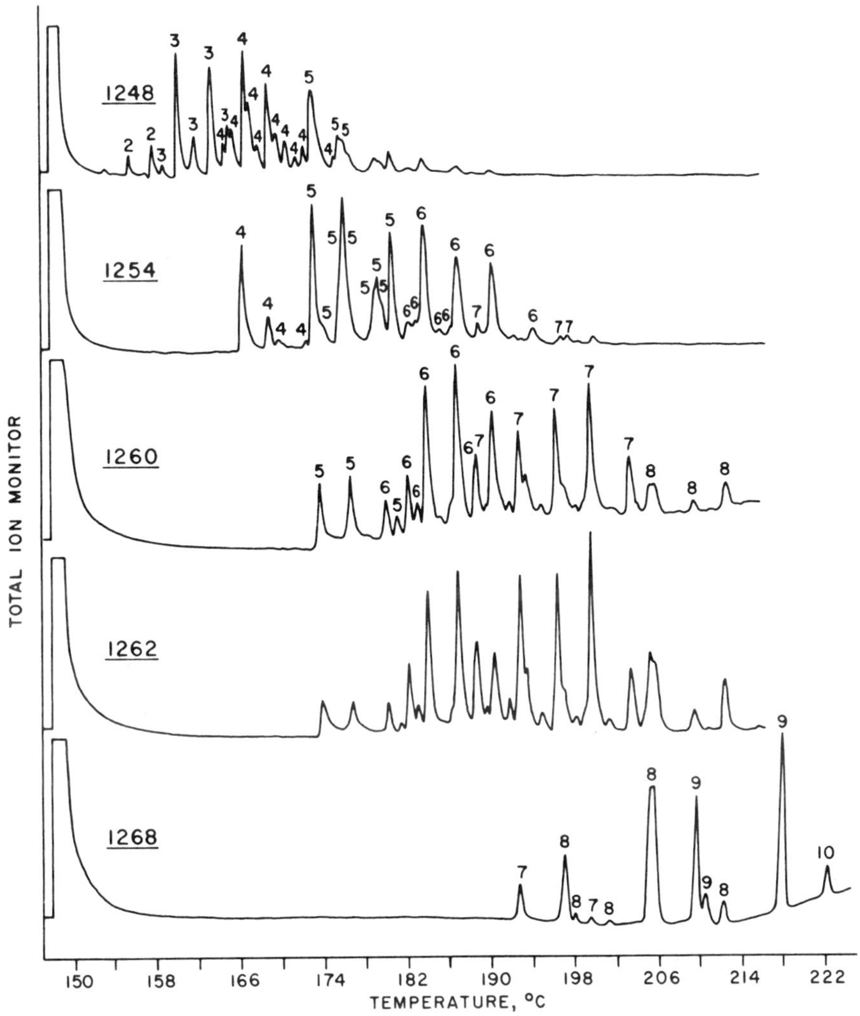


FIGURE 14. Capillary column chromatograms of some Aroclors. Numbers above the peaks designate the number of chlorine atoms per biphenyl molecule. GC-conditions: 50′ x 0.020″ SE-30 SCOT column, temperature program rate 2°/min, helium pressure 9 lb/in$^2$. The mass spectrometer was a Perkin-Elmer 270 instrument. (From *J. Assoc. Offic. Anal. Chem.*, 54, 801, 1972. With permission.)

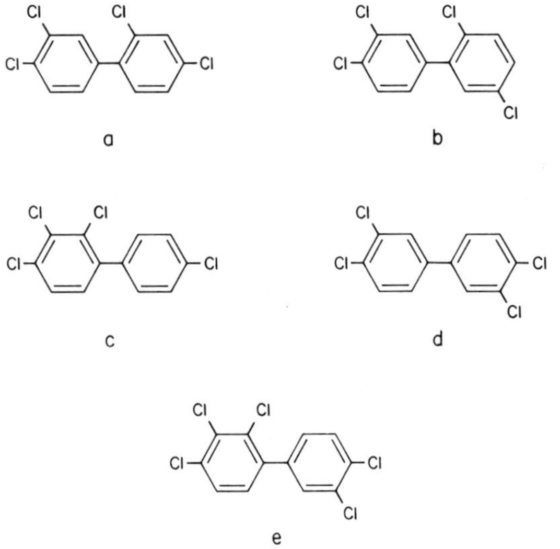

FIGURE 15. Structures of five major high-boiling components of Kanechlor-400.

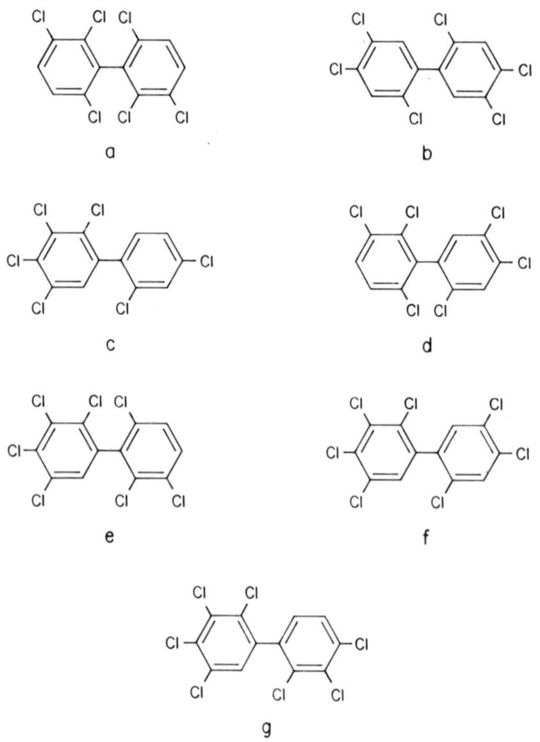

FIGURE 16. Structures of seven major components of Phenoclor DP6.

Figure 17 shows gas chromatograms of Aroclors 1221, 1232, 1242, 1248, and 1254; the retention times are marked to identify the peaks for Table 17. In this table, details of the structures and the mode of identification are given. Clarifying footnotes are given at the bottom of the table.

Webb and McCall have also synthesized a number of chlorobiphenyls that were subsequently found to be absent from the Aroclors investigated. These compounds and their retention times are listed in Table 18.

Sissons and Welti[35] have characterized the major constituents of Aroclor 1254 by 220 MHz proton magnetic resonance and mass spectrometry following separation by chromatography on alumina columns and preparative gas chromatography. The retention indices of these isolated compounds together with those of 40 synthesized chlorobiphenyls were used to predict the structures of the remaining constituents of Aroclor 1254 as well as those of Aroclors 1242 and 1260.

Figures 18, 19, and 20 show high resolution gas chromatograms of Aroclors 1254, 1242, and 1260. The column was a Perkin Elmer 50 ft x 0.02 in I.D. support coated open tubular (SCOT) Apiezon® L column fitted into a Pye® Model 104 gas chromatograph with a flame ionization detector. The oven temperature was kept at 205° and the carrier gas was helium at a flow of 2.5 ml/min.

Structures that were determined by 220 MHz NMR on isolated samples are drawn onto the chromatogram of Aroclor 1254 (Figure 18; see Table 19). Structures of all other components were predicted from retention indices and a complete list with relevant data is given in Table 20. Figure 19 and Table 21 show the chromatogram and predicted structures for Aroclor 1242 and the corresponding data for Aroclor 1260 are given in Figure 20 and Table 22.

The most common substitution patterns for the major chlorobiphenyl components found in PCB preparations are summarized in Figure 21.

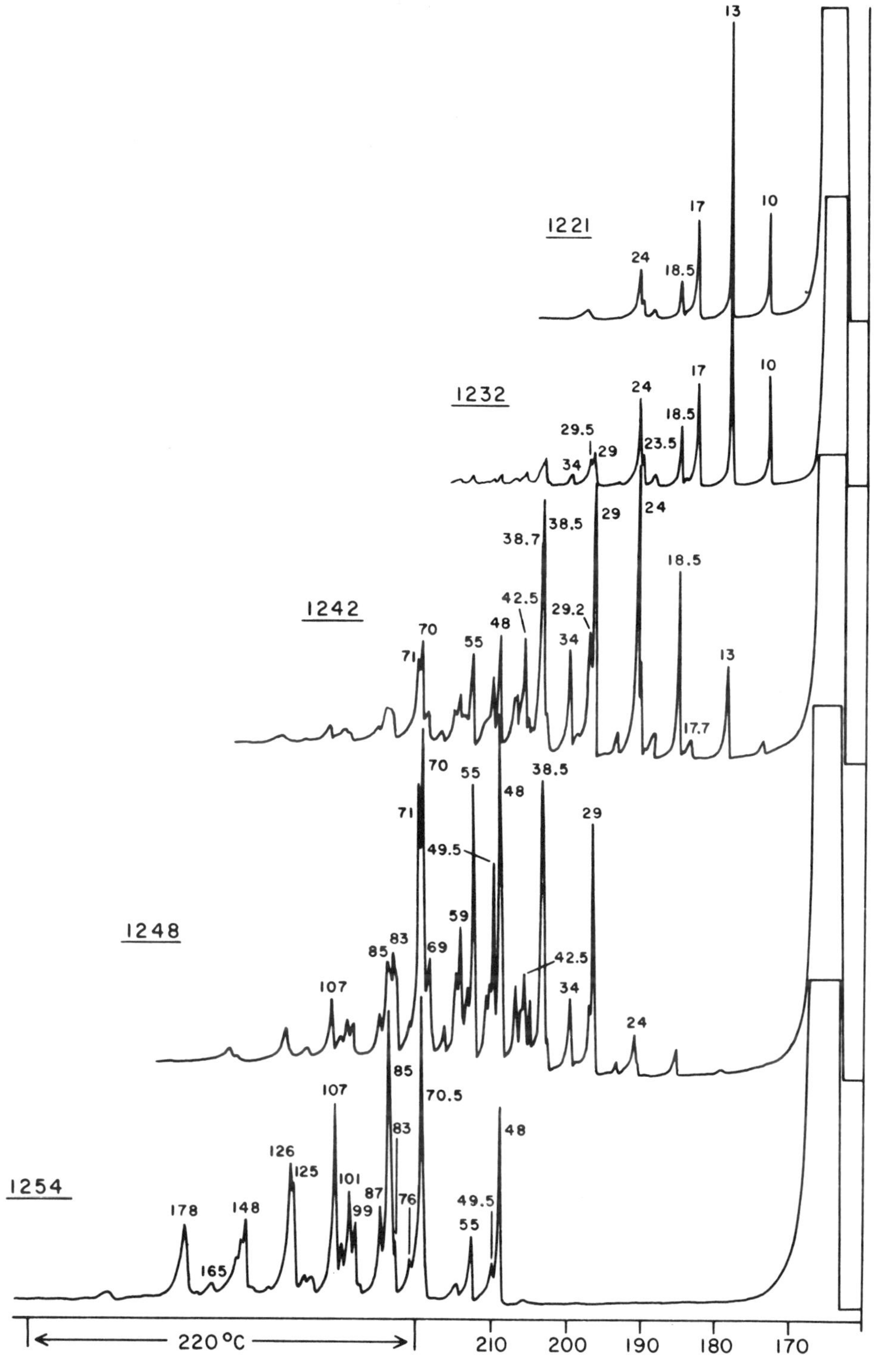


FIGURE 17. Capillary column chromatograms of some Aroclors. Peak identification numbers correspond to retention times of peaks relative to p,p′-DDT where both materials were chromatographed at 190°. Peaks correspond to retention times in Table 17. (From *J. Assoc. Offic. Anal. Chem.*, 55, 746, 1972. With permission.)

TABLE 17

**Identities and GC-retention Times of Chlorobiphenyls**

| Retention time (*p,p′*-DDE = 1.0) | Synthetic chlorobiphenyl | Aroclor 1221 | 1232 | 1242 | 1248 | 1254 |
|---|---|---|---|---|---|---|
| 0.10 | Biphenyl | X, ir | X | X | X | |
| 0.13 | 2 | X, ir | X | X | X | |
| 0.169 | 3 | X, R | | | | |
| 0.17 | 4 | X, ir | X | | | |
| 0.177 | | | | X | | |
| 0.185 | 2,2′ | X, ir | X | X, ir | X | |
| 0.213 | | X | X | X | | |
| 0.216 | | X | X | X | | |
| 0.235 | 2,3′ | X, ir | X | X | X | |
| 0.24 | 2,4′ | X, ir | X | X, ir | X | |
| 0.26 | 2,2′,6 | | X | X, R | X | |
| 0.29 | 2,2′,5 | | X | X, ir | X | |
| 0.292 | 2,2′,4 | X | X | X, ir | X | |
| 0.295 | 4,4′ | X, ir | X | | | |
| 0.34 | 2,2′,3 | | X | X, ir$^{+}$ | X | |
| 0.38 | 2,3′,4 | | X | X, ir | X | |
| 0.385 | 2,4,4′ | | X | X, ir | X | |
| 0.387 | 2,4′,5 | | X | X, ir | X | |
| 0.415 | | | | X | X | |
| 0.425 | 2′,3,4 | | | X, ir | X | |
| 0.43 | | | | X | X | |
| 0.44 | | | | X | X | |
| 0.445 | | | | | X | |
| 0.48 | 2,2′,5,5′ | | | X, ir | X | X, ir |
| 0.495 | 2,2′,4,5′ | | | X, ir | X | X, ir |
| 0.505 | 2,2′,4,4′ | | | X, R | X | |
| 0.515 | | | | X | X | |
| 0.55 | 2,2′,3,5′ | | | X | X | X, ir |
| 0.57 | | | | X | X | |
| 0.59 | 2,2′,3,6,6′ | | | X | X | X, R |
| 0.60 | | | | X | X | |
| 0.64 | | | | X | X | |
| 0.69 | | | | X | X | |
| 0.695 | | | | | X | X |
| 0.70 | 2,3′,4′,5 | | | X, ir$^{+}$ | X | |
| 0.705 | 2,2′,3,5′,6 | | | | | X, ir |
| 0.71 | 2,3′,4,4′ | | | X, ir$^{+}$ | X | |
| 0.72 | | | | X | X | |
| 0.76 | 2,2′,3,4′,6 | | | | | X, ir |
| 0.82 | | | | | X | |
| 0.83 | 2,2′,3,3′,6 | | | | X | X, ir |
| 0.84 | | | | | X | |
| 0.85 | 2,2′,4,5,5′ | | | | X | X, ir |
| 0.87 | 2,2′,4,4′,5 | | | | X | X, ir |
| 0.96 | | | | | | X |
| 0.99 | 2,2′,3′,4,5 | | | | X | X, ir |
| 1.01 | | | | | X | X |
| 1.04 | | | | | X | X |
| 1.07 | 2,3,3′,4′,6 | | | | X | X |
| 1.15 | | | | | | X |
| 1.17 | | | | | X | X |
| 1.19 | | | | | | X |
| 1.25 | 2,2′,3,4′,5′,6 | | | | | X, ir |

TABLE 17 (Continued)

**Identities and GC-retention Times of Chlorobiphenyls**

| Retention time ($p,p'$-DDE = 1.0) | Synthetic chlorobiphenyl | Aroclor 1221 | 1232 | 1242 | 1248 | 1254 |
|---|---|---|---|---|---|---|
| 1.26 | 2,3′,4,4′,5 | | | | X | X, ir |
| 1.35 | | | | | | X |
| 1.47 | | | | | | X |
| 1.48 | 2,2′,4,4′,5,5′ | | | | | X, ir |
| 1.49 | | | | | | X |
| 1.52 | | | | | | X |
| 1.65 | | | | | | X |
| 1.78 | | | | | | X |

X indicates a chromatographic peak at the specified retention time, (see Figure 17).
ir indicates the infrared spectrum and retention time has been matched with that of the synthetic compound.
R indicates that the retention time matches with that of the synthetic compound.
+ indicates that more than one compound is present.

TABLE 18

**Synthetic Chlorobiphenyls Found to be Absent from Aroclors 1221 → 1254**

| Chlorobiphenyl | Retention time* |
|---|---|
| 2,3,4′ | 44 |
| 3,4,4′ | 59.5 |
| 2,3′,5 | 36 |
| 3,3′,5 | 46 |
| 2,3′,6 | 32 |
| 2,2′,3,4′ | 57 |
| 2,3,3′,5′ | 68 |
| 2,2′,3,6′ | 47 |
| 2,2′,3,3′,5 | 93 |
| 2,2′,3,4′,5 | 85 |
| 2,2′,3,5,5′ | 83 |
| 2,2′,3,5,6′ | 68 |
| 2,3,3′,4′,5 | 121 |
| 2,2′,4,5,6′ | 70.5 |
| 2,3′,4,5,5′ | 104 |
| 2,2′,4,4′,6 | 65 |
| 2,2′,4,4′,6,6′ | 78 |
| 2,2′,3,3′,5,5′ | 138 |
| 2,2′,3,4′,5,5′ | 142 |
| 2,2′,3,3′,5,6′ | 116 |

*Chromatographic conditions are identical to those described in Figure 17.

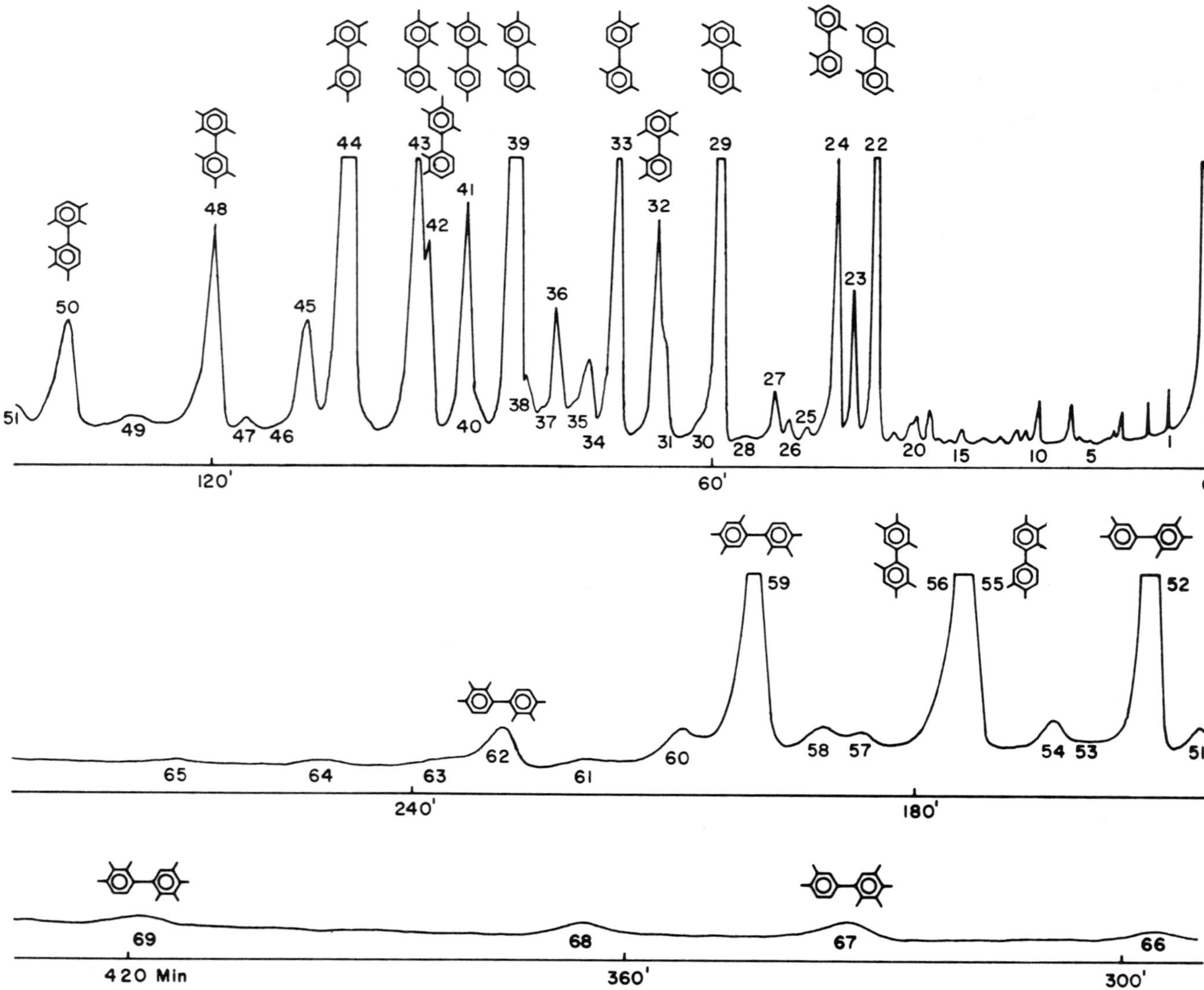


FIGURE 18. Capillary column chromatogram of Aroclor 1254. GC-conditions 50′ x 0.02″ Apiezon L SCOT column equipped with a flame ionization detector, the oven temperature was 205° and helium flow rate 2.5 ml/min. (From *J. Chromatogr.*, 60, 15, 1971. With permission.)

TABLE 19

**The Retention Indexes and Structures of the Major PCB Constituents in Aroclor 1254 as Determined by NMR**

| Peak no. | R.I. | NMR determined structure | Alternate predictions: Structure | Alternate predictions: R.I. |
|---|---|---|---|---|
| 22 | 1994 | 2,2′,5,5′ | 2,2′,3,5 | 1996 |
| 23 | 2010 | | 2,2′,4,5 | 2012 |
| | | | 2,2′,4,5′ | 2010 |
| 24 | 2022 | 2,2′,3,5′ | 2,3′,4,6 | 2021 |
| | | | 2,2′,4,4′ | 2027 |
| | | | 2,2′,3,6,6′ | 2017 |
| 29 | 2089 | 2,2′,3,5′,6 | 2,2′,3,5,6′ | 2092 |
| 32 | 2119 | 2,2′,3,3′,6 | | |
| 33 | 2136 | 2,3′,4′,5 | | |
| 36 | 2159 | | 2,2′,3,5,5′ | 2159 |
| 39 | 2175 | 2,2′,4,5,5′ | 2,2′,3,4′,5 | 2175 |
| | | | 2,3′,4,5′,6 | 2175 |
| | | | 2,2′,3,4,6,6′ | 2175 |
| | | | 2,2′,3,3′,6,6′ | 2172 |
| 41 | 2191 | 2,2′,4,4′,5 | 2,2′,3,3′,5 | 2189 |
| 42 | 2203 | 2,2′,3′,4,5 | | |
| 43 | 2207 | 2,2′,3,4,5′ | | |
| 44 | 2228 | 2,3,3′,4′,6 | 2,2′,3,4,4′ | 2226 |
| 45 | 2238 | | 2,2′,3,3′,4 | 2240 |
| | | | 2′,3,4,5,6′ | 2240 |
| | | | 3,3′,4,5′ | 2238 |
| | | | 2,2′,3,4′,5,6′ | 2240 |
| 48 | 2264 | 2,2′,3′,4,5,6′ | 2,2′,3,4,4′,6 | 2263 |
| 50 | 2299 | 2,2′,3,3′,4,6′ | | |
| 52 | 2321 | 2,3′,4,4′,5 | | |
| 55 | 2356 | 2,3,3′,4,4′ | | |
| 56 | 2356 | 2,2′,4,4′,5,5′ | 2,2′,3,4,5,5′ | 2360 |
| | | | 2,3,3′,4,5′,6 | 2357 |
| 59 | 2390 | 2,2′,3,4,4′,5′ | 2,2′,3,3′,4,5 | 2391 |
| | | | 2,3′,4,4′,5′,6 | 2388 |
| | | | 2,2′,3,3′,5,5′,6 | 2390 |

TABLE 20

**The Retention Indexes and Predicted Structures of the Minor Peaks in Aroclor 1254**

| Peak No. | R.I. | Structure | Predicted R.I. |
|---|---|---|---|
| 1 | 1490 | Biphenyl | 1491[b] |
| 2 | 1579 | 2 | 1577[b] |
| 3 | 1672 | 2,2′ | 1669[b] |
| 4 | 1688 | 4 | 1687[b] |
| 5 | 1750 | 2,5′ | 1754 |
| 6 | 1758 | 2,4 | 1756[b] |
| 7 | 1763 | 2,3′ | 1770 |
| 8 | 1774 | 2,2′,6 | 1765 |
| 9 | 1782 | 2,4′ | 1782[b] |
| 10 | 1833 | 2,2′,5 | 1831 |
| 11 | 1849 | 2,2′,4 | 1848 |
| 12 | 1863 | 2,2′,3 | 1864 |
| | | 2,3′,6 | 1866 |
| 13 | 1879 | 2,4′,6 | 1878 |
| 14 | 1896 | 4,4′ | 1894[b] |
| 15 | 1922 | 2,2′,4,6 | 1924[b] |
| 16 | 1935 | 2′,3,5 | 1932 |
| | | 2,2′,3,6 | 1932 |
| | | 2,3′,5 | 1932 |
| 17 | 1943 | 2,4′,5 | 1944 |
| 18 | 1952 | 2,3′,4 | 1949 |
| 19 | 1963 | 2,3,3′ | 1963 |
| | | 2,4,4′ | 1962[b] |
| 20 | 1966 | 2,3,3′ | 1963 |
| | | 2,4,4′ | 1962[b] |
| 21 | 1980 | 2′,3,4 | 1975 |
| 25 | 2040 | 2,3,4′,6 | 2041 |
| | | 2,2′,3,4′ | 2041 |
| 26 | 2051 | 2,2′,3,4 | 2047 |
| | | 2,2′,3,3′ | 2055 |
| 27 | 2058 | 2,2′,3,3′ | 2055 |
| 28 | 2072 | 2,3′,4′,6 | 2071 |
| 30 | 2097 | 2,3,3′,5 | 2097 |
| | | 2,2′,4,4′,6 | 2097 |
| | | 2,3′,5,5′ | 2094 |
| | | 2,2′,3,5,6′ | 2092 |
| 31 | 2115 | 2,2′,3,3′,6 | 2120 |
| | | 2,2′,3′,4,6 | 2112 |
| 34 | 2146 | 2′,3,4,5 | 2144 |
| | | 2,3,3′,4 | 2148 |
| 35 | 2152 | 2,3,3′,4 | 2148 |
| | | 2,3′,4,4′ | 2154 |
| 37 | 2164 | 2,3,3′,4′ | 2168 |
| | | 2,2′,3,4′,6,6′ | 2160 |
| 38 | 2169 | 2,3,3′,4′ | 2168 |
| | | 2,2′,3,3′,6,6′ | 2173 |
| 40 | 2186 | 2,2′,3,3′,5 | 2183 |
| | | 2,3,3′,5′,6 | 2185 |
| 46 | 2246 | 2,3′,4,5 | 2245 |
| | | 2,2′,3,4′,5,6 | 2241 |
| | | 2,2′,3,4,5′,6 | 2247 |
| | | 2,2′,3,3′,5,6′ | 2248 |
| 47 | 2254 | 3,4,4′,5 | 2257 |
| | | 2,2′,3,3′,5,6 | 2256 |
| | | 2,3,3′,5,5′ | 2259 |
| | | 2,2′,4,4′,5,6′ | 2256 |
| 49 | 2283 | 2,2′,3,3′,4,6 | 2278 |
| | | 2,2′,3,4,5,6′ | 2283 |
| | | 3,3′,4,4′ | 2282[b] |
| 51[a] | | 2,2′,3,3′,5,6,6′ | 2303 |
| 51 | 2310 | 2′,3,4,5,5′ | 2307 |
| | | 2,3,3′,4,5′ | 2310 |
| 53[a] | | 2,2′,3,3′,4,6,6′ | 2329 |
| 53 | 2335 | 2,3,3′,4,5 | 2338 |
| 54 | 2340 | 2,3,3′,5,5′,6 | 2335 |
| | | 2,2′,3,4′,5,5′ | 2340[b] |
| 57 | 2372 | 2,2′,3,4,4′,5 | 2376 |
| | | 2,2′,3,3′,4,5′ | 2374 |
| 58 | 2376 | 2,2′,3,4,4′,5 | 2376 |
| | | 2,3,3′,4′,5,6 | 2379 |
| | | 2,2′,3,3′,4,5′ | 2374 |
| 60 | 2400 | 2,3,3′,4,4′,6 | 2401 |
| | | 2,3,3′,4′,5′,6 | 2396 |
| 61 | 2413 | 2,2′,3,3′,4,5′,6 | 2412 |
| 62 | 2425 | 2,2′,3,3′,4,4′ | NMR |
| 63 | 2433 | 2,2′,3,3′,4,5,6′ | 2439 |
| | | 2,2′,3,4,4′,5′,6 | 2428 |
| | | 2,2′,3,4,4′,5,6′ | 2430 |
| 64 | 2451 | 2,2′,3,3′,4′,5,6 | 2450 |
| 65 | 2466 | 2,2′,3,3′,4,4′,6 | 2462 |
| 66 | 2489 | 2,3′,4,4′,5,5′ | 2488[b] |
| 67 | 2519 | 2,3,3′,4,4′,5 | NMR |
| 68 | 2543 | 2,2′,3,4,4′,5,5′ | 2541 |
| | | 2,3,3′,4′,5,5′,6 | 2548 |
| 69 | 2577 | 2,2′,3,3′,4,4′,5 | NMR |

[a]Small peaks.
[b]R.I. of synthesized compound.

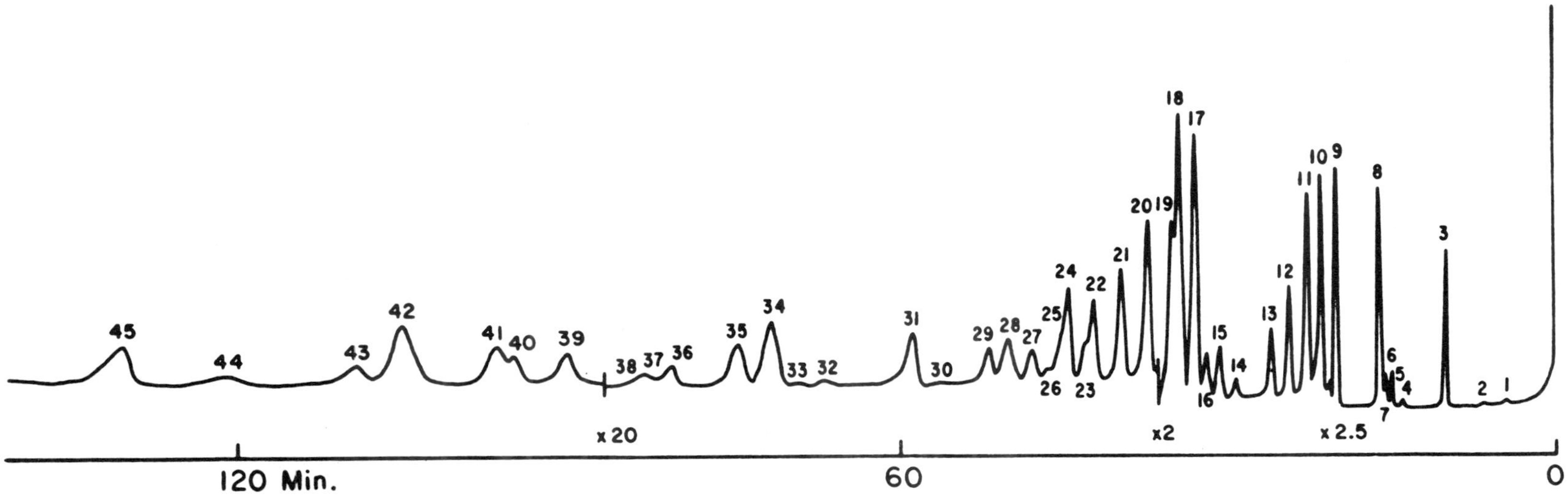


FIGURE 19. Capillary column chromatogram of Aroclor 1242. GC-conditions 50′ x 0.02″ Apiezon L SCOT column equipped with a flame ionization detector, the oven temperature was 205° and helium flow rate 2.5 ml/min. (From *J. Chromatogr.*, 60, 15, 1971. With permission.)

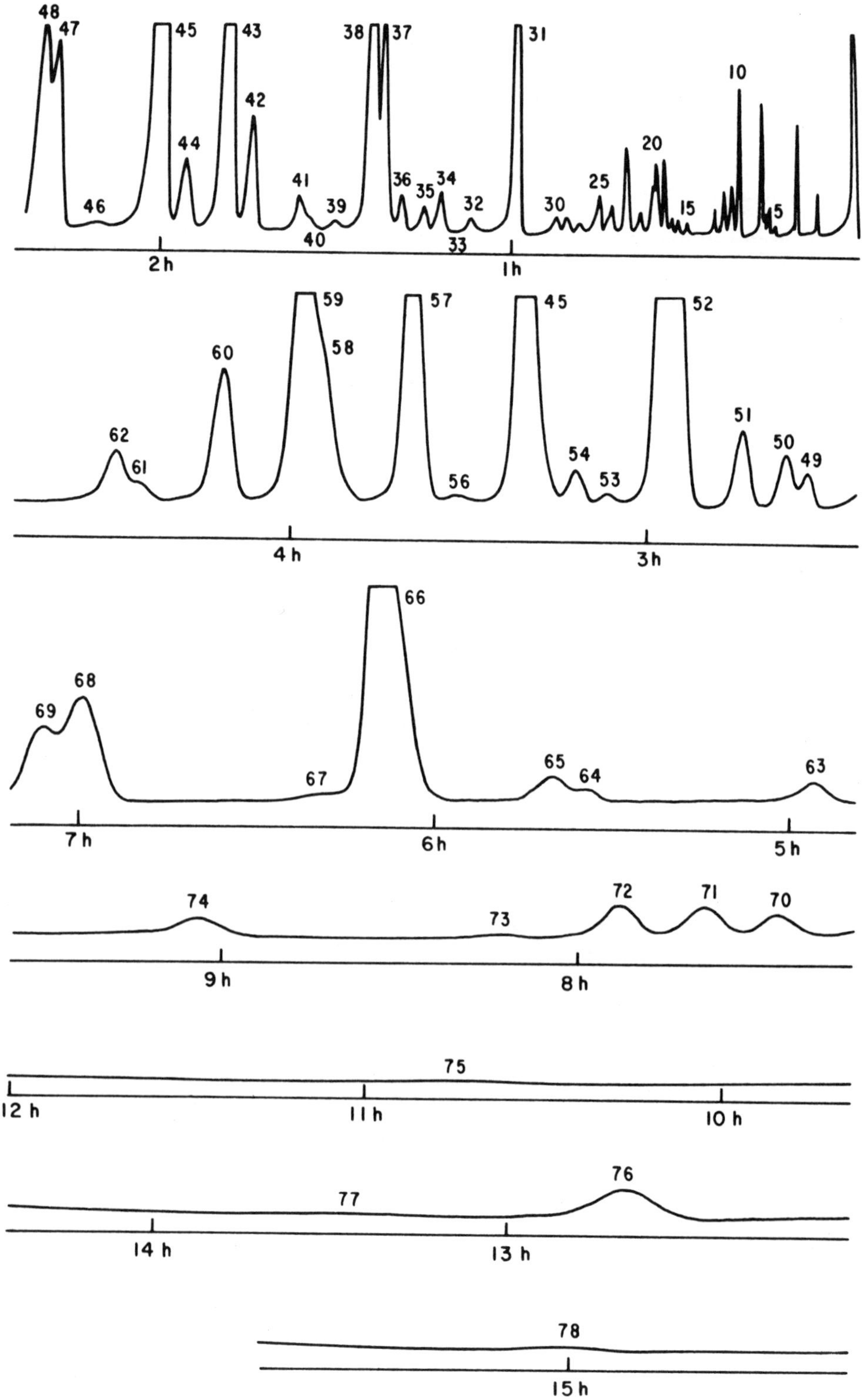


FIGURE 20. Capillary column chromatogram of Aroclor 1260. GC-conditions 50′ x 0.02″ Apiezon L SCOT column equipped with a flame ionization detector, the oven temperature was 205° and helium flow rate 2.5 ml/min. (From *J. Chromatogr.*, 60, 15, 1971. With permission.) 33

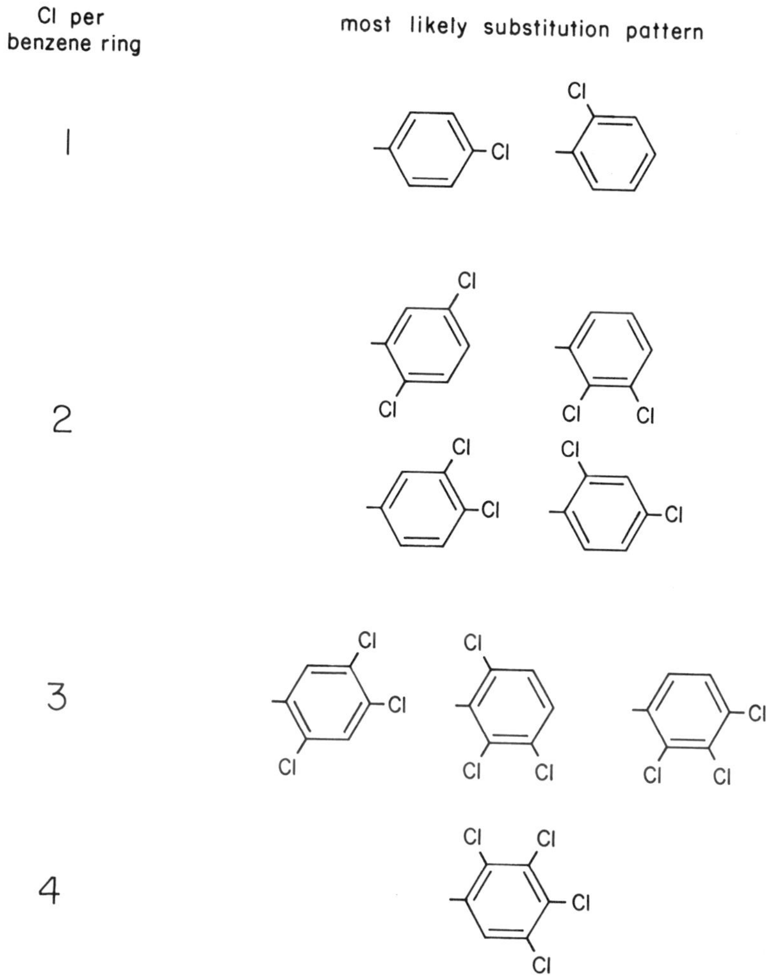


FIGURE 21. The most common substitution patterns for the chlorobiphenyls found in PCB preparations. Only one phenyl-ring is shown. The most abundant tetrachlorobiphenyls, for example, are those from the dichlorophenyl-moieties shown. One di- and one trichlorophenyl- would give most abundant penta-chlorobiphenyls etc.

TABLE 21

**The Retention Indexes and Predicted Structures of Aroclor 1242 Peaks**

| Peak no. | R.I. | Structure | Predicted R.I. | Peak no. | R.I. | Structure | Predicted R.I. |
|---|---|---|---|---|---|---|---|
| 1 | 1490 | Biphenyl | 1491[b] | 26 | 2037 | 2,3,4',6 | 2041 |
| 2 | 1580 | 2 | 1577[b] | 27 | 2041 | 2,2',3,4' | 2041 |
| 3 | 1672 | 2,2' | 1669[b] | 28 | 2052 | 2,2',3,4 | 2047 |
| 4 | 1750 | 2,5 | 1744 | | | 2,2',3,3' | 2055 |
| 5 | 1758 | 2,4 | 1756[b] | 29 | 2061 | | |
| 6 | 1765 | 2,3' | 1770 | 30 | 2080 | 3,3',4 | 2076 |
| 7 | 1774 | 2,2',6 | 1765 | 31 | 2092 | 3,4,4' | 2088 |
| 8 | 1781 | 2,4' | 1782[b] | | | 2,3',5,5' | 2094 |
| 9 | 1833 | 2,2',5 | 1831 | | | 2,2',4,4',6 | 2097 |
| 10 | 1849 | 2,2',4 | 1848 | | | 2,2',3,5',6[a] | 2089 |
| 11 | 1863 | 2,2',3 | 1864 | 32 | 2121 | 2,3,3',5' | 2125 |
| | | 2,3',6 | 1866 | | | 2,4,4',5 | 2125 |
| 12 | 1879 | 2,4',6 | 1878 | 33 | 2129 | 2,3,3',5' | 2125 |
| 13 | 1896 | 4,4' | 1894[b] | 34 | 2138 | 2,3',4',5 | 2138 |
| 14 | 1923 | 2,2',4,6 | 1924[b] | 35 | 2148 | 2',3,4,5 | 2144 |
| | | 2,2',5,6' | 1927 | | | 2,3,3',4 | 2148 |
| 15 | 1935 | 2',3,5 | 1932 | 36 | 2167 | 2,3,4,4' | 2160 |
| | | 2,3',5 | 1932 | 37 | 2174 | 2,3,3',4' | 2168 |
| 16 | 1944 | 2,4',5 | 1944 | 38 | 2178 | 2,2',3,4',5 | 2175 |
| 17 | 1951 | 2,3',4 | 1949 | | | 2,2',4,5,5'[a] | 2175[b] |
| 18 | 1961 | 2,3,3' | 1963 | | | 2,2',3,4,6,6' | 2175 |
| | | 2,4,4' | 1962[b] | | | 2,3,3',5',6 | 2183 |
| 19 | 1966 | 2,3,3' | 1963 | | | 2,3',4,5',6 | 2175 |
| | | 2,4,4' | 1962 | 39 | 2195 | 2,2',4,4',5[a] | 2191 |
| 20 | 1980 | 2',3,4 | 1975 | 40 | 2207 | 2,2',3',4,5 | 2206[b] |
| 21 | 1996 | 2,2',3,5 | 1996 | | | 2,2',3,4,5[a] | 2210 |
| | | 2,2',5,5'[a] | 1994[b] | 41 | 2212 | 2,2',3,4,5' | 2210 |
| 22 | 2011 | 2,2',4,5 | 2012 | 42 | 2232 | 2,2',3,4,4' | 2226 |
| | | 2,2',4,5' | 2010 | | | 2,3,3',4',6[a] | 2228 |
| 23 | 2015 | 2,2',4,5 | 2012 | 43 | 2242 | 2,2',3,4',5,6 | 2241 |
| 24 | 2024 | 2,3,3',6 | 2029 | | | 2,2',3,4,5',6 | 2247 |
| | | 2,3',4,6 | 2021 | | | 2,2',3,4',5,6' | 2240 |
| 25 | 2028 | 2,3',4,6 | 2029 | 44 | 2268 | 2,2',3,4,4',6 | 2263 |
| | | 2,2',3,5' | 2025 | | | 2,2',3,4',5',6[a] | 2264[b] |
| | | 2,2',4,4' | 2027[b] | 45 | 2287 | 2,2',3,4,5,6' | 2283 |
| | | 2,3',5',6 | 2028 | | | 3,3',4,4' | 2282[b] |
| | | 2,4,4',6 | 2033[b] | | | 2,2',3,4,4',6' | 2291 |

[a]Structures found in Aroclor 1254 by NMR determinations.

[b]R.I. of synthesized compound or NMR standard.

TABLE 22

**The Retention Indexes and Predicted Structures of Aroclor 1260 Peaks.***

| Peak no. | R.I. | Structure | Predicted R.I. |
|---|---|---|---|
| 22 | 1994 | 2,2′,3,5 | 1996 |
| | | 2,2′,5,5′ | 1994[c] |
| 23 | 2009 | 2,2′,4,5 | 2012 |
| | | 2,2′,4,5′ | 2010 |
| 24 | 2013 | 2,2′,4,5 | 2012 |
| | | 2,2′,4,5′ | 2010 |
| 25 | 2022 | 2,3′,4,6 | 2021 |
| | | 2,2′,3,5′ | 2025 |
| | | 2,2′,4,4′ | 2027[c] |
| 27 | 2030 | 3,3′,5 | 2033 |
| | | 2,3,3′,6 | 2029 |
| | | 2,4,4′,6 | 2033[c] |
| | | 2,2′,4,4′ | 2027[c] |
| | | 2,3′,5′,6 | 2028 |
| 28 | 2040 | 2,3,4′,6 | 2041 |
| | | 2,2′,3,4′ | 2041 |
| 29 | 2050 | 2,2′,3,4 | 2047 |
| | | 2,2′,3,3′ | 2055 |
| 30 | 2058 | 2,2′,3,3′ | 2055 |
| 31 | 2086 | 2,2′,3,5′,6[a] | 2089 |
| | | 2,2′,4,5′,6 | 2081 |
| 32 | 2117 | 2,3′,4,5 | 2114 |
| | | 2,2′,3,3′,6 | 2114 |
| 33 | 2125 | 2,4,4′,5 | 2125 |
| | | 2,3,3′,5 | 2125 |
| 34 | 2135 | 2,3′,4′,5[a] | 2138 |
| 35 | 2146 | 2′,3,4,5 | 2144 |
| | | 2,3,3′,4 | 2148 |
| 36 | 2158 | 2,2′,3,5,5′ | 2159 |
| 37 | 2168 | 2,2′,3,3′,6,6′ | 2173 |
| 38 | 2174 | 2,2′,3,4′,5 | 2175 |
| | | 2,2′,4,5,5′ | 2175[c] |
| | | 2,3′,4,5′,6 | 2175 |
| 39 | 2192 | 2,2′,3,3′,5 | 2189 |
| | | 2,2′,4,4′,5[a] | 2191 |
| 40 | 2205 | 2,2′,3′,4,5 | 2206[c] |
| 41 | 2209 | 2,2′,3,4,5′[a] | 2210 |
| 42 | 2229 | 2,2′,3,4,4′ | 2226 |
| | | 2,3,3′,4′,6 | 2228 |
| 43 | 2239 | 2,2′,3,4′,5,6 | 2241 |
| | | 2,2′,3,4′,5,6′ | 2240 |
| 44 | 2254 | 2,2′,3,3′,5,6 | 2256 |
| | | 2,2′,4,4′,5,6′ | 2256 |
| 45 | 2264 | 2,2′,3,4,4′,6 | 2263 |
| | | 2,2′,3,4′,5′,6 | 2264 |
| 46 | 2283 | 2,2′,3,3′,4,6 | 2278 |
| | | 2,2′,3,4,5,6′ | 2283 |
| | | 3,3′,4,4′ | 2282[c] |
| 47 | 2296 | 2,2′,3,3′,4,6[a] | 2299 |
| 48 | 2300 | 2,2′,3,3′,5,6,6′ | 2303 |
| | | 2,2′,3,4′,5,6,6′ | 2295 |
| 49 | 2321 | 2′,3,4,4′,5 | 2323 |
| | | 2,3,4,4′,5[a] | 2319 |
| 50 | 2326 | 2,2′,3,3′,4,6,6′ | 2329 |
| | | 2,2′,3,4,4′,6,6′ | 2321 |
| 51 | 2340 | 2,3,3′,5,5′,6 | 2335 |
| | | 2,2′,3,4′,5,5′ | 2340[c] |

TABLE 22 (Continued)

**The Retention Indexes and Predicted Structures of Aroclor 1260 Peaks.***

| Peak no. | R.I. | Structure | Predicted R.I. |
|---|---|---|---|
| 52 | | 2,2′,3,4,5,5′ | 2360 |
| | | 2,3,3′,4,5′,6 | 2357 |
| | | 2,2′,4,4′,5,5′[a] | 2356[c] |
| 53 | 2372 | 2,2′,3,4,4′,5 | 2376 |
| | | 2,2′,3,3′,4′,5 | 2374 |
| 54 | 2379 | 2,2′,3,4,4′,5 | 2376 |
| | | 2,3,3′,4′,5,6 | 2379 |
| | | 2,2′,3,3′,4,5′ | 2374 |
| 55 | 2390 | 2,2′,3,3′,4,5 | 2391 |
| | | 2,2′,3,4,4′,5′[a] | 2391[c] |
| | | 2,3′,4,4′,5′,6 | 2388 |
| | | 2,2′,3,3′,5,5′,6 | 2390 |
| 56 | 2402 | 2,3,3′,4,4′,6 | 2401 |
| 60 | 2445 | 2,2′,3,3′,4′,5,6 | 2440 |
| | | 2,2′,3,3′,4,5,6′ | 2439 |
| 61 | 2460 | 2,2′,3,3′,4,4′,6 | 2462 |
| 62 | 2464 | 2,2′,3,3′,4,5′,6,6′ | 2464 |
| 63 | 2486 | 2,2′,3,3′,4,5,6,6′ | 2489 |
| | | 2,2′,3,4,4′,5,6,6′ | 2481 |
| | | 2,2′,3,3′,4,4′,6,6′ | 2490[c] |
| | | 2,3′,4,4′,5,5′ | 2488[c] |
| 64 | 2518 | 2,3,3′,4,4′,5 | 2516 |
| 65 | 2522 | 2,2′,3,3′,4,5,5′ | 2525 |
| 66 | 2542 | 2,2′,3,4,4′,5,5′ | 2541 |
| 67 | 2550 | 2,3,3′,4′,5,5′,6 | 2548 |
| 68 | 2575 | 2,2′,3,3′,4,4′,5[a] | 2575 |
| | | 2,3,3′,4,4′,5′,6 | 2570 |
| 69 | 2578 | 2,2′,3,3′,4,5,5′,6 | 2577 |
| | | 2,2′,3,3′,4,5,5′,6′ | 2581 |
| 70 | 2587 | 2,2′,3,3′,4,5,5′,6′ | 2581 |
| 71 | 2597 | 2,2′,3,4,4′,5,5′,6 | 2593 |
| 72 | 2605 | 2,2′,3,3′,4,4′,5,6′ | 2603 |
| 73 | 2618 | 2,2′,3,3′,4,5,5′,6,6′ | 2624 |
| 74 | 2640 | 2,2′,3,3′,4,4′,5,6 | 2627 |
| b | | 2,2′,3,3′,4,4′,5,6,6′ | 2650 |
| 75 | 2682 | 2,3,3′,4,4′,5,5′ | 2685 |
| 76 | 2724 | 2,2′,3,3′,4,4′,5,5′ | 2726 |
| 77 | 2738 | 2,3,3′,4,4′,5,5′,6 | 2738 |
| 78 | 2770 | 2,2′,3,3′,4,4′,5,5′,6 | 2768 |
| b | | 2,2′,3,3′,4,4′,5,5′,6,6′ | 2810[c] |

*Peaks 2 to 21 in Aroclor 1254 chromatogram were also present in Aroclor 1260.
[a]Structures found in Aroclor 1254 by NMR determinations.
[b]Peaks observed on mass spectrometer oscilloscope, but not on GLC.
[c]R.I. of synthesized compound.

# REFERENCES


1. **Armour, J. A.,** *J. Chromatogr.,* 72, 275, 1972.
2. **Bagley, G. E., Reichel, W. L., and Cromartie, E.,** *J. Assoc. Offic. Anal. Chem.,* 53, 251, 1970.
3. Bayer Co., *Clophen,* 1968.
4. **Beaven, G. H., De La Mare, P.B.D., Hassan, M., Johnson, E. A., and Klassen, N. V.,** *J. Chem. Soc.,* 2749, 1961.
5. **Beraha, L. and Powell, D.,** *Phytopathology,* 43, 383, 1953.
6. **Biros, F. J., Walker, A. C., and Medbury, A.,** *Bull. Environ. Contam. Toxicol.,* 5, 317, 1970.
7. **Broadhurst, M. G.,** *Environ. Health Perspect.,* 1, 81, 1972.
8. **Cook, J. W.,** *Environ. Health Perspect.,* 1, 3, 1972.
9. **Deonier, C. C., Jones, H. A., and Incho, H. H.,** *J. Econ. Entomol.,* 39, 459, 1946.
10. **Eichelberger, J.,** *PCB Newsletter,* No. 3, July 1971.
11. **Fishbein, L.,** *J. Chromatogr.,* 68, 345, 1972.
12. **Freed, V. H.,** Personal communication (manuscript), 1971.
13. **Fuhremann, T. W. and Lichtenstein, E. P.,** *Toxicol. Appl. Pharmacol.,* 22, 628, 1972.
14. **Hattula, M. L. and Karlog, O.,** *Dan. Tidsskr. Farm.,* 45, 259, 1971.
15. **Helminen, M.,** PCB Conference, National Environmental Protection Board, Stockholm, 1970.
16. **Hoover, T. B.,** *PCB Newsletter,* No. 3, July 1971.
17. **Hornstein, I. and Sullivan, W. N.,** *J. Econ. Entomol.,* 46, 937, 1953.
18. **Hubbard, H. L.,** Chlorinated biphenyls and related compounds, in *Kirk-Ohmers Encyclopedia of Chemical Technology,* Vol. 5, 2nd ed., Interscience, New York 1964, 289.
19. **Hutzinger, O., Safe, S., and Zitko, V.,** unpublished observations.
20. **Isono, N.,** *Jishu Koza,* 1(1), 60, 1970; 1(4), 58, 1970. *Critical News,* No. 1, 1971.
21. **Jenkins, R. L.,** U.S. Patent 1,892,397 and 1,892,398 (1932); *Chem. Abstr.,* 27, 1897, 1933.
22. **Johnson, A. N.,** U.S. Patent 2,742,508 (1956); *Chem. Abstr.,* 50, 16848, 1956.
23. **Lichtenstein, E. P., Schulz, K. R., Fuhremann, T. W., and Liang, T. T.,** *J. Econ. Entomol.,* 62, 761, 1969.
24. **MacMullin, R. B.,** *Chem. Eng. Prog.,* 44, 183, 1948.
25. *Monsanto Tech. Bull.,* 0/PL-306A, Aroclor plasticizers; *Monsanto Tech. Bull.,* PL-321, Aroclor 1221, 1232, 1242 for polyvinyl/acetate – emulsion adhesives.
26. Monsanto, Presentation to the Interdepartmental Task Force on PCB's, Washington, D.C., 1972.
27. Monsanto, *Chem. Eng. News,* p. 15, Dec. 6, 1971.
28. **Murata, T., Takahashi, S., and Tachikawa, R.,** *Shimadzu Review,* 28, 29, 1971.
29. **Ohno, I., Suga, K., and Mogi, K.,** Japanese Patent 6264 (1957); *Chem. Abstr.,* 52, 11937, 1958.
30. Panel on Hazardous Trace Substances, Polychlorinated Biphenyls – environmental impact, *Environ. Res.,* 5, 249, 1972.
31. **Plapp, F. W., Jr.,** *Environ. Entomol.,* 1, 580, 1972.
32. **Racine, J.,** French Patent, 1,206,352, 1958; *Chem. Abstr.,* 56, 1807, 1962.
33. **Risbec, J.,** *Phytiatr. Phytopharm. Spec. Issue,* 513, 1954.
34. **Saeki, S., Tsutsui, A., Oguri, K., Yoshimura, H., and Hamana, M.,** *Fukuoka Igaku Zasshi,* 62, 20, 1971; *Chem. Abstr.,* 74, 146294, 1971; National Research Council of Canada, Technical Translation, Ottawa, 1972.
35. **Sissons, D. and Welti, D.,** *J. Chromatogr.,* 60, 15 1971.
36. **Stalling, D. L. and Huckins, J. N.,** *J. Assoc. Offic. Anal. Chem.,* 54, 801, 1971.
37. **Tas, A. C. and Kleipool, R. J. C.,** *Bull. Environ. Contam. Toxicol.,* 8, 32, 1972.
38. **Tas, A. C. and de Vos, R. H.,** *Environ. Sci. Technol.,* 5, 1216, 1971.
39. **Thruston, A.,** *PCB Newsletter,* No. 3, July 1971.
40. **Tsao, C. H., Sullivan, W. N., and Hornstein, I.,** *J. Econ. Entomol.,* 46, 882, 1953.
41. **Tsunoda, Y. and Muratu, A.,** Japanese Patent 6718 (1954); *Chem. Abstr.,* 50, 7867, 1956.
42. **Vollmann, H.,** Chlorierte Diphenyle and Polyphenyle in *Ullmanns Enzyklopädie der Technischen Chemie,* Vol. 5, Urban und Schwarzenberg, Berlin, 1956, 475.
43. **Vos, J. G., Koeman, J. H., van der Maas, H. L., Ten Noever de Brauw, M. C., and de Vos, R. H.,** *Food Cosmet. Toxicol.,* 8, 625, 1970.
44. **Wallnöfer, P. R., Königer, M., and Hutzinger O.,** *Analabs Research Notes,* in press.
45. **Webb, R. G. and McCall, A. C.,** *J. Assoc. Offic. Anal. Chem.,* 55, 746, 1972.
46. **Widmark, G.,** The OECD Study of Analysis of PCB, Report from the Institute of Analytical Chemistry, Stockholm, (manuscript), 1968.
47. **Willis, D. E. and Addison, R. F.,** *J. Fish. Res. Board. Can.,* 29, 592, 1972.
48. **Zal'kind, Y. S. and Belikova, M. V.,** *Zhurnal obshchei khimii,* 8, 1918, 1938; *Chem. Abstr.,* 33, 5833, 1939, National Research Council of Canada, Technical Translation, Ottawa, 1970.
49. **Zitko, V.,** *Bull. Environ. Contam. Toxicol.,* 5, 279, 1970.
50. **Zitko, V.,** *Fish. Res. Board. Can. Tech. Rep.,* No. 1083, 1970.

Chapter 3

# SYNTHESIS OF CHLOROBIPHENYLS

In this chapter, biphenyl syntheses are discussed which have been used for the preparation of chlorobiphenyls including labeled compounds.

The presentation throughout is from a preparative point of view. Mechanisms are usually not discussed and classification is strictly according to starting material (e.g., diazonium compounds, sulfonyl halides, etc.). Reactions with similar mechanisms can therefore be found in different subsections (e.g., diazoacetate reaction and nitroso acetylamine reaction).

The following symbols and abbreviations are used in the equations throughout the chapter: Δ(heat); ≡(equivalent); u(uniformly labeled); hν(light, u.v.); n(normal); t(tertiary); Ph(phenyl); Ac or Acet(acetyl, acetate); Bu or But(butyl); THF(tetrahydrofuran); DMSO(dimethyl sulfoxide); DMF(dimethylformamide).

## Methods for the Synthesis of Chlorobiphenyls

### *1. Formation of Chlorobiphenyls by Phenylation of Aromatic Substrates (Usually Involving Free Radical Mechanisms)*

These methods are well suited for the preparation of unsymmetrical chlorobiphenyls. Isolation and purification of products are greatly simplified when the substrate can be chosen to give only one isomer, e.g.,

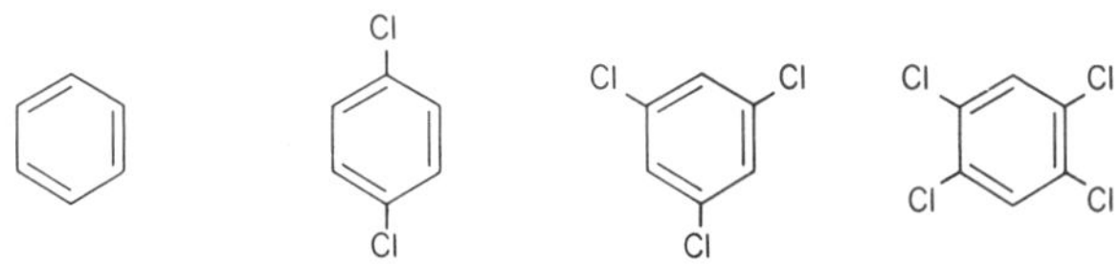

A number of free radical arylating systems are compared, particularly from a theoretical point of view in Reference 80. Mechanistic and preparative aspects of most of the homolytic aromatic arylations described in this section are discussed in References 7, 56, and 196.

#### *a. Arylation Via Aroyl Peroxides and Carboxylic Acids*

Aroyl peroxides have been popular sources of aryl radicals for mechanistic studies investigating substituent effects and isomer distribution in homolytic arylation reactions.

Decompositions of aroyl peroxides in appropriate substrates are also useful synthetic reactions in which high yields (e.g., 80%) of chlorobiphenyls have been obtained,[79] particularly in the presence of electron acceptors.[74,79,82]

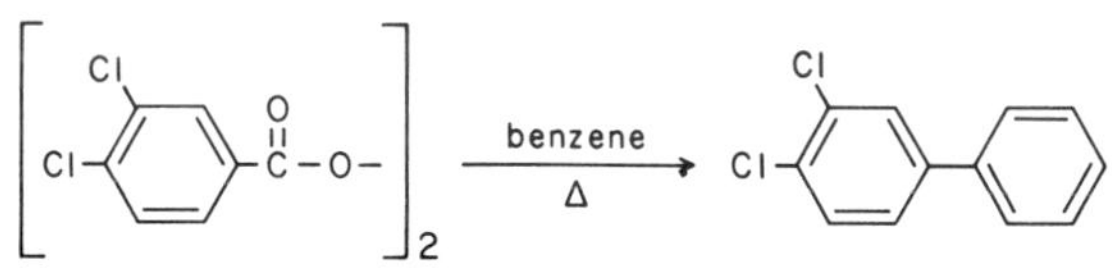

A number of simple chlorobiphenyls have been prepared by this reaction usually during the course of mechanistic studies.[6,46,59,79,82,102,147]

A method for the one-step homolytic decarboxylation of aromatic carboxylic acids has been described and this source of aryl radicals has been used for the preparation of several chlorobiphenyls.[173]

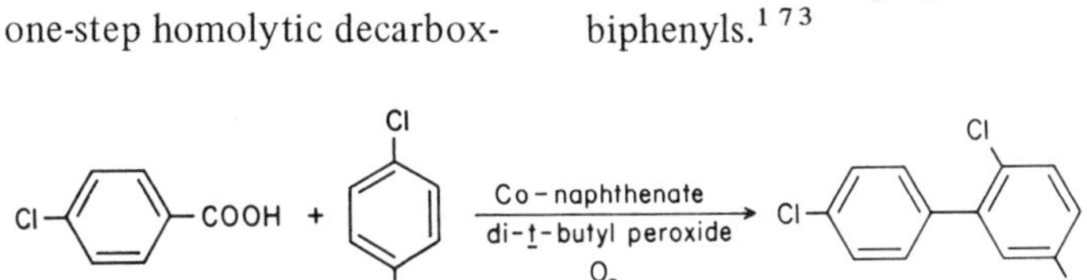

Mono- and dichlorobiphenyls have been prepared using new phenylating agents: iodosobenzene dibenzoate,[102] lead tetrabenzoate,[81,102] and silver iodide (bromide) dibenzoate.[28]

*b. Arylation Via Diazonium Salts (Gomberg-Bachmann, Meerwein and Related Reactions)*

This reaction in its various modifications is one of the most useful and widely used synthetic routes leading to unsymmetrical chlorobiphenyls. In its original version,[73] an aromatic amine is diazotized in the usual manner (for further information on diazotation see section on Sandmeyer reaction) and the diazonium salt solution mixed with the liquid aromatic substrate and sodium hydroxide (intermediate formation of diazonium hydroxides) or sodium acetate (intermediate formation of diazonium acetates). The literature on these reactions was thoroughly reviewed in 1944.[9] Recently, these reactions were used for the preparation of chlorobiphenyls.[95,191] Steric[191] and electronic[190] effects in the formation of chlorobiphenyls by these reactions have been investigated as have been isomer distributions generally.[6]

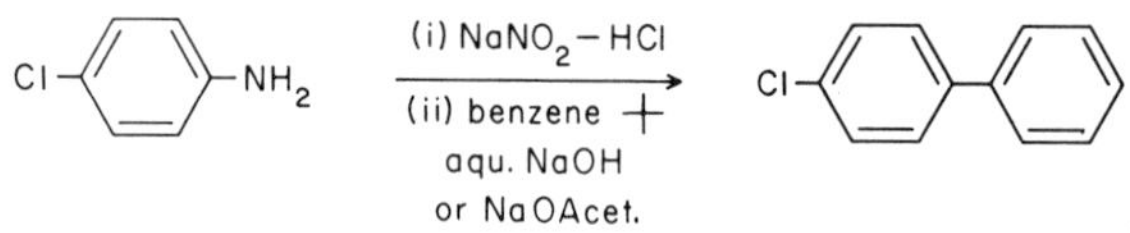

Contrary to the heterogeneous conditions prevailing in the reactions described above, decomposition of phenyldiazonium salts by cupric chloride (sometimes referred to as Meerwein reaction) is carried out in homogeneous solution (aqueous acetone[57,58]). A number of biphenyls including 2,4-dichlorobiphenyl (29% yield) were prepared by this procedure.

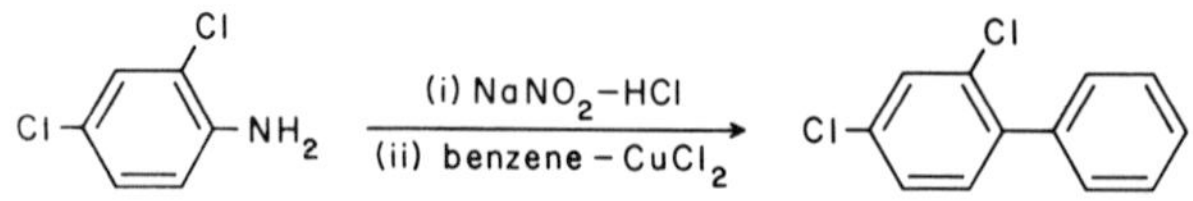

Solid diazonium salts[159] such as aryldiazonium chloride or zinc double salts have been used for arylations of this type.[9] In the last few years, the stable diazonium fluoroborates have been introduced for this purpose. Phenyldiazonium tetrafluoroborate, when heated with one equivalent of pyridine, provides a convenient source of aryl radicals. 4-Chlorobiphenyl[1] and a number of other substituted biphenyls[1,2] have been prepared by this method.

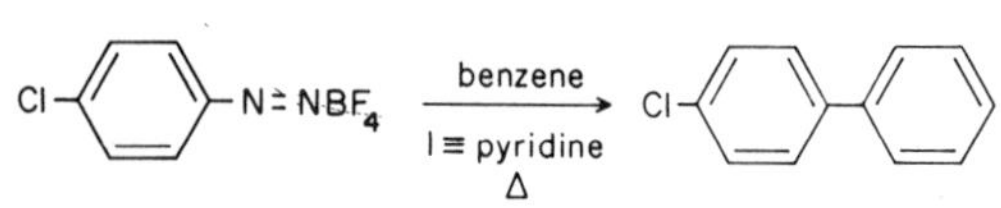

Sulfolan was found to be a convenient solvent to effect homogenous conditions. A number of biphenyls including 4-chlorobiphenyl have been prepared by decomposing phenyldiazonium tetrafluoroborates with copper and water or hydrated salts in the presence of aromatic substrate.[33]

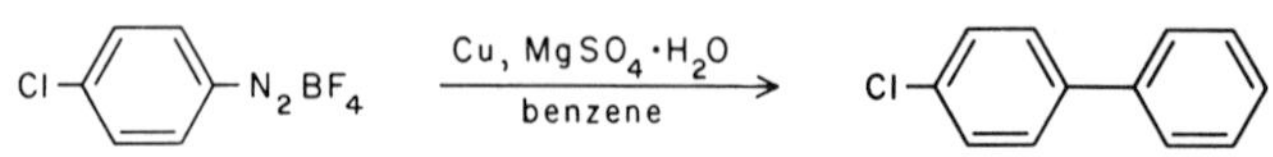

The isomeric monochlorobiphenyls were prepared by the decomposition of benzenediazonium tetrafluoroborate in a solution of chlorobenzene in acetone or dimethylsulfoxide[109] or chlorobenzene in dimethylsulfoxide in the presence of sodium nitrite.[107] Under these latter conditions, tetrafluoroborates decompose instantaneously with evolution of nitrogen. Preparations of tetrafluoroborates are described in Reference 155.

In a very convenient one-step synthesis, an aromatic amine, an organic nitrite (e.g., pentyl nitrite; the diazotizing agent), and excess substrate are heated to give good yields (typically: 50%) of biphenyls.[30,168] A number of biphenyls substituted with four and five chlorine atoms have been prepared by this method[156] as have been a large number of chlorobiphenyls including highly chlorinated compounds.[24]

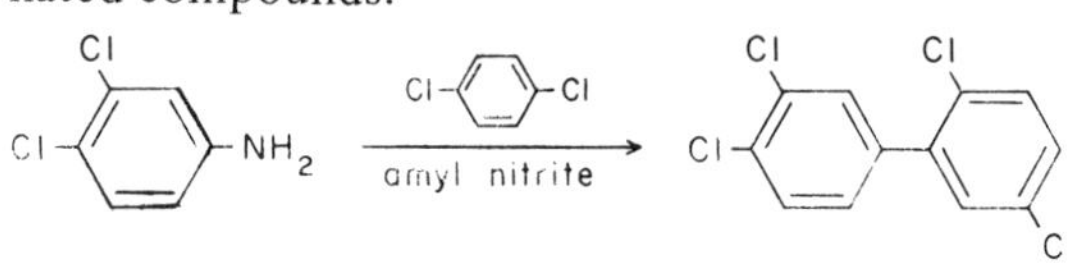

Deaminated products and (in the presence of carbon tetrachloride as solvent) formation of aryl chlorides are possible side reactions.[168]

*c. Arylation Via Acylarylnitrosamines*

Acylarylnitrosamines can conveniently be prepared from acylamines such as chloroacetanilides and nitrosylsulfuric acid. The decomposition of these compounds in hot aryl substrate gives good yield of chlorobiphenyls.[3,134]

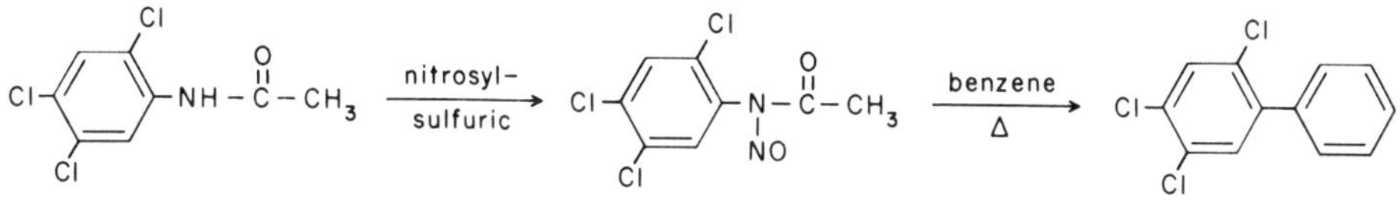


A similar reaction involving the treatment of acylamines with alkylnitrite-trifluoroacetic anhydride also yields biphenyls but the intermediate nitroso compounds were not isolated.[148] In a patent, the preparation of biaryls including 4-chlorobiphenyl from acetanilides is described.[124] The agents used for nitrosation were amyl nitrite, $N_2O_4$, $NO_3$, and NOCl. Part of a review by Bachmann and Hoffman[9] deals with the preparation of biphenyls via acylarylnitrosamines.

*d. Arylation Via Diazoaminobenzenes (Aryltriazenes)*

Diazoaminobenzenes, which can be prepared from diazo compounds and amines decompose in the presence of aryl substrate and an acid catalyst or aluminum chloride to give biphenyls.[78,199,200]

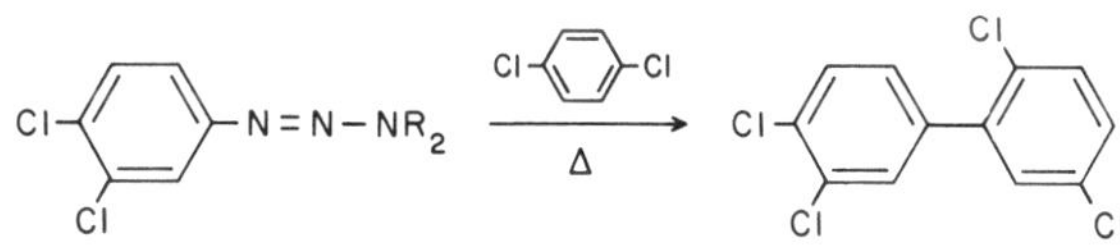


*e. Arylation Via Phenylhydrazines*

Oxidation of phenylhydrazines with a silver oxide and a variety of other oxidizing agents in the presence of aryl substrate leads to the formation of biphenyls.[77]

*f. Arylation Via Arylsulfonyl Halides*

Metal salt-catalyzed thermal decomposition of aryl sulfonyl halides in suitable substrates provide a useful method for the preparation of chlorobiphenyls which seems to have received little attention.[12,118]

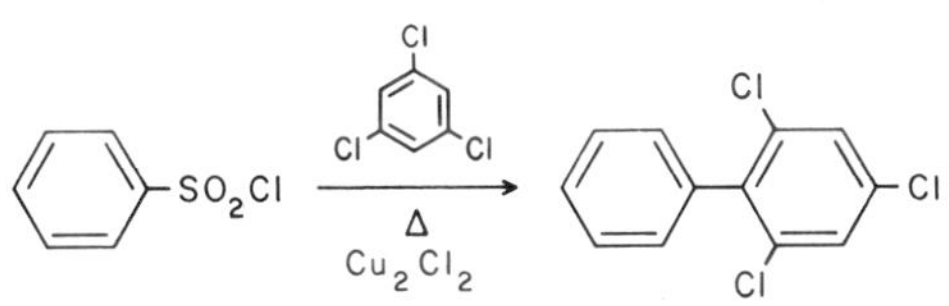


*g. Arylation by Photolysis of Arylthallium Ditrifluoroacetates*

Unsymmetrical biaryl can be prepared in good yield by photolysis of the readily accessible arylthallium ditrifluoroacetates.[181] Solvent is the substrate to be arylated, the wavelength is 300 nm, and a typical reaction time is 18 hr.

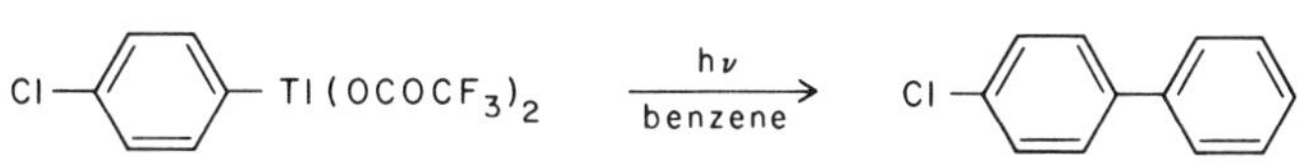


*h. Arylation Via Phenylazotriphenylmethane and Phenylazo-p-Tolylsulfone*

Phenylazotriphenylmethane when decomposed in chlorobenzene gives a mixture of chlorobiphenyl isomers.[89,108]

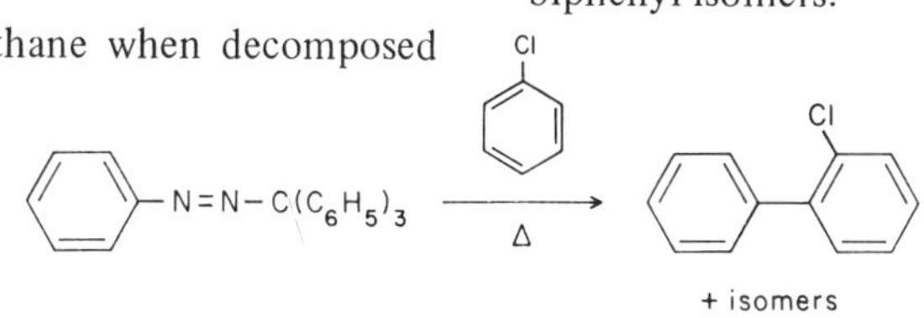

Phenylazo-*p*-tolysulfone can be used similarly.[106] However, from the distribution of isomers, a phenyl cation mechanism and not phenyl radicals was suggested.

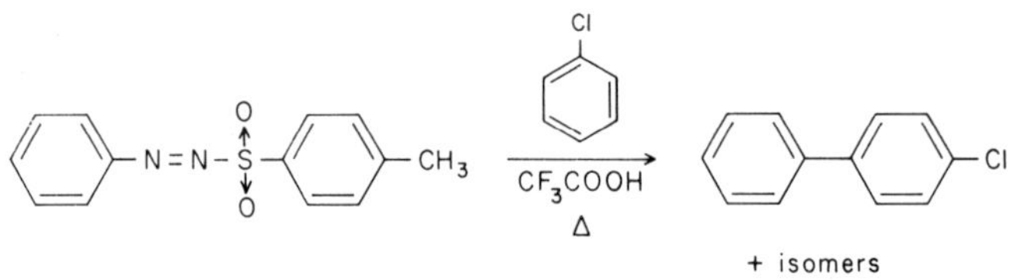

*2. Formation of Chlorobiphenyls by Other Aryl-condensation Reactions*

***a. Copper Mediated Condensations of Aryl Halides (Ullmann and Related Reactions)***

The condensation of two molecules of aromatic halides in the presence of finely divided copper is one of the most useful methods for the synthesis of biaryls. It is by far the most widely used preparative procedure leading to symmetrical biphenyls (including chlorobiphenyls) since its discovery by Ullmann in 1904.[183]

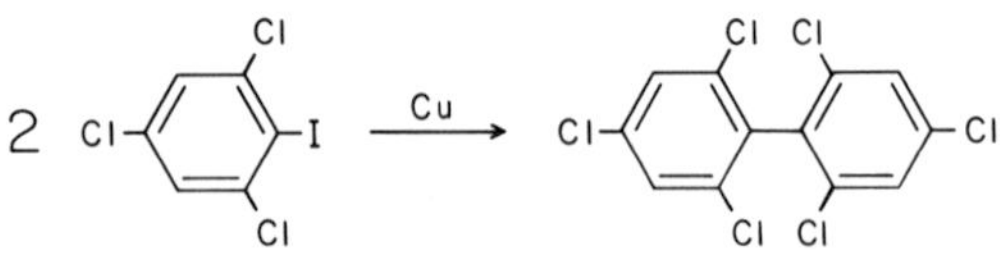

The reactivity of halides in this reaction is I > Br > Cl. Aromatic chloro compounds react only if activated by one or more nitro groups, ortho or para to the chlorine atom. Aromatic iodo compounds, which can readily be prepared from amines via diazonium salts, are usually the compounds of choice. Copper(I)iodide in the mixture can speed the reaction with arylbromides probably by halogen-halogen exchange.[20] Several equivalents of copper are usually mixed or successively introduced in the mixture with the halide and heated for several hours, commenced usually at 200° or higher. The quality of the finely divided copper is often critical in determining yields.[63] Improved yields have been reported using copper activated by treatment with complexing agents such as EDTA.[117] In the authors' laboratory, activation of commercial copper bronze with iodine, acetone, and hydrochloric acid[69,104] was found satisfactory. Improved yields were reported for a number of Ullmann reactions when the aryl halide and copper were diluted with solvents[63] (e.g., dimethylformamide[1,111] or sulfolan[1,92]) and reactions carried out at lower temperatures and prolonged times. The Ullmann reaction can also be used for the preparation of unsymmetrical biaryls. Generally, a mixture of two different aryl halides is used and three products (two symmetrical and one unsymmetrical) are obtained which may be difficult to separate.

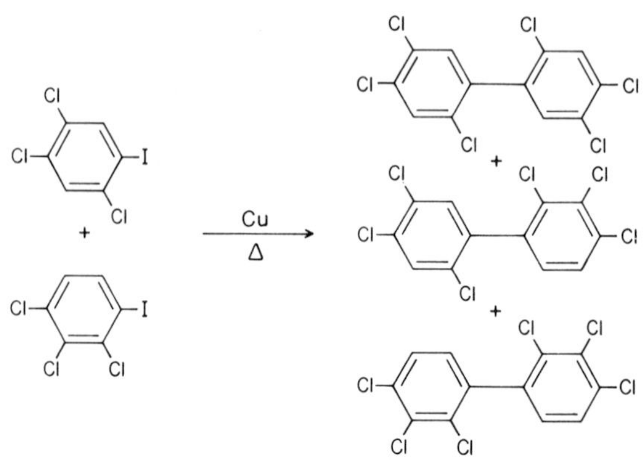

When one bromo- and one iodo-component are used, yields of unsymmetrical product can sometimes be increased.[63] All aspects of the Ullmann reaction were thoroughly reviewed in 1946,[62] 1964,[63] and in a recent article.[64] A review on copper promoted reactions deals also with Ullmann reactions, particularly mechanistic aspects.[10] The preparation of a number of symmetrical and unsymmetrical polychlorobiphenyls by the Ullmann reaction has recently been reported.[24,95,178-180,188]

In reactions related to the Ullmann biaryl

synthesis phenyl copper derivatives[11,37,144] and 2-furyl copper[184] couple with organic halides (*cf.* References 35, 55, and 141).

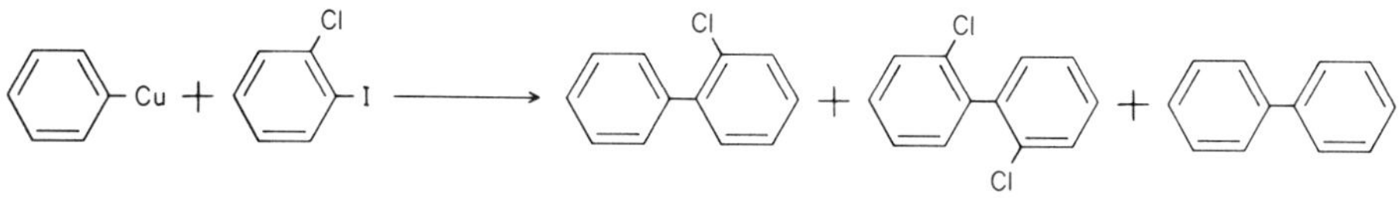

This reaction, which can give yields up to 90% with certain compounds,[35] seems particularly suited for the preparation of unsymmetrical biphenyls. Other biphenyl synthesis in which organo-copper intermediates are postulated are the decarboxylation of nitroaromatic acids in the presence of copper (I) oxide and iodobenzene,[21,140] and the coupling of nitrobenzenes with iodobenzene in the presence of copper (I) oxide.[20,21] Another biaryl synthesis involving organo-copper intermediates is described in 2g.

*b. Condensation Via Grignard Reagents*

2,2′4,4′-Tetrachlorobiphenyl was obtained by the action of cupric chloride on 2,4-dichlorophenylmagnesium iodide.[75]

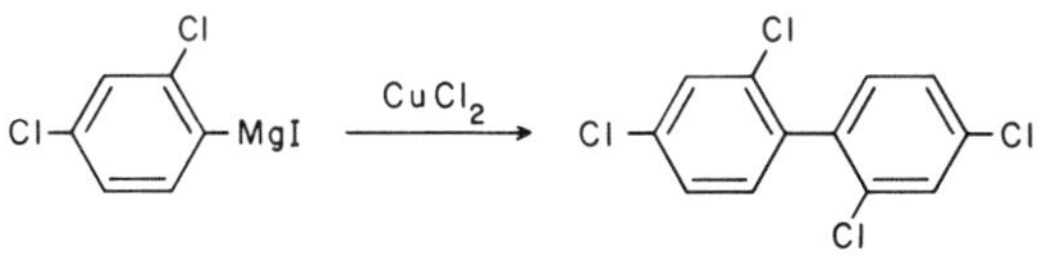

In a similar reaction, 4,4′-dichlorobiphenyl was obtained from the corresponding Grignard reagent and ethyl trichloroacetate.[101] Recently, it was shown that high yields of chlorobiphenyls can be obtained when thallium (I) bromide is used as the coupling reagent.[125,126] Ortho substituted biaryls cannot be prepared by this procedure.

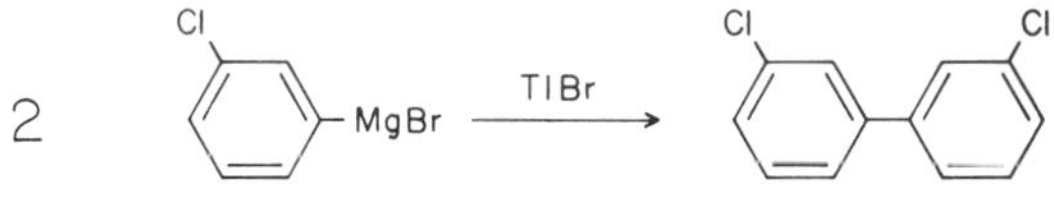

Grignard reagents have been used also for the synthesis of biphenyls not containing chlorine. Usually, metal salts are the coupling reagents[47,164] and good yields of biphenyls were reported by the action of diphenoquinones on Grignard reagents.[154] The coupling of some Grignard reagents with cobaltous chloride and transition metals generally to give biaryls (no chlorobiphenyls) is discussed in Reference 135 and 182a. Older examples of the preparation of biphenyls via Grignard reagents are discussed in Reference 103. Grignard reagents are also used in a new synthesis for unsymmetrical biphenyls (2g).

*c. Condensation Via Organolithium Intermediates*

The preparation of 4,4′-dichlorobiphenyl by the action of *n*-butyllithium on *p*-chlorobromobenzene in tetrahydrofuran in a yield of less than 20% is described in the literature.[20]

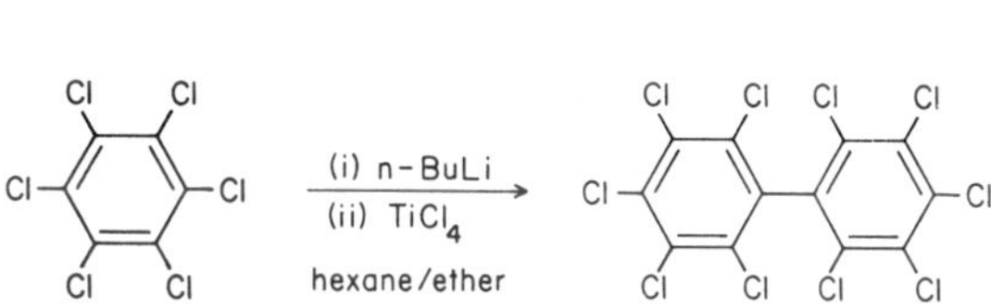

Using a similar procedure, decachlorobiphenyl was obtained from hexachlorobenzene *n*-butyllithium and titanium tetrachloride[42] in small yield.

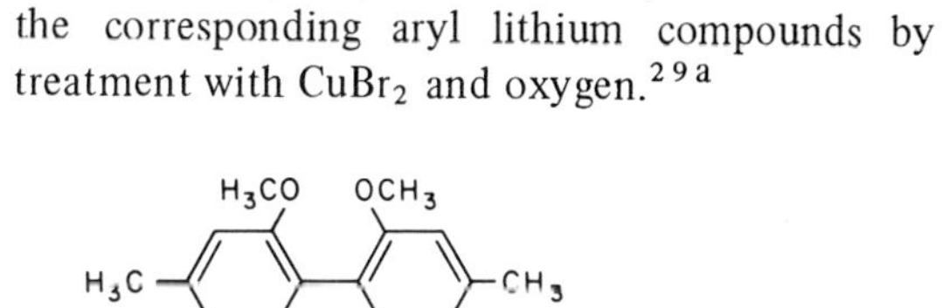

Methoxybiphenyls were recently obtained from the corresponding aryl lithium compounds by treatment with $CuBr_2$ and oxygen.[29a]

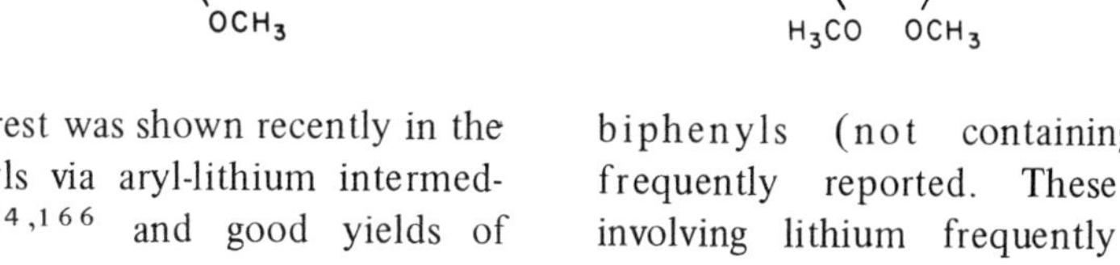

Considerable interest was shown recently in the preparation of biaryls via aryl-lithium intermediates[39-41,91,154,164,166] and good yields of biphenyls (not containing chlorine) were frequently reported. These biaryl syntheses involving lithium frequently involve dehydrobenzene intermediates.[86] It should be pointed out that organolithium reagents pose a potential explosion hazard.[27]

### d. Condensation Via Diels-Alder Reaction

Chlorobiphenyls can be obtained from chlorobicyclo(2:2:1)-heptadiene derivatives. The latter are prepared by Diels-Alder reaction from 5,5-disubstituted tetrachlorocyclopentadienes and phenylacetylene or styrene derivatives.

The synthetic approaches[87,123] to 7,7-dialkoxybicyclo-(2:2:1)-hepta-2,5-diene derivatives, which are thermally labile and decompose to give the corresponding biphenyl derivative, are shown below.

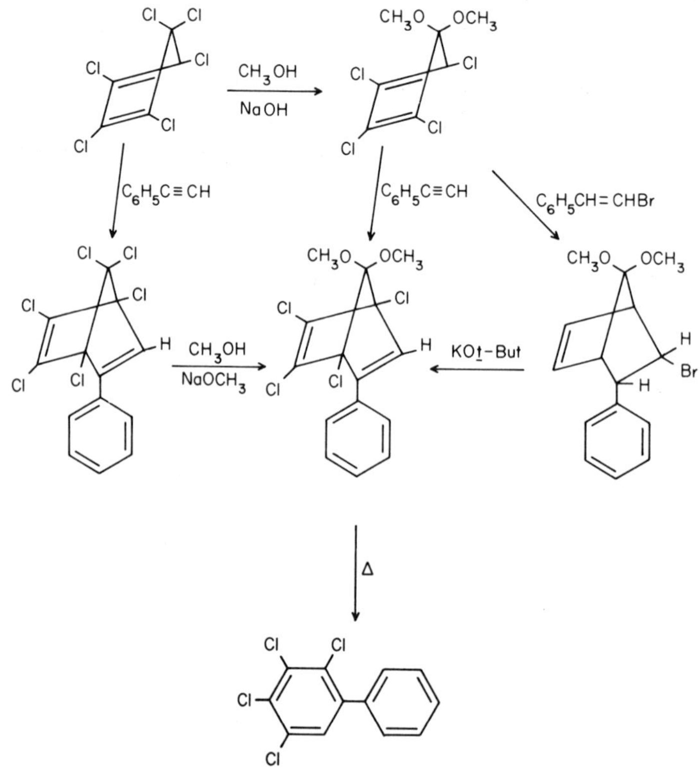


The reaction of styrenes with chlorocyclopentadiene derivatives gives adducts which result in 1,2-dihydrochlorobiphenyls which can be aromatized to give chlorobiphenyls.[122]

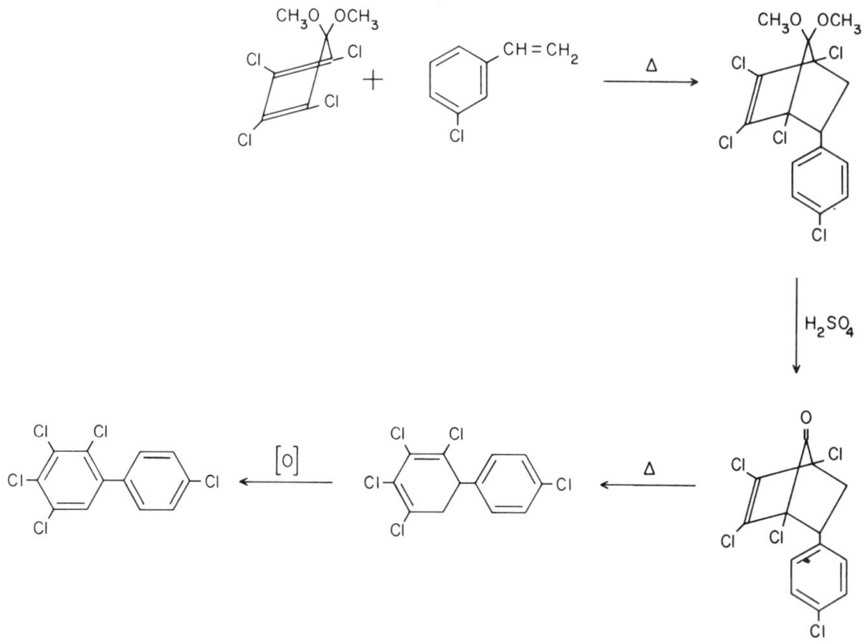

*e. Condensation of Phenylsulfinates*

A patent recently described the preparation of isomerically pure biphenyls, including 4,4′-dichlorobiphenyl, by the palladium catalyzed decomposition of phenylsulfinic acids salts.[114]

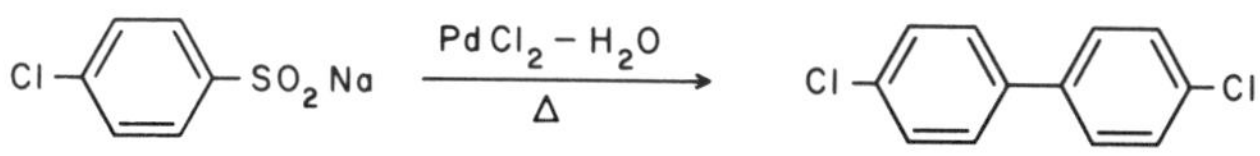

*f. Decomposition of Diaroyl Peroxides and Acid Anhydrides*

The preparation or 4,4′-di- and 2,2′,4,4′-tetrachlorobiphenyl by thermal decomposition of the corresponding benzoyl peroxides is described in the literature.[66] In the authors' laboratory, however, only poor yields were obtained and other unexpected chlorobiphenyls were observed as by-products.

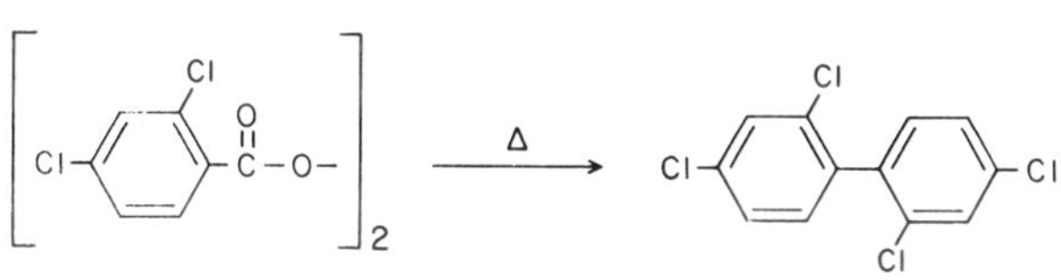

4-Chlorobenzoic anhydride has been converted to 4,4-dichlorobiphenyl in 3% yield using chlorotris(triphenylphosphine)rhodium as a catalyst during the thermal decomposition.[23]

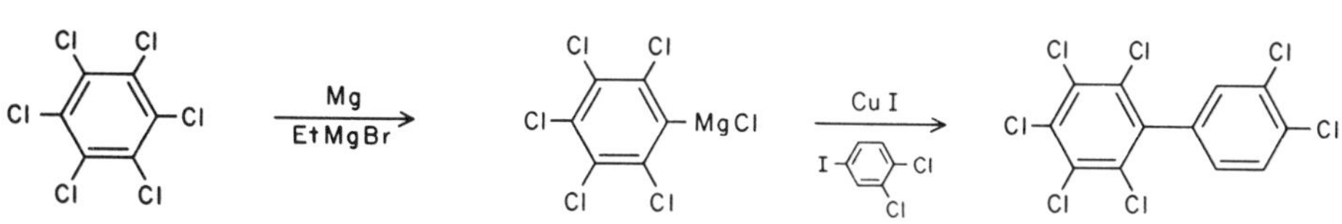

*g. Copper (I) Iodide Mediated Condensation of Grignard Reagents with Aryl Halides*

Several unsymmetrical chlorobiphenyls were recently prepared in 20 to 50% yield from pentachlorophenylmagnesium chloride and mono- and dichloroiodobenzene in the presence of copper (I)iodide.[116] In the absence of CuI, no biphenyls are formed, which indicates pentachlorophenyl copper to be the species responsible for the coupling.

### 3. *Formation of Chlorobiphenyls by Direct Substitution on the Preformed Biphenyl System*

#### *a. Chlorinations (H → Cl and -CH = CH- → -CH(Cl)-CH(Cl)-)*

**Substitution by chlorine**—Direct chlorination of biphenyl in the presence of a catalyst is the industrial process used for the preparation of PCB formulations. Mixtures containing more than two atoms of chlorine per biphenyl molecule cannot be easily separated into the individual components (for example by distillation) and preparation of such compounds by chlorination of biphenyl is therefore not feasible. The reaction of molten biphenyl with one equivalent of chlorine, in the presence of iron filings, gave after distillation the following amounts of chlorobiphenyls:[98] 31.6% 2-chlorobiphenyl; 25.7% 4-chlorobiphenyl; and 1.9% dichlorobiphenyls. In the presence of clay catalysts (e.g., bentonite) a 70% yield of ortho isomer was reported.[161] Mechanistic studies on the isomer distribution in monochlorobiphenyls on chlorination of biphenyl under a variety of conditions were reported.[15,49,53,192] Some of the results are given below:

| Chlorination process | Reference | Isomer distribution % o- | m- | p- |
|---|---|---|---|---|
| $Cl_2$, acetic acid, 25° (electrophilic) | 51 | 53 | 0 | 47 |
| Vapor phase chlorination, 350–450° | 53 | 35–38 | 47–52 | 14–16 |
| Catalyzed ($AlCl_3$-HCl) equilibrium of isomers at 160° | 193 | 3 | 64 | 33 |
| Statistic distribution | | 40 | 40 | 20 |

Early reports of formation of *m*-isomer in the electrophilic chlorination process[192] were later shown to be due to thermal decomposition of polychloro-adducts which are also formed during the chlorination.[49]

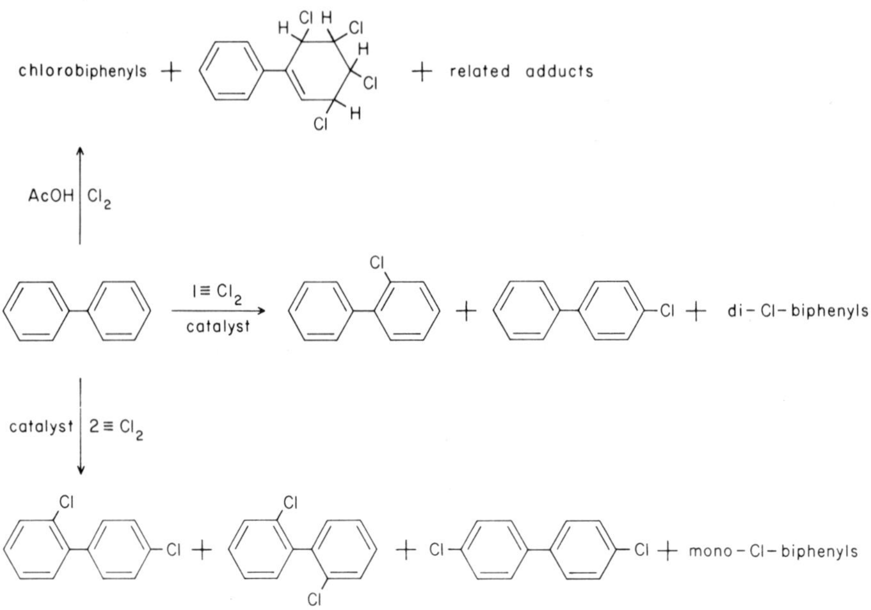


The chlorination of biphenyl with two equivalents of chlorine in the presence of ferric chloride catalyst yielded a mixture containing mainly dichlorobiphenyls.[43] 4,4′-Dichlorobiphenyl and 2,4′-dichlorobiphenyl were easily separated from the mixture and 2,2′-dichlorobiphenyl was collected in a fraction together with monochlorobiphenyls.

Chlorination of biphenyl with liquid chlorine in the presence of ferric chloride yielded a mixture of nona- and decachlorobiphenyls.[115] Metal chlorides are known to chlorinate aromatic compounds.[112] Under vigorous conditions (i.e., high temperatures) decachlorobiphenyl was obtained from biphenyl or low-chlorinated biphenyls with antimony pentachloride[129,189] or titanium tetrachloride.[158] An analytical method which involves conversion of all chlorobiphenyl compounds into a single derivative (decachlorobiphenyl) is based on this reaction[18,96] and antimony pentachloride alone or in the presence of catalytic amounts of iodine as well as a sulfuryl chloride-sulfur monochloride-aluminum chloride mixture (BMC-reagent) have been used.

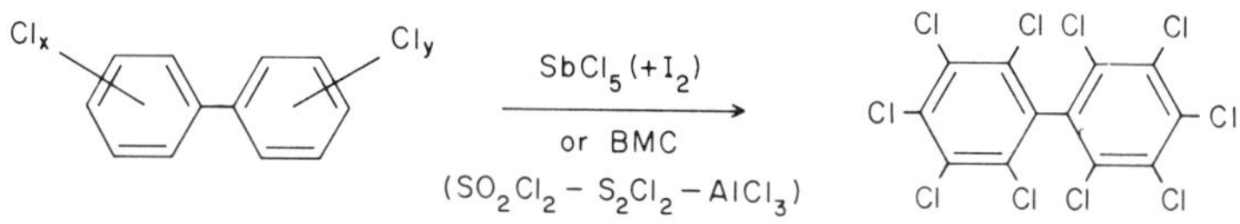

Because of the strongly ortho-para directing and activating influence of the amino group in electrophilic aromatic substitution reactions, aminobiphenyls (and amino-chlorobiphenyls) are suitable starting materials for specific chlorination to give chlorobiphenylamines (aminochlorobiphenyls) which can easily be converted to highly substituted chlorobiphenyls. Chlorination of *N,N′*-diacetylbenzidine in acetic acid solution for example yields *N,N′*-diacetyl-3,3′,5,5′-tetrachlorobenzidine.[185] Recently, this reaction was more closely examined and several by-products (mainly those formed by excessive chlorination) were characterized.[8] When chlorination of 2,2′,5,5′-tetrachlorobenzidine was carried out at −30° in dichloromethane solution, 2,2′,3,3′,5,5′-hexachlorobenzidine was obtained. No protection of the amino groups was necessary in this reaction, but similarly to above, a compound could be separated from the crude reaction product whose mass spectrum corresponded to that of a heptachlorobenzidine.[92] Various aspects of the chlorination of aromatic compounds are discussed in References 26, 50, 65, 143, and 175.

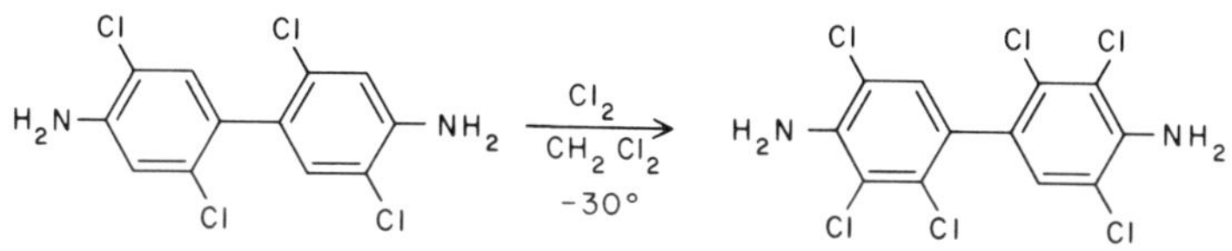

**Addition of chlorine** – Addition of chlorine to biphenyl in carbon tetrachloride, under the influence of daylight leads to the formation of two isomeric hexachlorocyclohexylbenzenes ($C_{12}H_{10}Cl_6$). The structures of these compounds were proven by NMR and X-ray data.[52,54]!.

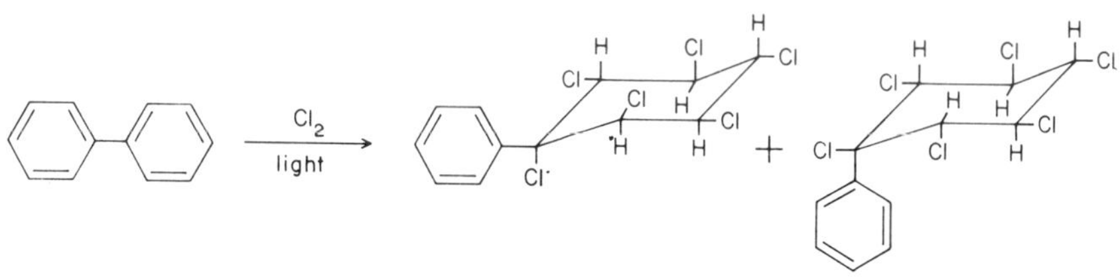

Even under conditions of electrophilic substitution, some addition of chlorine takes place. For example, in the chlorination of biphenyl with chlorine in acetic acid more than 10% of the chlorine used is consumed for the formation of tetrachlorides.[15,49] As far as the authors are aware, addition products of this type have not been found in commercial PCB preparations. If at all formed in the industrial chlorination process, these thermally labile adducts would probably decompose to chlorobiphenyls in the work-up procedure. When an excess of chlorine is passed into a hot solution of 3,5-dichloro-1-phenylcyclohexa-2,4-diene in chloroform, 3,5-dichlorobiphenyl and a small quantity of 2,3,5-trichlorobiphenyl are formed. This reaction was explained to proceed via intermediate formation of chlorine addition compounds and subsequent dehydrochlorination.[84]

*b. Sandmeyer Reaction ($NH_2 \rightarrow Cl$)*

Replacement of amino group by chlorine in chlorobiphenylamines (aminochlorobiphenyls) (see Chapter 4) is a convenient route to highly chlorinated biphenyls.[95,157,185] Of the most common reactions involved in this replacement (diazotations[36,88,151,159] and Sandmeyer reactions[44,88,159,174]) are well-established procedures and reviews on a large number of reáction conditions and variations are available.

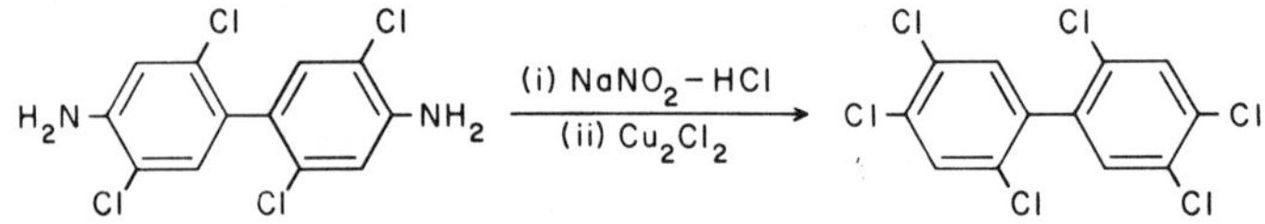

Some more recent developments in the Sandmeyer procedure lead to good yields and offer simplicity of procedure. The reaction of solid biphenyldiazonium fluoroborates (*cf.* 155) with ferric chloride in dimethyl sulfoxide solution was reported to give chlorobiphenyls[105] in good yield.

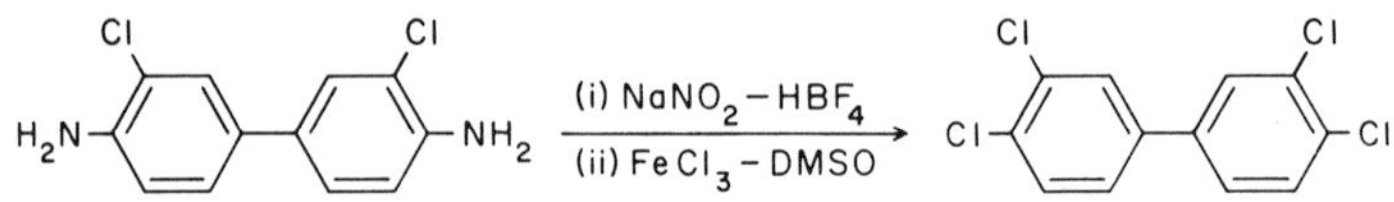

In a convenient one-step procedure, the aromatic amine in acetonitrile is reacted with one equivalent of cupric chloride under an atmosphere of NO to give yields of over 80%.[25] Moderate yields of arylchlorides are obtained when amines are reacted with pentylnitrite in the presence of chloroform or carbon tetrachloride.[32]

*c. Removal of Amino Groups (Reduction of Diazonium Salts) ($NH_2 \rightarrow H$)*

For each chlorobiphenylamine (aminochlorobiphenyl) in hand, two different chlorobiphenyls can easily be prepared, one by replacing the amino group with chlorine (Sandmeyer reaction *cf.* 3b) and one by replacing the amino group with hydrogen (reduction of diazonium salts).[157,185] The most commonly used reducing agents are ethanol and hypophosphorous acid.[110,151,159] A simple, one-pot deamination

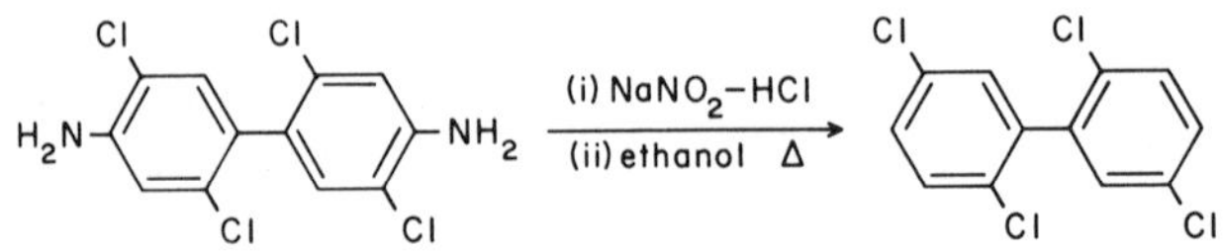

of aromatic amines involving aprotic diazotization with pentyl nitrite in boiling tetrahydrofuran has recently been described.[31]

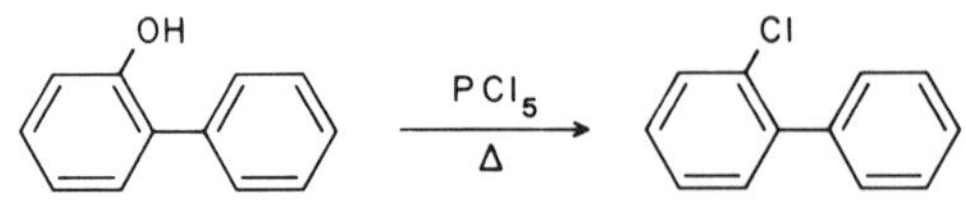

*d. Replacement of Hydroxy Groups by Chlorine ($OH \rightarrow Cl$)*

Although sometimes considered a "textbook reaction" of little preparative value, the preparation of chlorobiphenyls by the action of phosphorus pentachloride on the corresponding hydroxybiphenyl has been known for some time[162] and yields of over 90% have been claimed in this reaction.[139] A new development in this reaction is the use of triaryloxyphosphorus dichloride in the place of phosphorus pentachloride.[38]

3,5-Dichlorobiphenyl and 2,3,5-trichlorobiphenyl have been prepared by a related reaction. A small quantity of 3,5-dichlorobiphenyl is formed during the action of phosphorus pentachloride on phenyldihydroresorcinol. The main product of the reaction is 3,5-dichloro-1-phenylcyclohexa-2,4-diene.

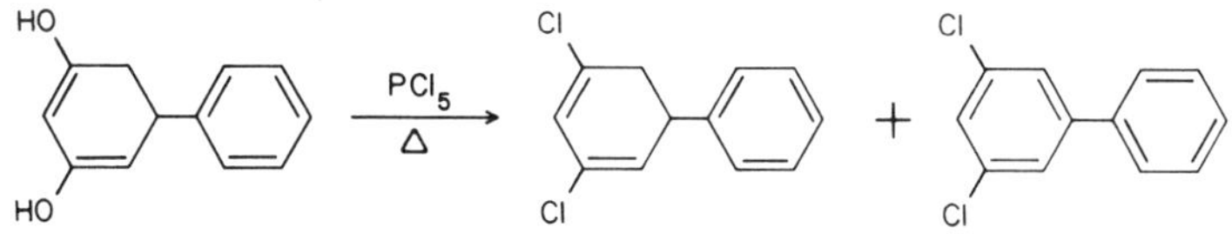

When chlorine is passed through a hot solution of this compound 3,5-dichlorobiphenyl and a smaller quantity of 2,3,5-trichlorobiphenyl is formed.[84]

*e. Replacement of Nitro Groups by Chlorine ($NO_2$ → Cl)*

Dichloro-[119] and trichlorobiphenyls[153] were obtained when the corresponding nitro compounds were heated with thionyl chloride to approximately 200°. Although fair yields were observed, this method has not been used frequently, probably because of inconvenient experimental conditions (sealed tube) and side reactions (chlorination).

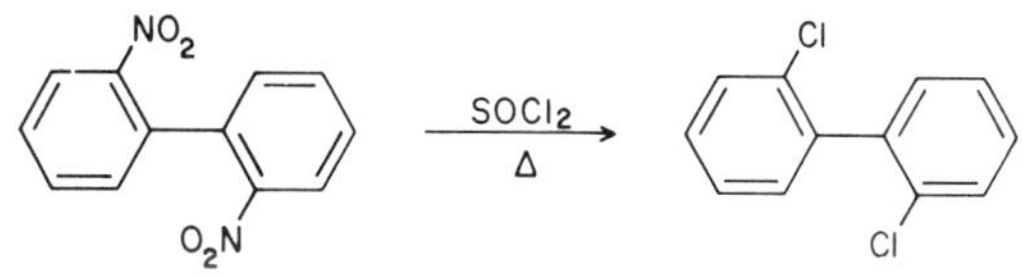

Recently, it was shown that certain nitronaphthalenes can be converted in good yield to their corresponding chloronaphthalene by a photosubstitution reaction.[68]

*f. Replacement of Sulfonyl Chloride and Sulfonic Acid Groups by Chlorine ($SO_2Cl$ → Cl; $SO_3H$ → Cl)*

A recent patent describes the preparation of simple chlorobiphenyls in yields of over 90% by heating biphenylsulfonic acids, their salts or acid chlorides with carbon tetrachloride or phosgene to 230 to 270° in an autoclave.[45]

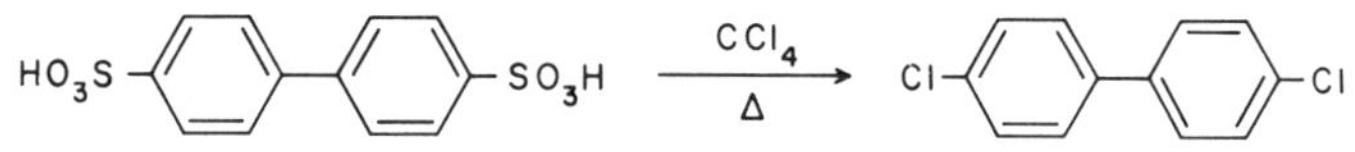

Similarly, the catalyzed or uncatalyzed thermal decomposition of biphenyl-2-sulfonyl chloride gave fair yields of 2-chlorobiphenyl.[138]

Recently, it was shown that aryl sulfonyl halides can be converted to their corresponding aryl halides with the aid of noble-metal catalysts.[22]

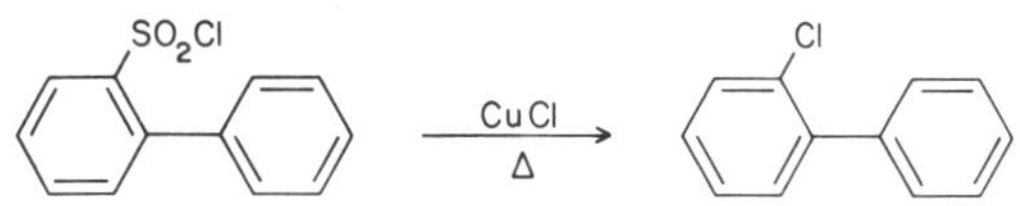

*g. Decarboxylations (COOH → H)*

Several chlorobiphenyls have been obtained from the corresponding chlorofluorenes by oxidation to 9-fluorenone derivatives, opening of the five-membered ring and decarboxylation of the resulting carboxylic acid.[165] Thermal decarboxylation of octachloro-4,4′-biphenyldicarboxylic acid

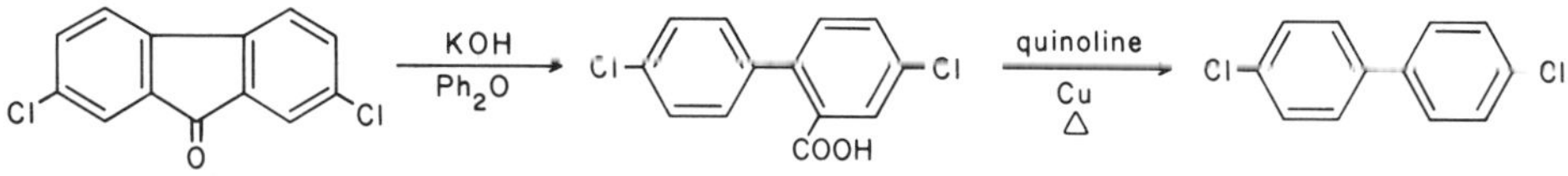

is also the final step in a preparation of 2,2′, 3,3′, 5,5′, 6,6′-octachlorobiphenyl.[14]

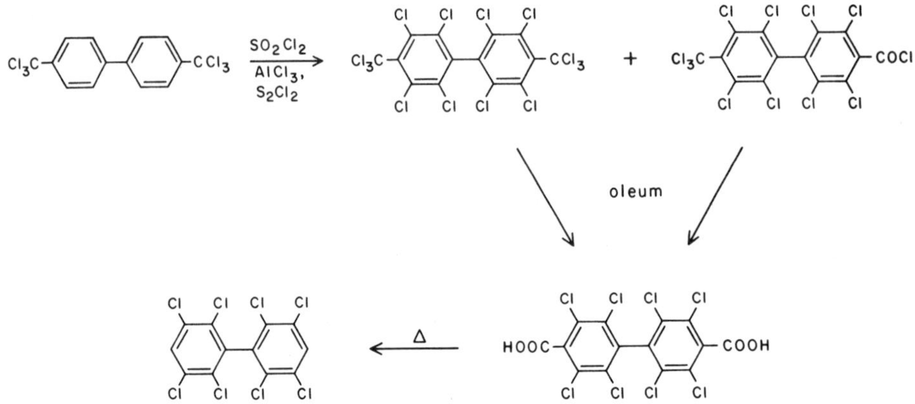

*h. Dechlorination of Decachlorobiphenyl by Reduction or Via Organometallic Derivatives*

One or two chlorine atoms are removed from decachlorobiphenyl when this compound is treated with lithium aluminum hydride, or when the corresponding Grignard or lithium derivatives are decomposed with water.[19] A mixture of starting material, nonachlorobiphenyl, and octachlorobiphenyl is obtained with one equivalent of the reagents mentioned, whereas octachlorobiphenyl is mainly formed on treatment with two equivalent of these reagents.

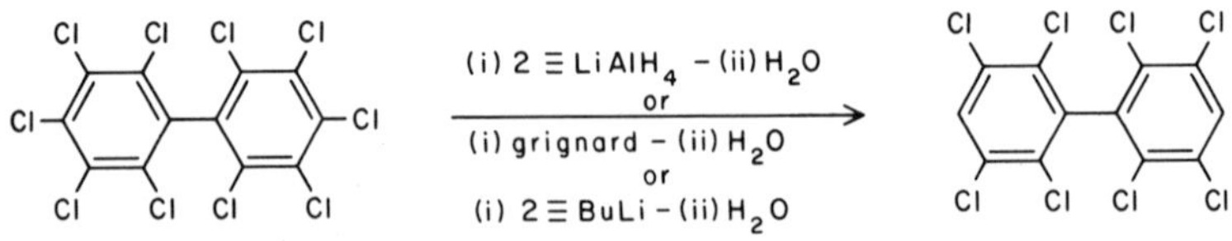

**Tables Describing Known Chlorobiphenyls**

The tables contain a brief summary of the melting points, the methods of preparation, and the more useful and accessible references on the synthesis and properties of all chlorobiphenyls described in the literature to date. A complete bibliography on chlorobiphenyls until 1949 can be found in Beilsteins Handbuch.[16] Generally, only those chlorobiphenyls are included in the tables for which data properly characterizing them (melting point, analysis, etc.) have been obtainable by the authors. The few exceptions are marked with an asterisk in the melting point column. The retention times of a number of chlorobiphenyls which are not included in the tables have been given.[188] Only one melting point (the highest one reported) is given if the variation is not more than 2 to 3°. Melting points deviating more than 2 to 3° are listed separately; a question mark is placed next to those that are most likely erroneous.

The chlorobiphenyls are arranged in the following order: (a) increasing chlorine content, (b) unprimed position numbers before primed, and (c) lower position numbers before higher. Among dichlorobiphenyls, for instance, the following order would apply: 2,3-, 2,4-, 2,2′-, 2,4′-. More information on nomenclature may be found in the introductory chapter.

The literature references and the synthetic routes listed for each compound are in no particular order. However, the melting point given is always taken from the reference on the same line.

The structures of the chlorobiphenyls shown in the tables are drawn to conform with the following numbering of positions:

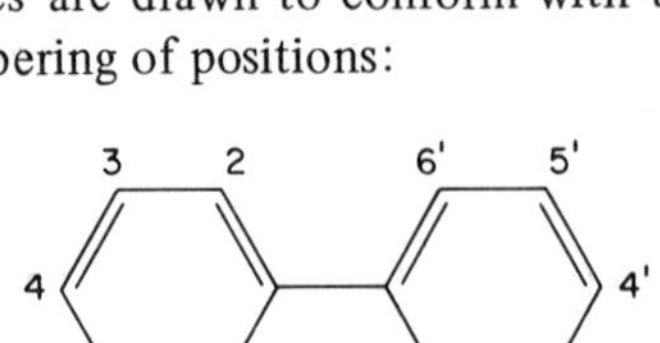


A number of chlorobiphenyls are now available commercially (Analabs).

| Compound | | | | |
|---|---|---|---|---|
| Name | Formula | M.P. °C | Synthetic route | Lit. ref. |
| Monochlorobiphenyls | | | | |
| 2- | 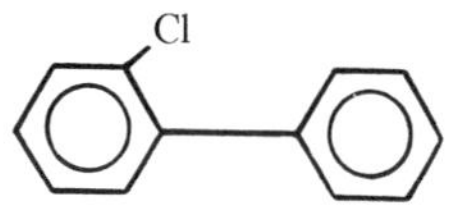 | 34 | 3-b | 15 |
| | | | | 6 |
| | | | | 105 |
| | | | 1-b | 60 |
| | | | 3-d | 139 |
| | | | 1-e | 77 |
| | | | 3-f | 138 |
| 3- | 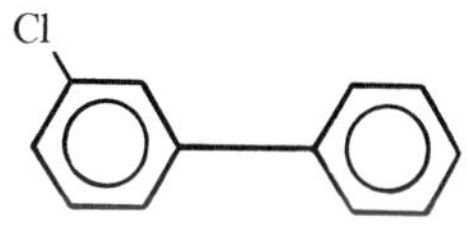 | | 3-b | 105 |
| | | Oil | 1-b | 60 |
| | | | | 73 |
| | | | | 6 |
| | | 16–17 | 1-b | 95 |
| 4- | 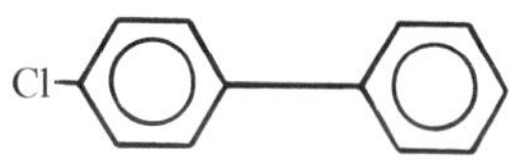 | 77.7 | 1-b | 73 |
| | | | | 60 |
| | | | | 30 |
| | | | | 6 |
| | | | | 1 |
| | | | 3-b | 167 |
| | | | 3-a | 98 |
| | | | 1-a | 147 |
| | | | 3-g | 165 |
| | | | 1-g | 181 |
| Dichlorobiphenyls | | | | |
| 2,3- | 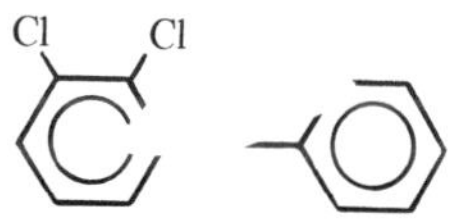 | 27.7–28.2 | 1-b | 191 |
| | | Oil | 3-b | 48 |
| | | | 1-d | 199 |
| 2,4- | 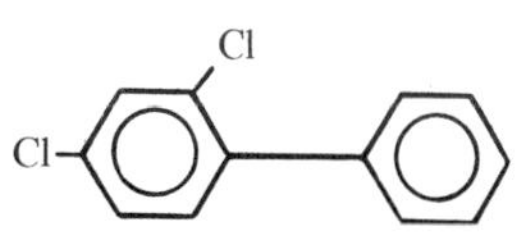 | 24.1–24.4 | 1-b | 191 |
| | | | | 57 |
| 2,5- | 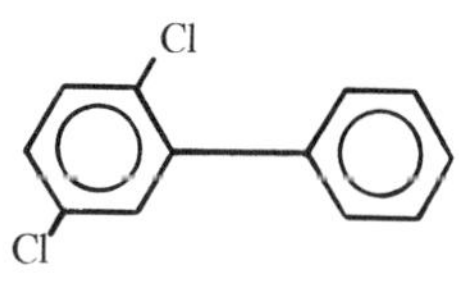 | Oil | 3-b | 160 |
| | | | | 48 |
| | | | 1-a | 6 |
| | | 22–23? | 1-b | 193 |

| Compound: Name | Compound: Formula | M.P. °C | Synthetic route | Lit. ref. |
|---|---|---|---|---|
| 2,6- | 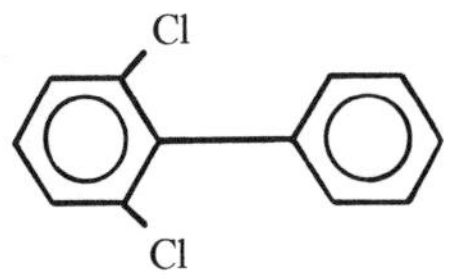 | 35–36 | 1-b | 95 |
| 3,4 | 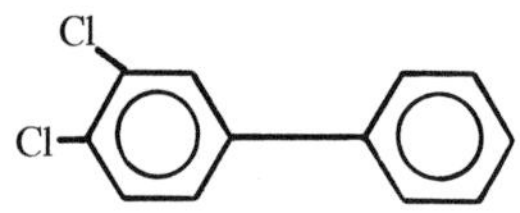 | 49–50 | 1-b | 191 |
| | | | 1-a | 79 |
| | | | 1-c | 3 |
| | | | 1-d | 200 |
| 3,5- | 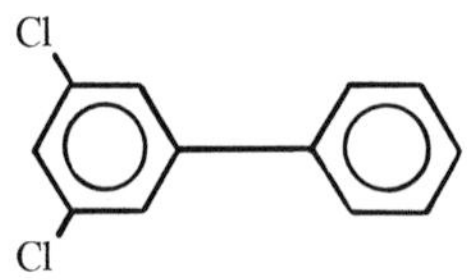 | 31–32 | 1-b | 191 |
| | | 36 | 3-b | 84 |
| | | | 3-d | |
| | | | 3-c | 160 |
| 2,2′- | 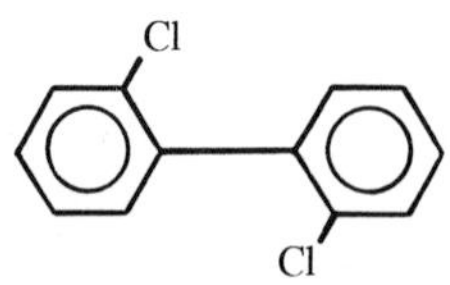 | 60.5 | 3-b | 163 |
| | | | 2-a | 76 |
| | | | 3-e | 119 |
| 2,3′- | 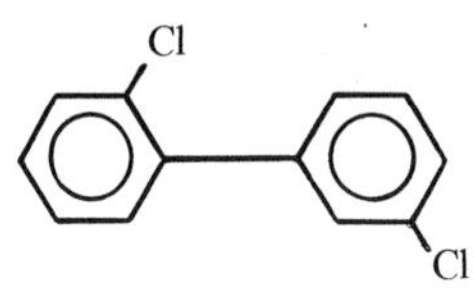 | Oil | 1-b | 23 |
| 2,4′- | 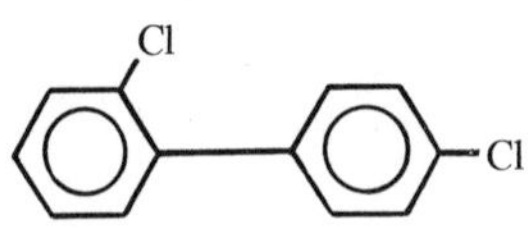 | 46 | 3-c | 67 |
| | | | 3-b | 67 |
| | | | | 132 |
| | | 44.5 | | 99 |
| | | 32 | 3-a | 43 |
| | | 41 | 1-b | 197 |
| 3,3′- | 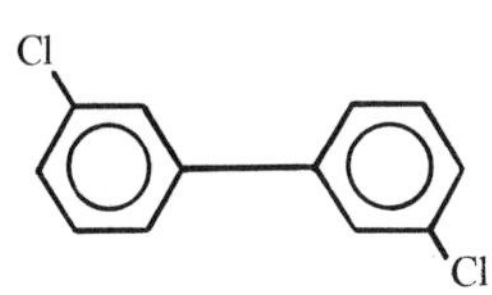 | 29 | 3-c | 34 |
| | | | 2-a | 183 |
| | | | 2-b | 126 |
| | | | 3-e | 119 |
| | | | 3-b | 95 |
| 3,4′- | 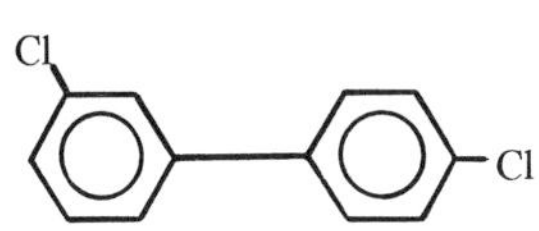 | Oil | 1-b | 23 |

| Compound: Name | Compound: Formula | M.P. °C | Synthetic route | Lit. ref. |
|---|---|---|---|---|
| 4,4′- | 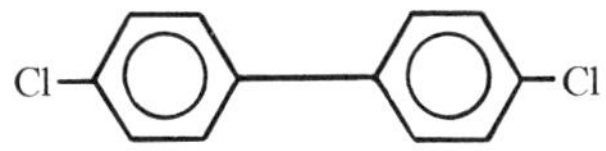 | 148–149 | 2-b | 125 |
| | | | 2-a | 183 |
| | | | 3-d | 162 |
| | | | 3-a | 167 |
| | | | | 162 |
| | | | 3-b | 85 |
| | | 149–150 | 2-c | 72 |
| | | | 2-e | 114 |
| | | 145–146 | 3-g | 165 |
| | | | 3-f | 45 |
| Trichlorobiphenyls | | | | |
| 2,3,4- | 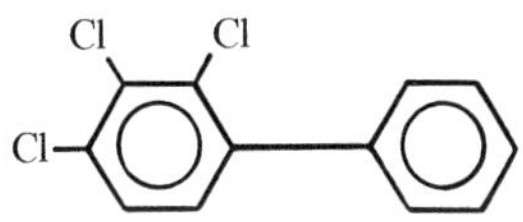 | 101–102 | 1-b | 24 |
| 2,3,5- | 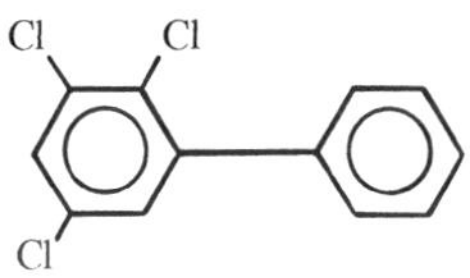 | 41 | 3-d | 84 |
| 2,3,6- | 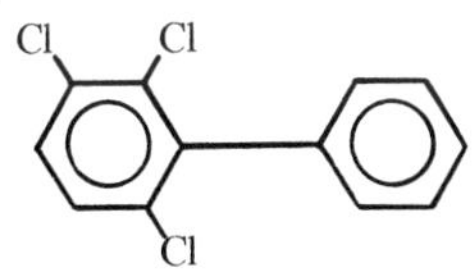 | * | 1-b | 145 |
| 2,4,5 | 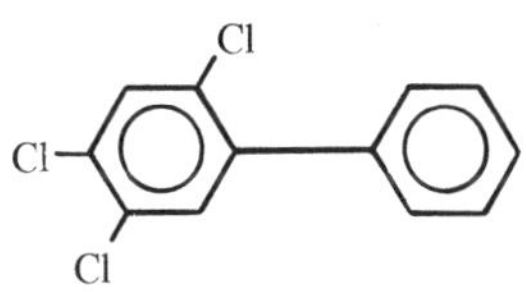 | 78–79 | 1-b | 71 |
| | | | 1-c | 3 |
| | | | 1-d | 199 |
| 2,4,6- | 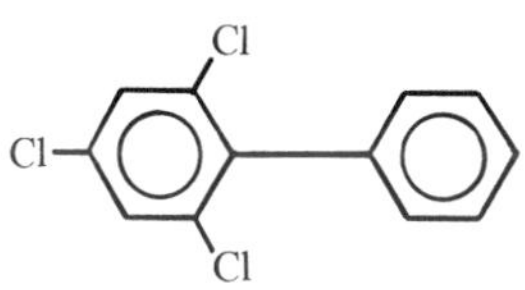 | 62.5 | 1-a | 6 |
| | | | 1-f | 12 |
| | | 46 | 1-b | 156 |
| 2,2′,3- | 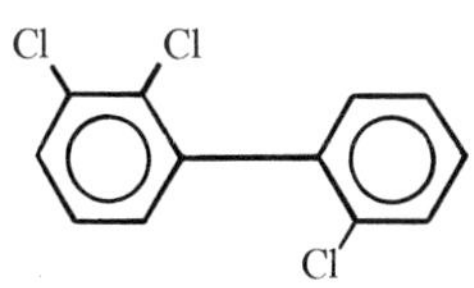 | 28.1–28.8 | 1-b | 191 |
| 2,3,4′- | 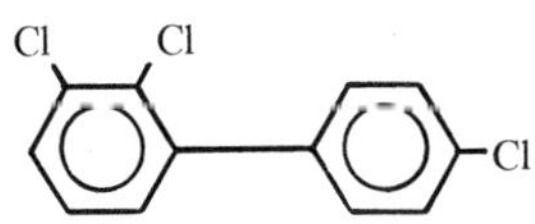 | 73–73.2 | 1-b | 191 |
| 2,4,4′- | 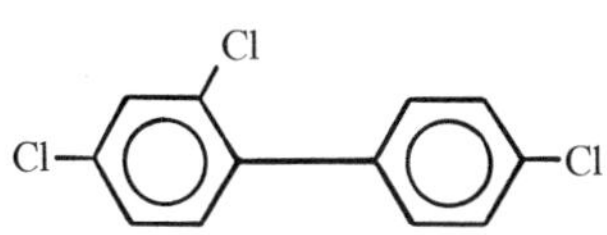 | 57-58 | 3-b | 95 |
| | | | 3-e | 153 |
| | | 206–207 | 3-g | 165 |

| Compound: Name | Compound: Formula | M.P. °C | Synthetic route | Lit. ref. |
|---|---|---|---|---|
| 2,2′,5- | 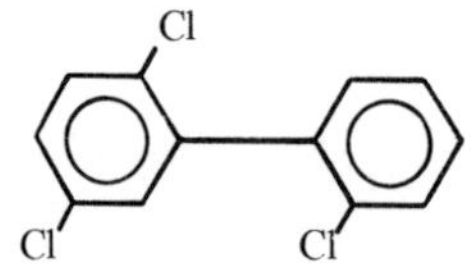 | 43–44 | 1-b | 24 |
| 2,3′,5- | 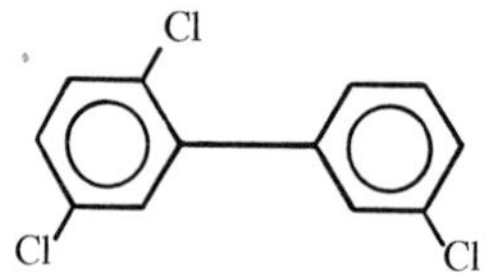 | 40–40.5 | 1-b | 24 |
| 2,4′,5- | 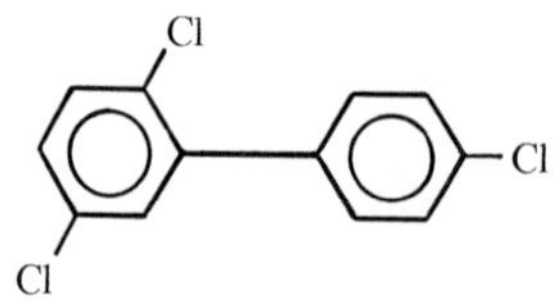 | 67<br><br>63.5–64.5 | 3-b<br>1-a<br>1-b | 17<br>173<br>24<br>188 |
| 2′,3,4- | 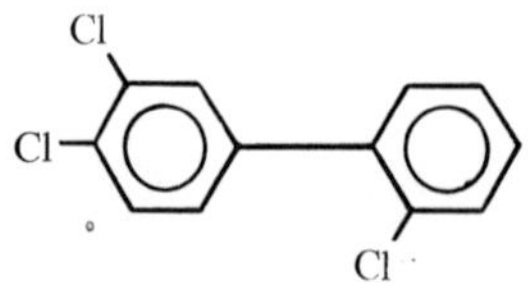 | 60.1–60.4<br>54<br>65–66 | 1-b<br>3-b | 191<br>17<br>120 |
| 3,4,4′- | 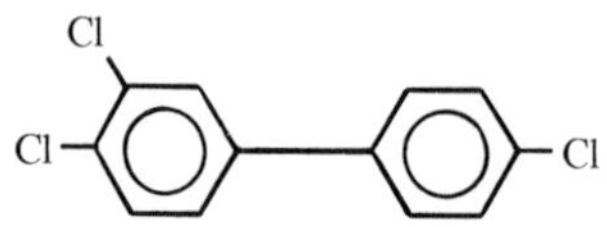 | 86.8–87.8 | 1-b | 191 |
| 2′,3,5- | 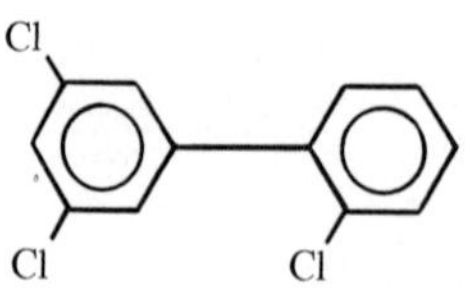 | 58 | 3-b | 84 |
| 3,4′,5- | 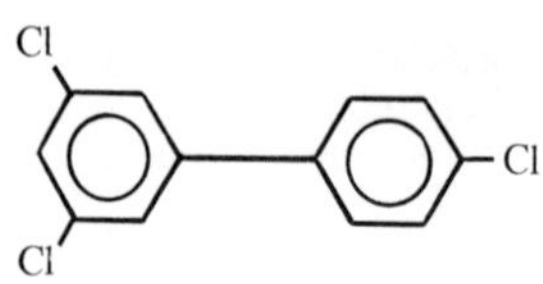 | 88 | 3-b | 84 |
| Tetrachlorobiphenyls | | | | |
| 2,3,4,5- | 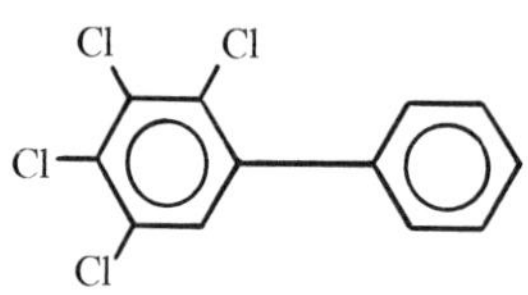 | 92–92.5 | 2-d<br><br><br>1-b | 122<br>123<br>87<br>95 |

| Compound | | M.P. °C | Synthetic route | Lit. ref. |
|---|---|---|---|---|
| Name | Formula | | | |
| 2,3,5,6- | 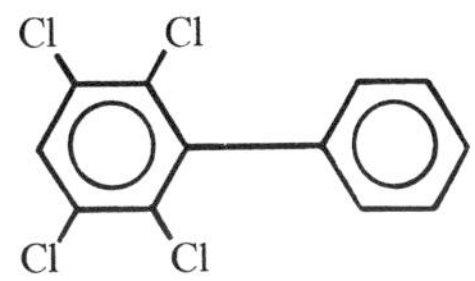 | 79 | 1-f<br>1-b | 12<br>95 |
| 2,3,4,4′- | 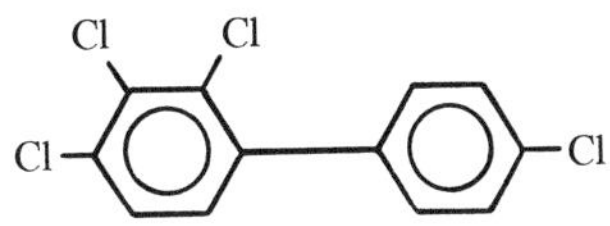 | 142 | 1-b | 156 |
| 2,4,4′,5- | 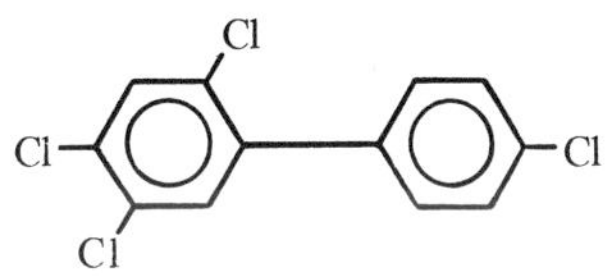 | 125 | 1-b | 156 |
| 2,2′,4,6- | 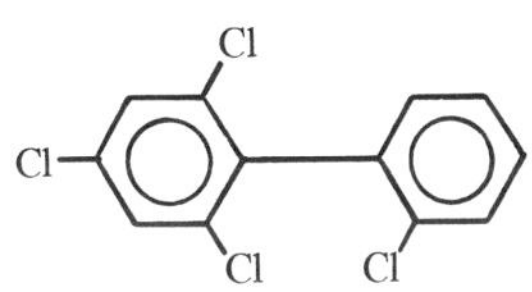 | * | 1-f | 118 |
| 2,4,4′,6- | 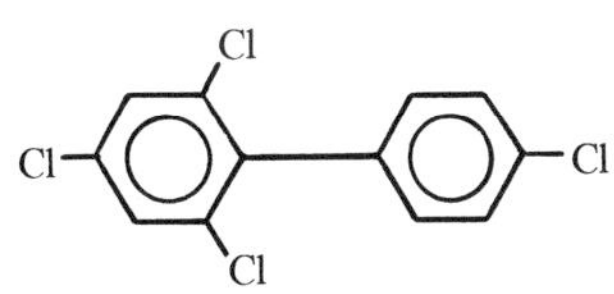 | * | 1-f | 118 |
| 2,2′,3,3′- | 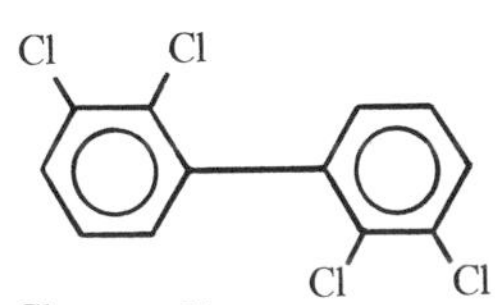 | 119.5–121.5 | 2-a | 24 |
| 2,2′,3,4′- | 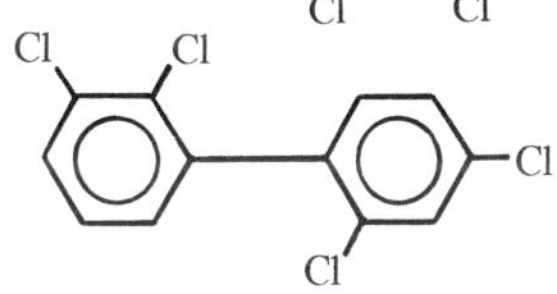 | 68–70 | 1-b,<br>2-a | 188 |
| 2,2′,3,5′- | 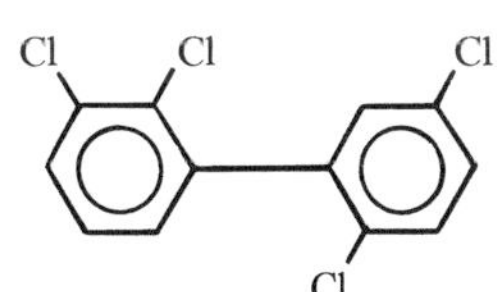 | 46.5–47 | 1-b | 24 |
| 2,2′,4,4′- | 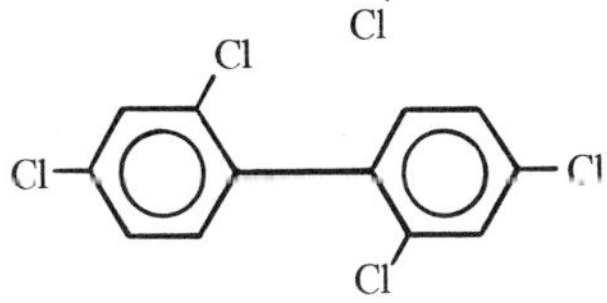 | 83<br>83<br>41<br>41–42 | 2-f<br>2-a<br>2-a<br>2-b<br>1-b | 66<br>183<br>95<br>75<br>156 |

| Compound Name | Formula | M.P. °C | Synthetic route | Lit. ref. |
|---|---|---|---|---|
| 2,2′,4,5′- | 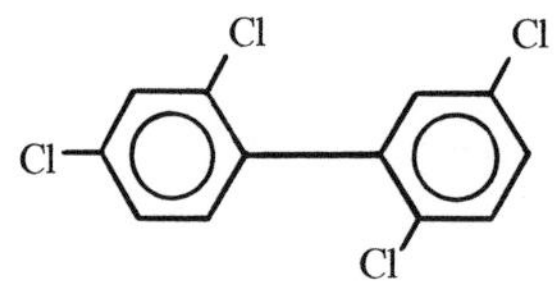 | 64–66 | 1-b | 24 |
| | | 66–68.5 | 1-b | 188 |
| 2,3′,4,4′- | 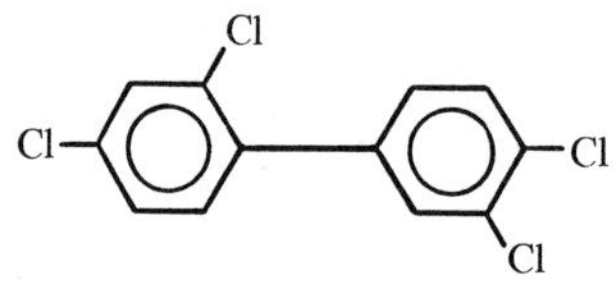 | 124 | 1-b | 156 |
| | | 127–127.5 | 1-b, | 188 |
| | | | 2-a | |
| | | | 3-b | 92 |
| 2,2′,5,5′- | 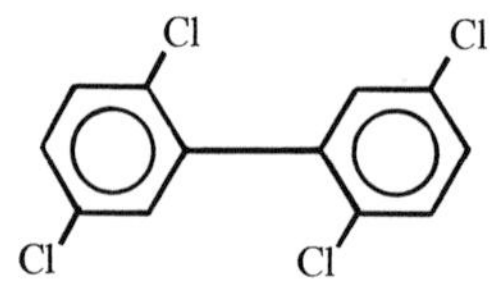 | 86.5–87 | 3-c | 157 |
| | | 87–89 | 2-a | 188 |
| 2,3′,4′,5- | 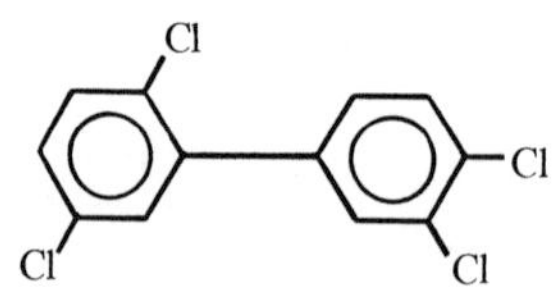 | 104 | 3-b | 17 |
| | | | 1-d | 200 |
| | | | 1-b | 156 |
| | | | | 188 |
| 2,2′,6,6′- | 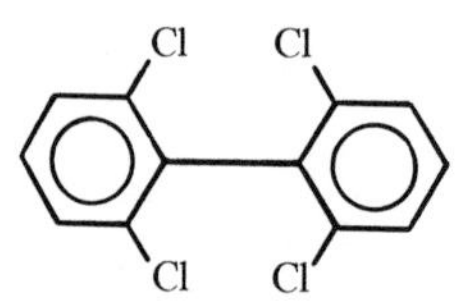 | 198 | 3-c | 185 |
| | | | 2-a | 95 |
| 3,3′,4,4′- | 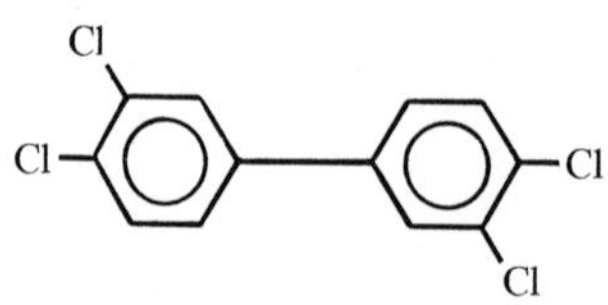 | 173 | 3-b | 185 |
| | | | | 95 |
| | | | 1-b | 156 |
| | | 182–184 | 2-a | 24 |
| 3,3′,5,5′- | 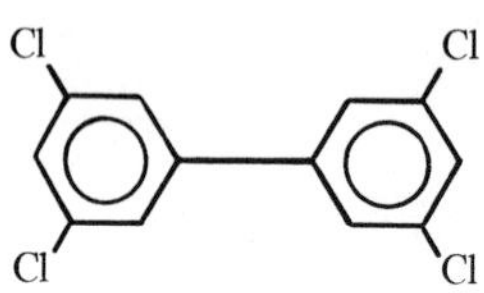 | 164 | 3-c | 185 |
| | | | 2-a | 95 |
| Pentachlorobiphenyls | | | | |
| 2,3,4,5,6- | 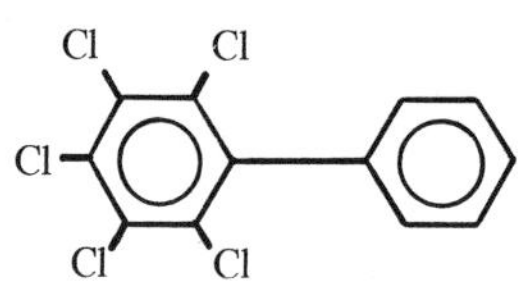 | 123 | 1-b | 95 |
| | | | 2-g | 116 |
| 2,3,4,4′,5- | 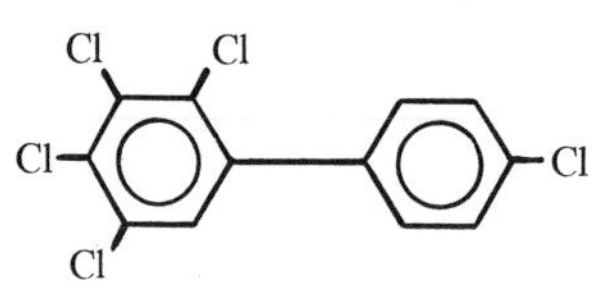 | 98–99 | 2-d | 122 |

| Compound Name | Formula | M.P. °C | Synthetic route | Lit. ref. |
|---|---|---|---|---|
| 2,2′,3,4,5′- | 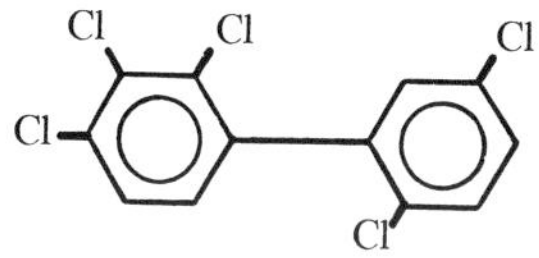 | 111.5–113 | 1-b | 24 |
| 2,3,3′,4,4′- | 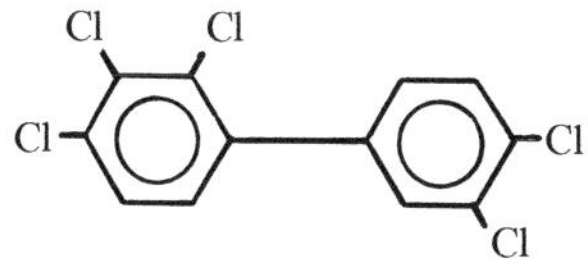 | 101–105 | 1-b | 156 |
| 2,2′,3′,4,5- | 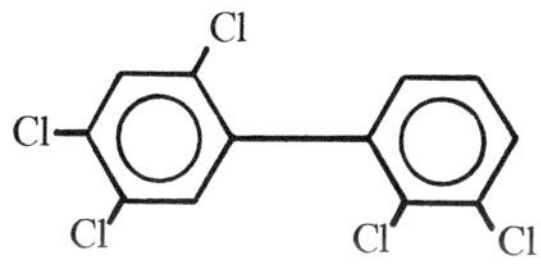 | 81–82– | 1-b<br>2-a | 145<br>24 |
| 2,2′,4,5,5′- | 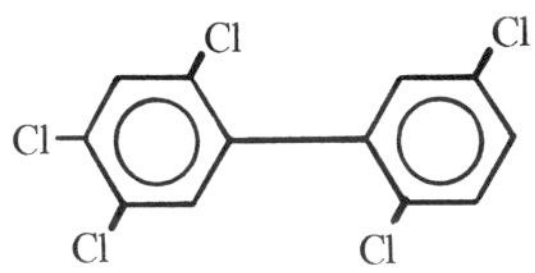 | 76.5–77.5 | 1-b | 24<br>188 |
| 2,3′,4,4′,5- | 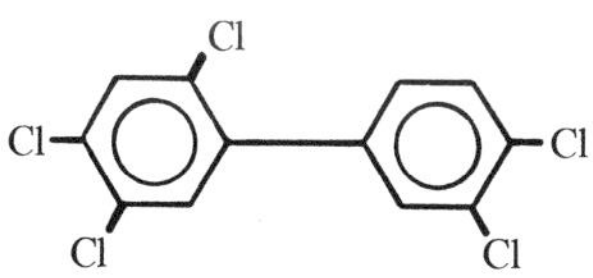 | 82*<br>105–107 | 1-b<br>3-b | 156<br>83 |
| Hexachlorobiphenyls | | | | |
| 2,2′,3,4,5,6- | 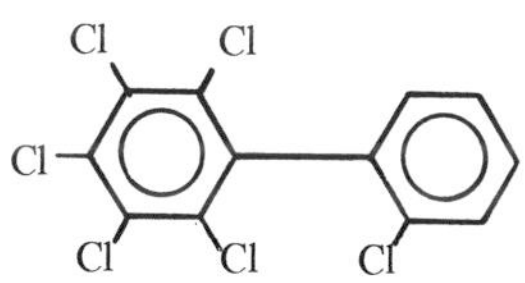 | 134–137 | 2-g | 116 |
| 2,3,3′,4,5,6- | 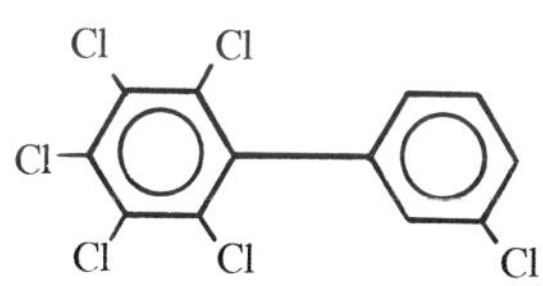 | 97–100 | 2-g | 116 |
| 2,3,4,4′,5,6- | 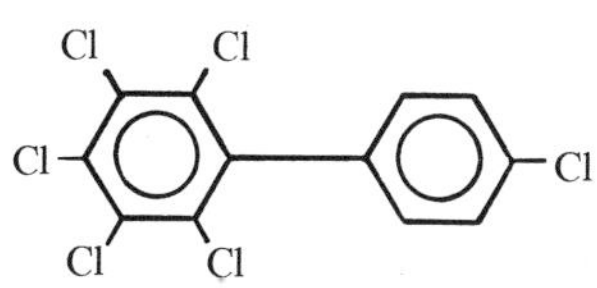 | 160–165 | 2-g | 116 |
| 2,2′,3,4,4′,5- | 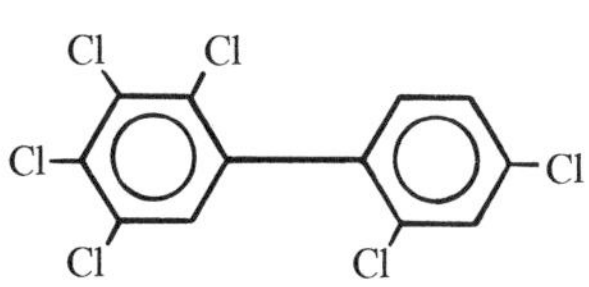 | 77–78 | 1-a | 157 |

*See last chapter.

| Compound: Name | Compound: Formula | M.P. °C | Synthetic route | Lit. ref. |
|---|---|---|---|---|
| 2,2′,3,5,5′,6- | 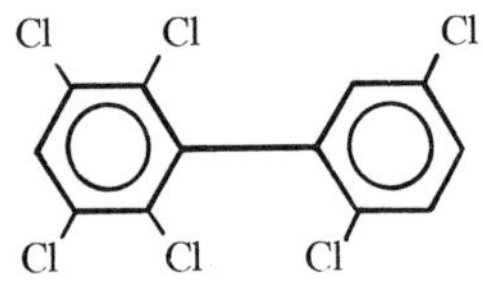 | 100–101 | 1-b | 24 |
| 2,2′,3,3′,4,4′- | 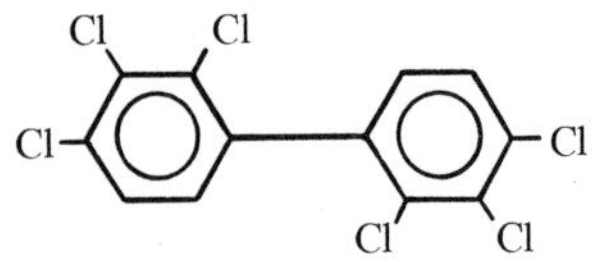 | 145.5–146.6<br>150–152 | 2-a<br>2-a | 178<br>24 |
| 2,2′,3,4,4′,5′- | 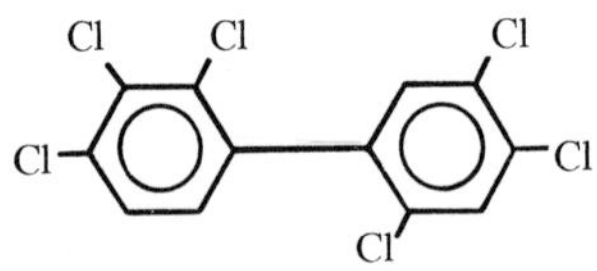 | 78.5–80 | 2-a | 178 |
| 2,2′,3,3′,5,5′- | 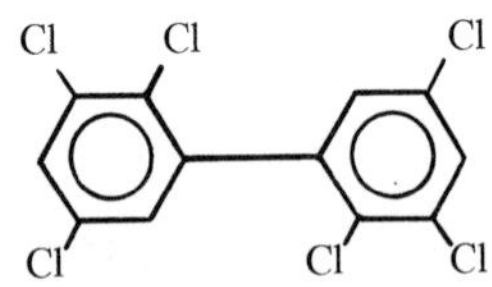 | 128–129 | 3-c | 157 |
| 2,2′,3,3′,6,6′- | 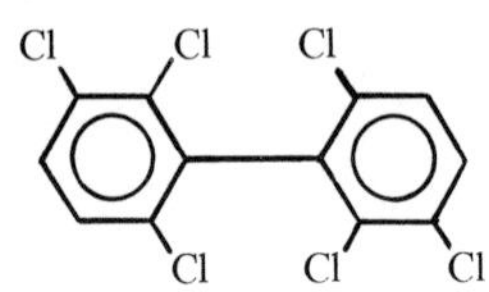 | 114–114.5 | 2-a | 179 |
| 2,2′,3,4′,5′,6- | 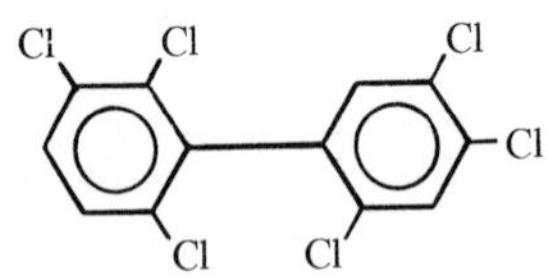 | Oil | 2-a | 179 |
| 2,2′,4,4′,5,5′- | 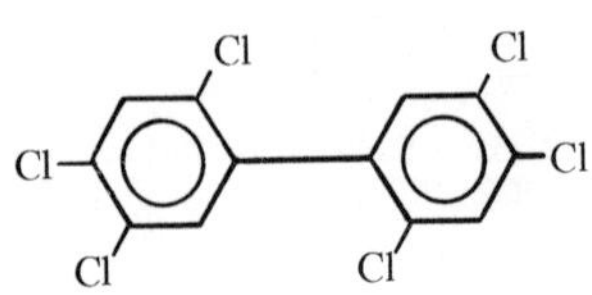 | 103–104<br><br><br>137–138? | 3-b<br>2-a | 157<br>178<br>188<br>150 |
| 2,2′,4,4′,5,6′- | 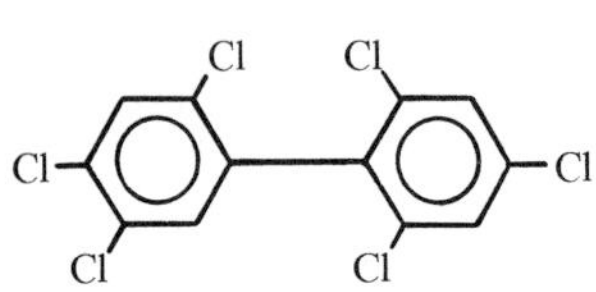 | Oil | 2-a | 180 |
| 2,3′,4,4′,5,5′- | 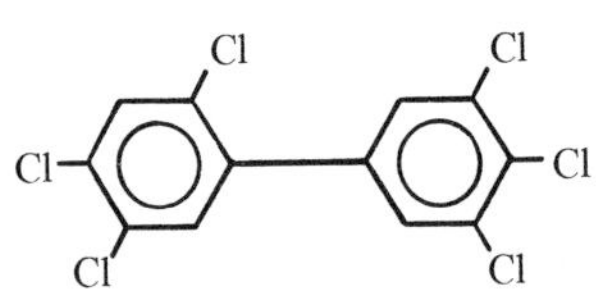 | * | 1-b | 145 |

| Compound | | | | |
|---|---|---|---|---|
| Name | Formula | M.P. °C | Synthetic route | Lit. ref. |
| 2,2′,4,4′,6,6′- | 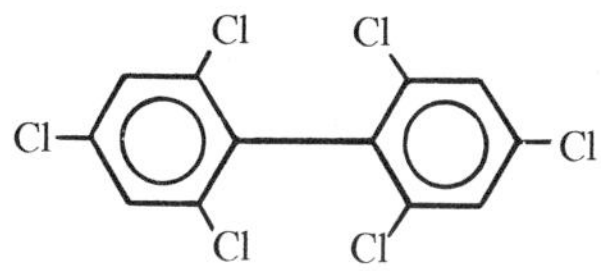 | 112.5 | 2-a<br>3-b | 183<br>185 |
| 3,3′,4,4′,5,5′- | 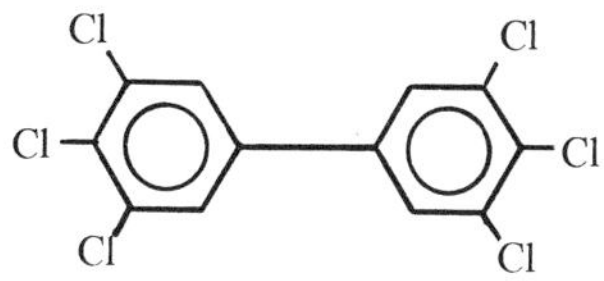 | 201–202 | 3-b | 185<br>95 |
| Heptachlorobiphenyls | | | | |
| 2,2′,3,4,5,5′,6- | 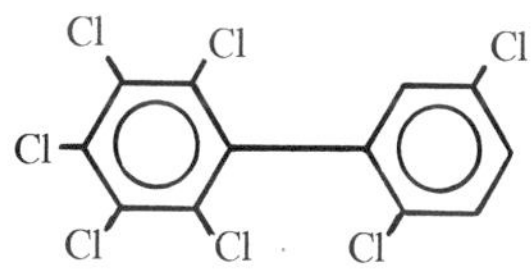 | 147–150 | 1-b | 24 |
| 2,3,3′,4,4′,5,6- | 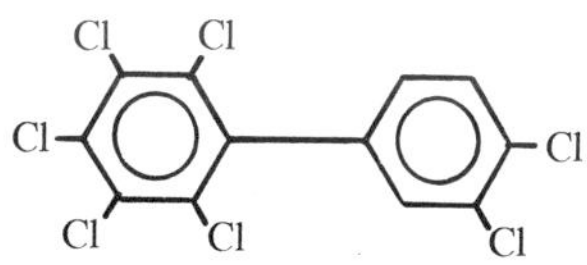 | 116–118 | 2-g | 116 |
| 2,2′,3,3′,4,4′,5- | 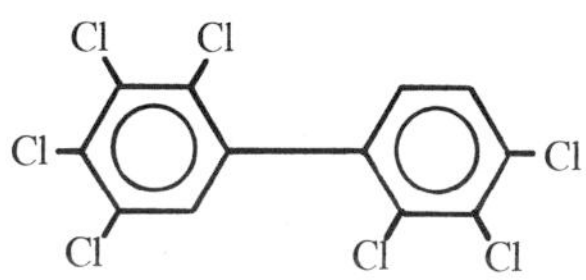 | 134.5–135.5 | 2-a | 178 |
| 2,2′,3,3′,4,5,6′- | 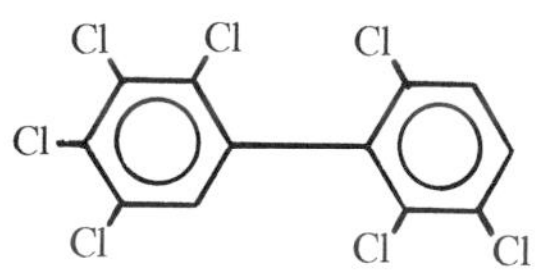 | 130.5–130.7 | 2-a | 179 |
| 2,2′,3,4,4′,5,5′- | 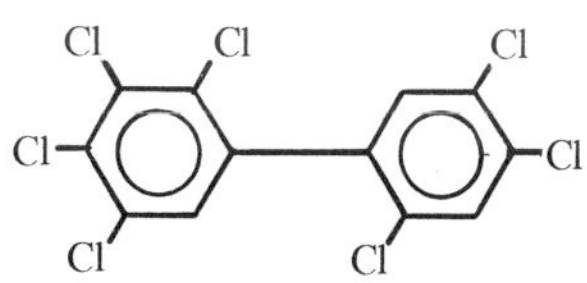 | 109–110 | 2-a | 178 |
| 2,2′,3,4,4′,5,6′- | 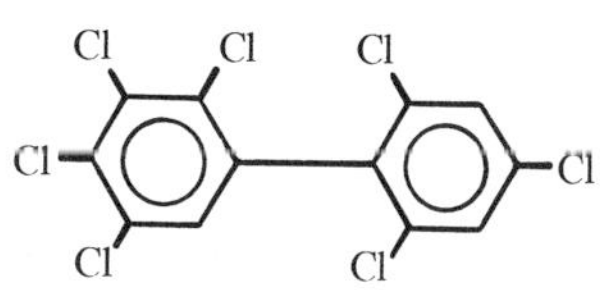 | 152–153 | 2-a | 180 |
| 2,3,3′,4,4′,5,5′- | 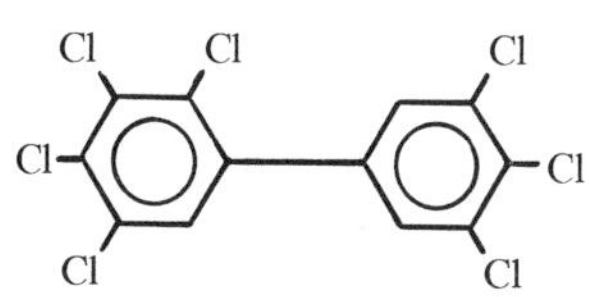 | 162–163<br>178 | 2-a<br>3-b | 180<br>92 |

| Compound: Name | Compound: Formula | M.P. °C | Synthetic route | Lit. ref. |
|---|---|---|---|---|
| Octachlorobiphenyls | | | | |
| 2,2′,3,3′,4,4′,5,5′- | 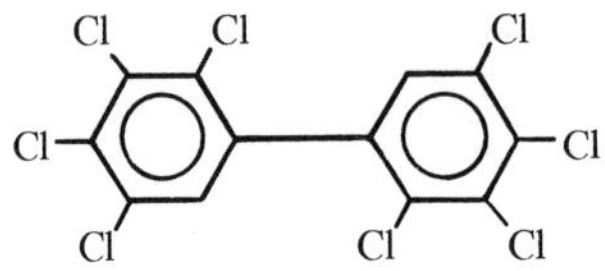 | 156–157 | 3-b | 157 |
| | | 152–153 | 2-a | 178 |
| | | 159–160 | 2-a | 19 |
| | | | 3-h | 19 |
| 2,2′,3,3′,4,4′,6,6′- | 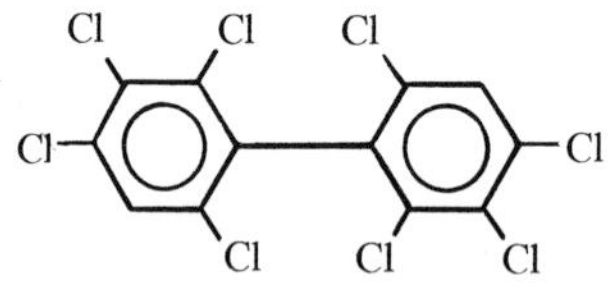 | 132 | 3-b | 95 |
| 2,2′,3,3′,5,5′,6,6′- | 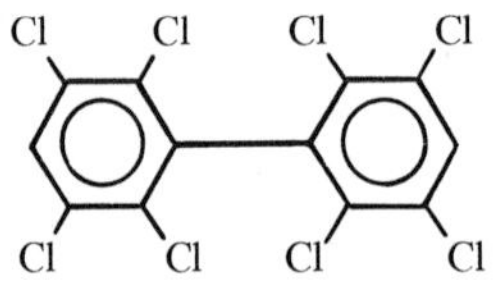 | 161 | 3-c | 185 |
| | | | 2-a | 95 |
| | | | 3-g | 14 |
| Nonachlorobiphenyls | | | | |
| 2,2′,3,3′,4,4′,5,5′,6- | 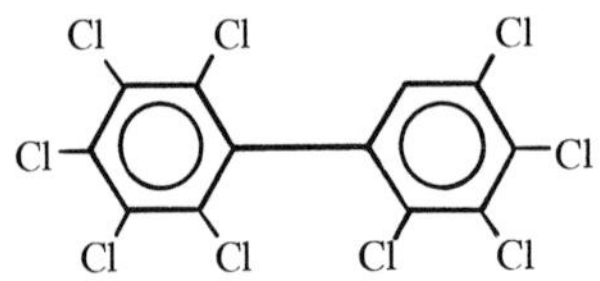 | 204.5–206.5 | 1-b | 24 |
| 2,2′,3,3′,4,5,5′,6,6′- | 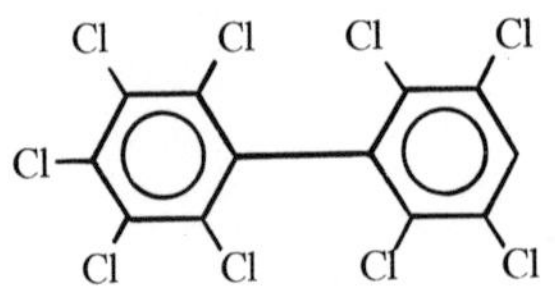 | * | 3-h | 19 |
| Decachlorobiphenyl | | | | |
| | 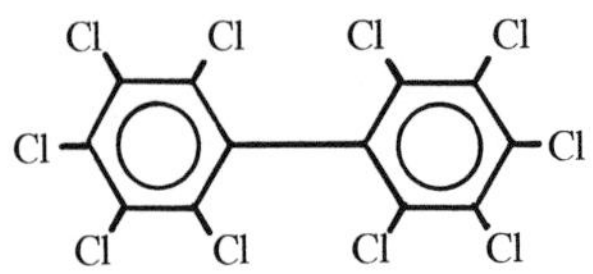 | 310 | 3-b | 185 |
| | | 305–6 | 3-a | 95 |
| | | | | 96 |
| | | | | 14 |
| | | | 2-c | 42 |

**Methods for the Synthesis of Labeled Chlorobiphenyls**

References on the synthesis of labeled molecules in general[137,170,195] and on specific isotopes such as carbon-14,[29,149] deuterium,[182] and tritium[61,194] are available.

Determinations of isotope distribution in labeled compounds are discussed in Reference 169. The "Isotope Index" provides useful information on the availability of specific compounds.[171]

*Chlorobiphenyls Labeled with Carbon-14*

For most studies on the metabolism, distribution, and accumulation of chlorobiphenyls in biological system, $^{14}C$ is the preferred label. $^{14}C$ has a long half-life (5700y) and its relatively intense (0.155 MeV) $\beta^-$ radiation allows for high counting efficiency. Most important, however, label is located in the carbon skeleton and initial metabolic alteration of the chlorobiphenyl molecule is not likely to remove the tagged atom. The removal of the label by metabolic action on the other hand is more likely with $^{36}Cl$ and particularly $^{3}H$. Although a limited number of carbon-14-labeled chlorobiphenyls are available commercially (Mallinckrodt), very few examples of their preparation have actually been described in the literature.

The three isomeric monochlorobiphenyls ($^{14}C$) were obtained in the course of mechanistic studies by the decomposition of *N*-nitrosoacetanilide (phenyl-$^{14}C$) in chlorobenzene.[130] 2,3′,4,4′-Tetrachlorobiphenyl ($^{14}C$) and 2,2′,3,4′-tetrachlorobiphenyl ($^{14}C$) (specific activity of mixture: 7.8 Ci/mol) were prepared by the reaction of diazotized 2,4-dichloroaniline ($^{14}C$) with 1,2-dichlorobenzene.[136] Analogously, 2,2′,3,4,4′,6′- and 2,3,′,4,4′,5′,6-hexachlorobiphenyl ($^{14}C$) were prepared from 2,4,6-trichloroaniline ($^{14}C$) and 1,2,3-trichlorobenzene.[176,186]

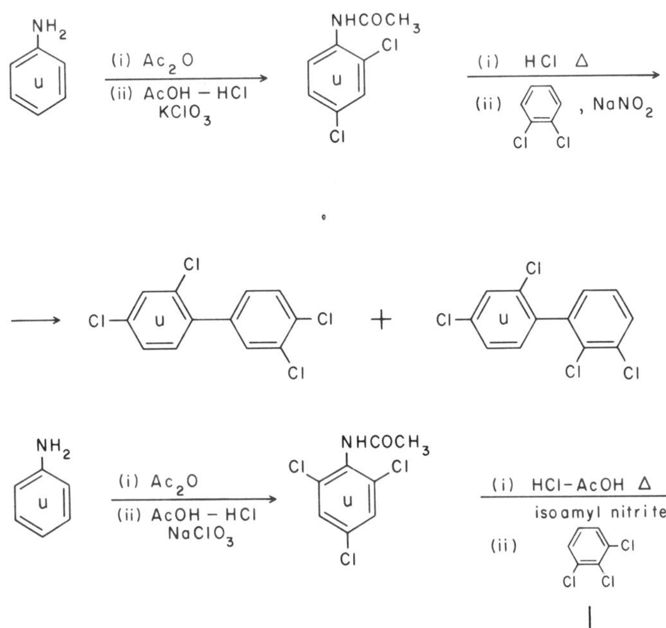


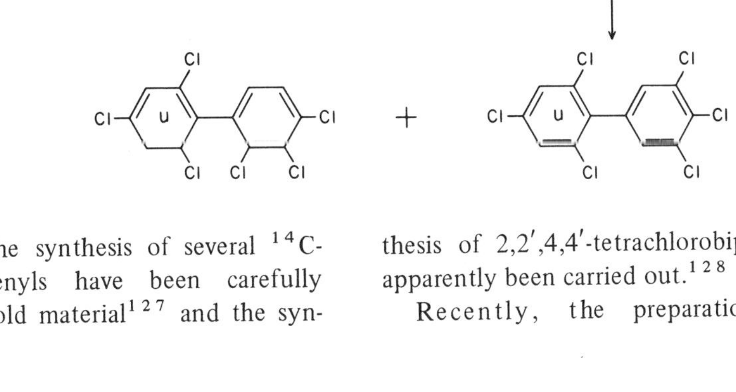


Conditions for the synthesis of several $^{14}C$-labeled chlorobiphenyls have been carefully worked out using cold material[127] and the synthesis of 2,2′,4,4′-tetrachlorobiphenyl ($^{14}C$) has apparently been carried out.[128]

Recently, the preparation of 3,3′,4,4′-

tetrachloroazobenzene ($^{14}C$) has been described.[177] 3,4-Dichloroaniline ($^{14}C$) was oxidized with pertrifluoroacetic acid to the corresponding nitro compound and subsequent reduction with zinc and aqueous sodium hydroxide gave the desired product. These types of compounds with a different chlorine substitution pattern (positions 4 and 4′ unsubstituted) would be useful intermediates for the synthesis of chlorobiphenyls.

Biphenyl ($^{14}C$) itself has been prepared by the nitrosamine reaction.[113] By Ullmann coupling[152,198] and by the reaction of phenylmagnesium bromide with cyclohexanone ($^{14}C$).[90] A good summary for synthetic methods leading to terphenyls-($^{14}C$) is given in Reference 100.

As far as the authors are aware, no biphenyl derivatives enriched with the carbon-13 isotope have been prepared.

*Chlorobiphenyls Labeled with Chlorine-36*

Chlorine-36 with its long half-life ($4 \times 10^5$ y) and relatively intense (0.716 MeV) $\beta^-$ radiation is a useful label since it is easily introduced in many instances into the chlorobiphenyl in the last step of a preparative procedure, assuring efficient use of the relatively inexpensive inorganic $^{36}Cl$. This isotope appears particularly useful if selective removal of chlorine atoms is to be studied.

Commercial Aroclor mixtures 1248 and 1254 were irradiated with neutrons to prepare $^{36}Cl$-labeled PCB mixtures.[172] Specific activities of 75 to 100 dpm per $\mu$g were obtained at flux densities of 4 to $5 \times 10^{13}$ neutrons/ $cm^3$ and 40 min irradiation time. The n,$\nu$ reaction is accompanied by an n,p reaction to yield sulfur-35 and correspondingly sulfur-containing chlorobiphenyls were detected in the mixtures. Irradiation induced polymerization decreased the yield of $^{36}Cl$-PCB to 10%.

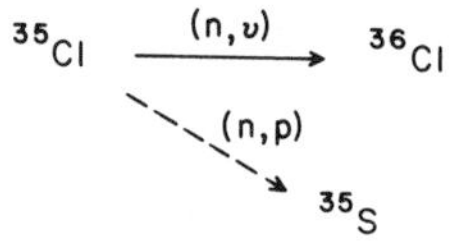


Several mono- and dichlorobiphenyls ($^{36}Cl$) were obtained during mechanistic studies on the chlorination of biphenyl (employing $^{36}Cl_2$)[15] and the decomposition of *N*-nitroso-*p*-chloroacetamide ($^{36}Cl$) in a number of substrates.[131-133]

2,2′,4,4′,5,5′-Hexachlorobiphenyl (4,4′-$^{36}Cl$; specific activity, $1.2 \times 10^{-3}$ $\mu$Ci/mM) and 3,3′,4,4′-tetrachlorobiphenyl (4,4′-$^{36}Cl$; specific activity, $1.4 \times 10^{-3}$ $\mu$Ci/mM) have been prepared from the corresponding tetrachlorobenzidine and dichlorobenzidine.[94] The bis-diazoniumfluoroborates were reacted with ferric chloride-$^{36}Cl$ in dimethyl sulfoxide, a Sandmeyer-type procedure[105] which was particularly suited to small scale preparations.

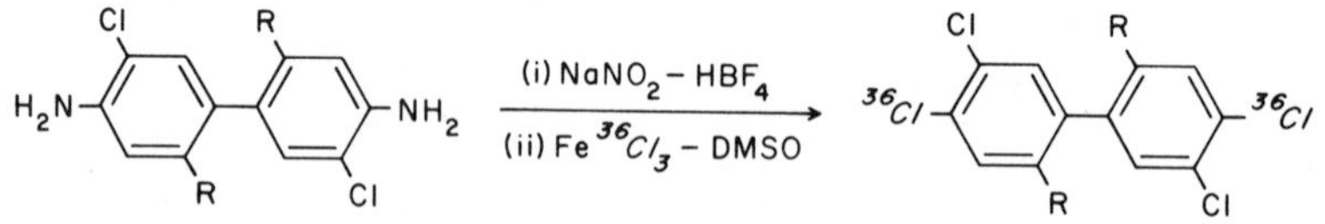


Since chlorobiphenylamines (aminochlorobiphenyls) are easily accessible compounds (Chapter 4) this method of labeling chlorobiphenyls with chlorine-36 seems to be generally useful.

*Chlorobiphenyls Labeled with Deuterium and Tritium*

The disadvantages of labeling with tritium are its relative weak (0.018) $\beta^-$ emission (half-life: 12.5y) and the fact that the activity can easily be lost for instance by metabolic substitution at the position of T-labeling and possibly also by exchange with the biological medium.

On the other hand, tritiated compounds are usually easy to prepare from inexpensive precursors ($T_2$, $T_2O$, $LiAlH_3T$). Tritium can easily be introduced into specific positions in the molecule which can greatly aid metabolic investigations.

Deuterium can serve as isotope to work out conditions for tritiation. Incorporation as well as position of labeling can be ascertained by NMR and mass spectrometry. Of course, deuterium can be used as stable-isotope label in its own right particularly when radioactive isotopes cannot be used because of health hazards.

A tritium-labeled PCB mixture corresponding to Kanechlor-400 was prepared by chlorination of specifically tritiated biphenyls with chlorine and ferric chloride catalyst. The tritiated biphenyls were prepared by the reduction of 2- and 3-bromobiphenyl with tritium gas and Pd-black and the reaction of the Grignard reagents[97] from 2- and 4-bromobiphenyl and tritiated water in tetrahydrofuran.

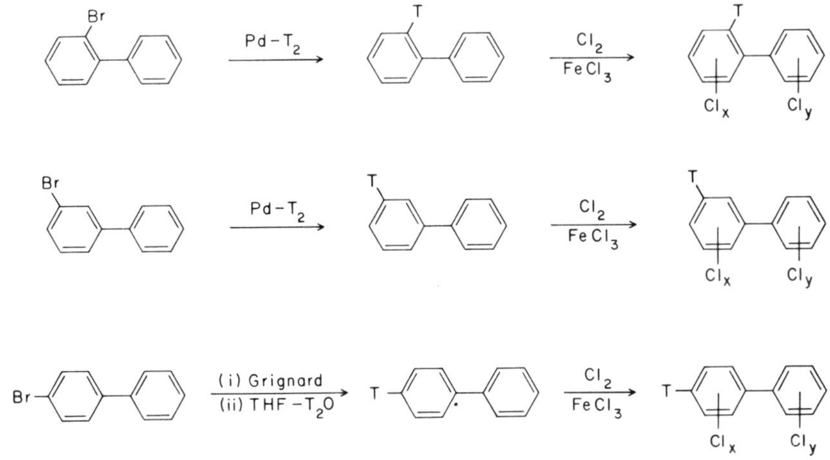

$x + y \sim 4$

Chlorobiphenyls with four chlorine atoms per molecule or less exchange readily with tritiated trifluoroacetic acid and platinum catalyst and high activities, particularly for mono- and dichlorobiphenyls (~1 mCi/mg) were obtained by this method.[187]

When highly active nonspecific, tritium-labeled chlorobiphenyls with Cl ⩾4 are required, the tritium exchange step is preferably carried out on an intermediate chlorobiphenylamine (aminochlorobiphenyl).[93] The amino group facilitates exchange, and for each tritiated aminochlorobiphenyl, two chlorobiphenyls can be obtained. For example, 2,2′,4,4′,5,5′-hexachlorobiphenyl and 2,2′,5,5′- tetrachlorobiphenyl (both of a specific activity of 1.9 mCi/m mol) were obtained from tritiated 2,2′,5,5′-tetrachlorobenzidine which was diluted 2.5:1 with cold material.[93] A similar acid catalyzed tritiation of a dimethylbiphenylamine was recently reported.[142]

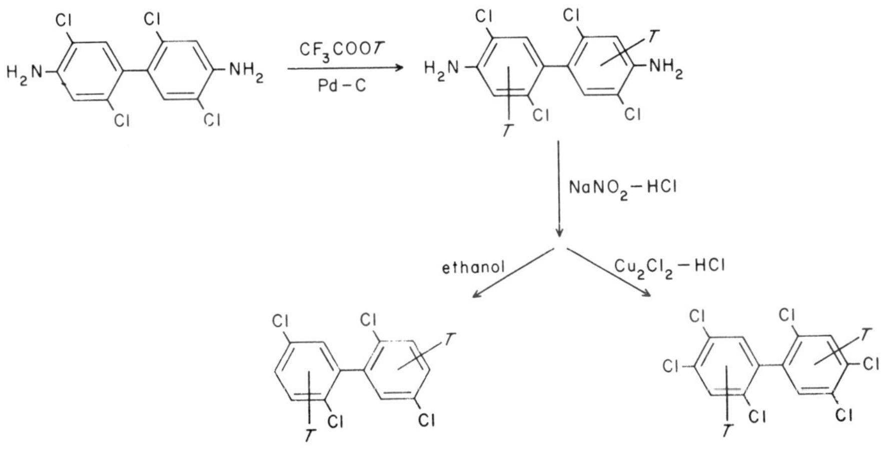

Several deuterated chlorobiphenyls were prepared[94] from the corresponding chloroiodobiphenyls by selective reduction of the iodine by lithium aluminum deuteride in isotopic purity of >98% (Table 1).

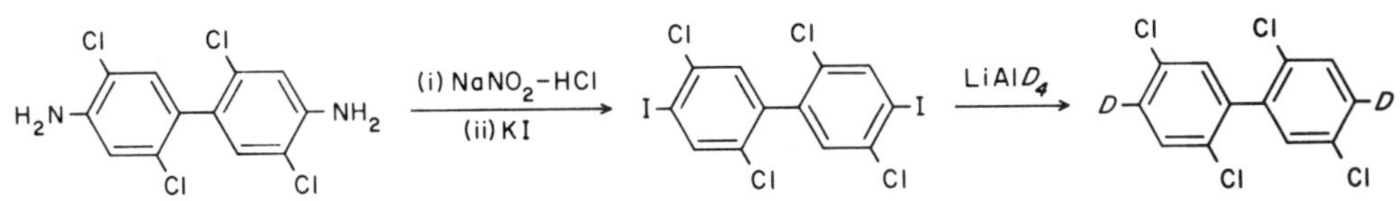

TABLE 1

**Preparation of Chlorobiphenyls (di-D)***

| Starting material | | Deuteriochlorobiphenyl | |
|---|---|---|---|
| Chloroiodobiphenyl | m.p.°C | Structure | $M^+$ *m/e* |
| 2,2′-dichloro-4,4′-diiodobiphenyl | 104–105° | 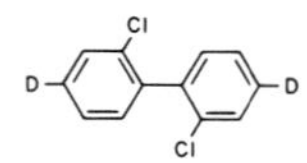 | 224 |
| 3,3′-dichloro-4,4′-diiodobiphenyl | 165 | 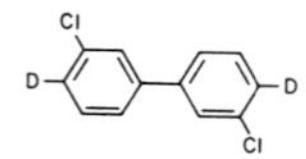 | 224 |
| 2,2′,5,5′-tetrachloro-4,4′-diiodobiphenyl | 119–120 | 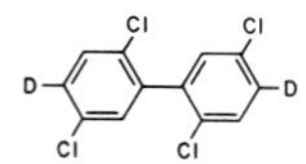 | 292 |
| 2,2′,4,4′,6,6′-hexachloro-3,3′-diiodobiphenyl | 166–167 | 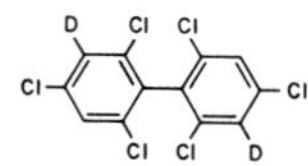 | 360 |

*From chloroiodobiphenyl (200 mg) and lithium aluminum deuteride (100 mg) in tetrahydrofuran (10 ml) stirred at room temperature for 20 hr.

Biphenyl has been labeled with deuterium or tritium.[4,5,13,69a,70,100,121,146] The most common methods for the preparation of these labeled biphenyls have been exchange with catalyst and/or acid, reduction of a halide with hydrogen isotope and palladium, decomposition of a Grignard reagent with $D_2O$ or $T_2O$ and deamination via diazonium salt and hypophosphoric acid.

## REFERENCES


1. **Abramovitch, R. A. and Koleoso, O. A.,** *J. Chem. Soc.,* (B), 779, 1969.
2. **Abramovitch, R. A. and Saha, J. G.,** *Tetrahedron,* 21, 3297, 1965.
3. **Ahmad, Y., Qureshi, M. I., and Baig, M. I.,** *Can. J. Chem.,* 45, 1539, 1967.
4. **Akawie, R. I.,** *J. Org. Chem.,* 26, 243, 1961.
5. **Akawie, R. I., Scarborough, J. M., and Burr, J. G.,** *J. Org. Chem.,* 24, 946, 1959.
6. **Augood, D. R., Hey, D. H., and Williams, G. H.,** *J. Chem. Soc.,* 44, 1953.
7. **Augood, D. R. and Williams, G. H.,** *Chem. Rev.,* 57, 123, 1957.
8. **Ayres, D. E.,** *Nature,* 240, 161, 1972.
9. **Bachmann, W. E. and Hoffman, R. A.,** *Org. React.,* 2, 224, 1944.
10. **Bacon, R. G. R. and Hill, H. A. O.,** *Quart. Rev.,* 19, 95, 1965.
11. **Bähr, G.,** Herstellung und Umwandlung von Organokupfer-Verbindungen, in *Methoden der Organischen Chemie* Vol. XIII/I, E. Müller Ed., Thieme, Stuttgart, 1970, 727.
12. **Bain, P. J., Lynch, E. R., McCall, E. B., and Roberts, R. J.,** *Proc. Chem. Soc.,* 186, 1962.
13. **Baker, R., Bott, R. W., and Eaborn, C.,** *J. Chem. Soc.,* 2136, 1963.
14. **Ballester, M.,** et al., *The Chlorocarbons,* Anthology 1, Office of Aerospace Research, ARL, 0167, (U.S. Clearinghouse No. AD 661099), 148, 1967.
15. **Beaven, G. H., de la Mare, P. B. D., Hassan, M., Johnson, E. A., and Klassen, N. V.,** *J. Chem. Soc.,* 2749, 1961.
16. **Beilstein, F. K.,** *Beilsteins Handbuch der Organischen Chemie,* Springer Verlag, Berlin, Vol. 5, Hauptwerk, 579; Ergänzungsband I, 272; Ergänzungsband II, 483; Ergänzungsband III, 1736.
17. **Bellavita, V.,** *Gazz. Chim. Ital.,* 65, 632, 1935.
18. **Berg. O. W., Diosady, P. L., and Rees, G. A. V.,** *Bull. Environ. Contam. Toxicol.,* 7, 338, 1972.
19. **Binns, F. and Suschitzky, H.** *J. Chem. Soc.,* (C), 1913, 1971.
20. **Björklund, C.,** *Acta Chem. Scand.,* 25, 2825, 1971.
21. **Björklund, C. and Nilsson, M.** *Tetrahedron Lett.,* 675, 1966.
22. **Blum, J. and Scharf, G.,** *J. Org. Chem.,* 35, 1895, 1970.
23. **Blum, J., Milstein, D., and Sasson, Y.,** *J. Org. Chem.,* 35, 3233, 1970.
24. **Bolgar, M.,** personal communication.
25. **Brackman, W. and Smit, P. J.,** *Rec. Trav. Chim.,* 85, 857, 1966.
26. **Braendlin, H. P. and McBee, E. T.,** in *Friedel-Crafts and Related Reactions,* Vol. III/2, Olah, G. A., Ed., Interscience, New York, 1964, 1517.
27. **Bretherick, L.,** *Chem. Ind. (Lond.),* 1017, 1971.
28. **Bryce-Smith, D. and Clarke, P.,** *J. Chem. Soc.,* 2264, 1956.
29. **Bübner, M. and Schmidt, H.,** *Die Synthese Kohlenstoff-14 Markierter Organischer Verbindungen,* Leipzig, 1966.

29a. **Buchi, G., Klaubert, D. H., Shank, R. C., Weinreb, S. M., and Wogan, G. N.,** *J. Org. Chem.,* 36, 1143, 1971.

30. **Cadogan, J. I. G.,** *J. Chem. Soc.,* 4257, 1962.
31. **Cadogan, J. I. G. and Molina, G. A.,** *J. Chem. Soc.,* Perkin I, 541, 1973.
32. **Cadogan, J. I. G., Roy, D. A., and Smith, D. M.,** *J. Chem. Soc.,* (C), 1249, 1966.
33. **Cadogan, J. I. G., Hibbert, P. G., Siddiqui, M. N. U., and Smith, D. M.,** *J. Chem. Soc.,* Perkin I, 2555, 1972.
34. **Cain, J. C.,** *J. Chem. Soc.,* 7, 1904.
35. **Cairncross, A. and Sheppard, W. A.,** *J. Am. Chem. Soc.,* 90, 2186, 1968.
36. **Challis, B. C. and Butler, A. R.,** Substitution at an amino group, in *The Chemistry of the Amino Group,* Patai, S., Ed., Interscience, New York, 1968, 277.
37. **Coates, G. E.,** *Organometallic Compounds,* Methuen, London, 1968.
38. **Coe, D. G., Rydon, H. N., and Tonge, B. L.,** *J. Chem. Soc.,* 323, 1957.
39. **Cohen, S. C. and Massey, A. C.,** *Chem. Commun.,* 457, 1966.
40. **Cohen, S. C. and Massey, A. C.,** *Tetrahedron Lett.,* 4393, 1966.

41. Cohen, S. C., Fenton, D. E., Shaw, D., and Massey, A. G., *J. Organometal. Chem.*, 8, 1, 1967.
42. Cohen, S. C., Fenton, D. E., Tomlinson, A. J., and Massey, A. G., *J. Organometal. Chem.*, 6, 301, 1966.
43. Coquillon, J., French Patent 1,316,510, 1963; *Chem. Abstr.*, 59, 6310, 1963.
44. Cowdrey, W. A. and Davies, D. S., *Quart. Rev.*, 6, 358, 1952.
45. Cumbo, C. C., German Patent, 1,804,695, 1969; *Chem. Abstr.*, 71, 49510, 1969.
46. Dannley, R. L. and Gregg, E. C., Jr., *J. Am. Chem. Soc.*, 76, 2997, 1954.
47. Davies, D. I., Done, J. N., and Hey, D. H., *J. Chem. Soc.*, (C), 2506, 1969.
48. DeCrauw, T., *Rec. Trav. Chim.*, 50, 753, 1931.
49. De la Mare, P. B. D. and Johnson, E. A., *J. Chem. Soc.*, 4076, 1963.
50. De la Mare, P. B. D. and Ridd, H. H., *Aromatic Substitution,* Butterworths, London, 1959.
51. De la Mare, P. B. D., Harris, D. M., and Hassan, M., *Chem. Ind. (Lond.)*, 1086, 1958.
52. DeMey, C. A., DeKok, A. J., Lugtenburg, J., and Romers, C., *Rec. Trav. Chim.*, 91, 383, 1972.
53. DeMey, C. A. and Kooyman, E. C., *Rec. Trav. Chim.*, 90, 1337, 1971.
54. DeMay, C. A. and Kooyman, E. C., *Rec. Trav. Chim.*, 91, 357, 1972.
55. DePasquale, R. J. and Tamborski, C., *J. Org. Chem.*, 34, 1736, 1969.
56. Dermer, O. C. and Edmison, M. T., *Chem. Rev.*, 57, 77, 1957.
57. Dickerman, S. C. and Weiss, K., *J. Org. Chem.*, 22, 1070, 1957.
58. Dickerman, S. C., Weiss, K., and Ingberman, A. K., *J. Org. Chem.*, 21, 380, 1956.
59. Eberhardt, M. and Eliel, E. L., *J. Org. Chem.*, 27, 2289, 1962.
60. Elks, J., Haworth, J. W., and Hey, D. H., *J. Chem. Soc.*, 1284, 1940.
61. Evans, E. A., *Tritium and its Compounds,* Butterworths, London, 1966.
62. Fanta, P. E., *Chem. Rev.*, 38, 139, 1946.
63. Fanta, P. E., *Chem. Rev.*, 64, 613, 1964.
64. Fanta, P. E., *Synthesis,* in press.
65. Feast, W. J. and Musgrave, W. K. R., in *Rodd's Chemistry of Carbon Compounds,* Vol. III, Part A, Coffey, S., Ed., Elsevier Publishing Co., Amsterdam, 1971, 241.
66. Fichter, F. and Adler, M., *Helv. Chim. Acta,* 9, 287, 1926.
67. Finzi, C. and Bellavita, V., *Gazz. Chim. Ital.*, 64, 335, 1934.
68. Frater, G. and Havinga, E., *Tetrahedron Lett.*, 4603, 1969.
69. Fuson, R. C. and Cleveland, E. A., *Org. Synthesis Collect.*, Vol. 3, 339, 1955.
69a. Garnett, J. L. and Hodges, R. J., *Chem. Commun.*, 1220, 1967.
70. Geiss, F., Sandroni, S., and Blech, G., *J. Labeled Comp.*, 3, 271, 1967.
71. General Aniline and Film Corp., U.S. Patent 2,280,504, 1939; *Chem. Abstr.*, 36, 5658, 1942.
72. Gilman, H. and Gaj, B. J., *J. Org. Chem.*, 22, 447, 1957.
73. Gomberg, M. and Bachman, W. E., *J. Am. Chem. Soc.*, 46, 2339, 1924.
74. Hall, C. D., *Chem. Ind. (Lond.)*, 384, 1965.
75. Hall, D. M. and Minhaj, F., *J. Chem. Soc.*, 4585, 1957.
76. Hampson, G. C., and Weissberger, A., *J. Am. Chem. Soc.*, 58, 2111, 1936.
77. Hardie, R. L. and Thompson, R. H., *J. Chem. Soc.*, 2512, 1957.
78. Hardie, R. L. and Thompson, R. H., *J. Chem. Soc.*, 1286, 1958.
79. Hey, D. H. and Perkins, M., *J. Org. Synthesis,* 49, 44, 1969.
80. Hey, D. H., Nechvatal, A. and Robinson, T. S., *J. Chem. Soc.*, 2892, 1951.
81. Hey, D. H., Stirling, C. J. M., and Williams, G. H., *J. Chem. Soc.*, 2747, 1954.
82. Hey, D. H., Liang, K. S. Y., Perkins, M. J., and Williams, G. H., *J. Chem. Soc.*, (C), 1153, 1967.
83. Hickmott, P. W. and Hudson, A. T., *J. Chem. Soc.*, (C), 762, 1971.
84. Hinkel, L. E. and Hey, D. H., *J. Chem. Soc.*, 2786, 1928.
85. Hodgson, H. H., *J. Chem. Soc.*, 745, 1946.
86. Hoffmann, R. W., *Dehydrobenzene and Cycloalkynes,* Verlag Chemie, Weinheim and Academic Press, New York, 1967.
87. Hoffmann, R. W. and Häuser, H., *Tetrahedron Lett.*, 197, 1964.
88. Holzach, K., *Die Aromatischen Diazoverbindungen,* Ferdinand Enke, Stuttgart, 1947.
89. Huisgen, R. and Grashey, R., *Ann. Chem.*, 607, 46, 1957.
90. Huisgen, R., Jakob, F., and Grashey, R., *Chem. Ber.*, 92, 2206, 1959.
91. Huisgen, R., Sauer, J., and Hauser, A., *Chem. Ber.*, 91, 2366, 1958.
92. Hutzinger, O., Unpublished results.
93. Hutzinger, O. and Safe, S., *Bull. Environ. Contam. Toxicol.*, 7, 374, 1972.
94. Hutzinger, O., Pothier, C., and Safe, S., *J. Assoc. Offic. Anal. Chem.*, 55, 753, 1972.

95. Hutzinger, O., Safe, S., and Zitko, V., *Bull. Environ. Contam. Toxicol.*, 6, 209, 1971.
96. Hutzinger, O., Safe, S., and Zitko, V., *Int. J. Environ. Anal. Chem.*, 2, 95, 1972.
97. Ikegami, S., Kawamoto, K., Kasida, Y., Akaboshi, S., Enogaki, K., and Baba, S., *Radioisotopes*, 20, 65, 1971.
98. Jenkins, R. L., McCollough, R., and Booth, C. F., *Ind. Eng. Chem.*, 22, 31, 1930.
99. Johnson, G. J., Ecobichon, D., and Hutzinger, O., *Toxicol. Appl. Pharmacol.*, in press.
100. Juppe, G., Otto, P. P. H. L., Hafferl, W., and Fischer, E., Die Synthese von Radiokohlenstoff und Tritium Markierten Di- und Terphenylen, in *Methods of Preparing and Storing Labelled Compounds*, Sirchis, J., Ed., Euratom, Brussels, 1968, 379.
101. Kaluszyner, A. and Reuter, S., *J. Am. Chem. Soc.*, 75, 5126, 1953.
102. Karelsky, M. and Pausacker, K. H., *Aust. J. Chem.*, 11, 39, 1958.
103. Kharasch, M. S. and Reinmuth, O., *Grignard Reactions of Nonmetallic Substances*, Prentice-Hall, New York, 1954.
104. Kleiderer, E. C. and Adams, R., *J. Am. Chem. Soc.*, 55, 4219, 1933.
105. Kobayashi, M. and Yamada, E., *Org. Prep. Proced.*, 1, 221, 1969.
106. Kobayashi, M., Minato, H., and Kobori, N., *Bull, Chem. Soc. Jap.*, 43, 219, 1970.
107. Kobayashi, M., Minato, H., Kobori, N., and Yamada, E., *Bull. Chem. Soc. Jap.*, 43, 1131, 1970.
108. Kobayashi, M., Minato, H., Wanatabe, N., and Kobori, N., *Bull. Chem. Soc. Jap.*, 43, 258, 1970.
109. Kobayashi, M., Minato, H., Yamada, E., and Kobori, N., *Bull. Chem. Soc. Jap.*, 43, 215, 1970.
110. Kornblum, N., *Org. React.*, 2, 262, 1944.
111. Kornblum, N. and Kendall, D. L., *J. Am. Chem. Soc.*, 74, 5782, 1952.
112. Kovacic, P., in *Friedel-Crafts and Related Reactions*, Vol. 4, Olah, G. A., Ed., Interscience, New York, 1965, 111.
113. Koyama, H., Morikawa, N., Migita, T., Ito, R., and Tsuchihashi, I., *Radioisotopes*, 9, 112, 1960; Extensive English summary on p. 115.
114. Krause, H. W., Selke, R., Mennenga, H., and Kreuzfeld, H. J., East German Patent 72,262, 1970; *Chem. Abstr.*, 73, 109510, 1970.
115. Lawlor, F. E., U.S. Patent, 2,608,532, 1952; *Chem. Abstr.*, 48, 3390, 1954.
116. Levy, L. A., *Synthesis*, 170, 1973.
117. Lewin, A. H., Zovko, M. J., Rosewater, W. H., and Cohen, T., *Chem. Commun.*, 80, 1967.
118. Lidgett, R. A., personal communication.
119. Mascarelli, L., Gatti, D., Jona, E., and Leoncini, V., *Gazz. Chim. Ital.*, 59, 867, 1929.
120. Mascarelli, L. and Gatti, D., *Gazz. Chim. Ital.*, 63, 654, 1933.
121. Matsuo, M., Matsui, E., Matsuo, T., and Kasida, Y., *Radioisotopes*, 17, 483, 1968; *Chem. Abstr.*, 70, 36766, 1969.
122. McBee, E. T., Diveley, W. R., and Burch, J. E., *J. Am. Chem. Soc.*, 77, 385, 1955.
123. McBee, E. T., Idol, J. D., Jr., and Roberts, C. W., *J. Am. Chem. Soc.*, 77, 6674, 1955.
124. McCall, E. B. and Blackman, E. J., British Patent 929,093, 1963; *Chem. Abstr.*, 59, 12704, 1963.
125. McKillop, A., Elsom, L. F., and Taylor, E. C., *J. Am. Chem. Soc.*, 90, 2423, 1968.
126. McKillop, A., Elsom, L. F., and Taylor, E. C., *Tetrahedron*, 26, 4041, 1970.
127. Melvas, B., *Nord. Hyg. Tidskr.*, 52, 40, 1971; *Chem. Abstr.*, 75, 88226, 1971; National Research Council of Canada, Technical Translation, Ottawa, 1972.
128. Melvas, B. and Brandt, I., PCB Conference II, Stockholm, December 1972.
129. Merz, V. and Weith, W., *Ber.*, 16, 2869, 1883.
130. Migita, T., *Yuki Gosei Kagaku Kyokai Shi.*, 19, 840, 1961; *Chem. Abstr.*, 56, 10008, 1962.
131. Migita, T., Ito, R., and Osamu, S., *Yuki Gosei Kagaku Kyokai Shi.*, 19, 613, 1961; *Chem. Abstr.*, 55, 24667, 1961.
132. Migita, T., Morikawa, N., and Simamura, O., *Bull. Chem. Soc. Jap.*, 36, 980, 1963.
133. Migita, T., Ito, R., Kazuo, T., and Osamu, S., *Yuki Gosei Kagaku Kyokai Shi.*, 19, 609, 1961; *Chem. Abstr.*, 55, 24667, 1961.
134. Miles, P. and Suschitzky, H., *Tetrahedron*, 18, 1369, 1962.
135. Morizur, J. P., *Bull. Soc. Fr.*, 1331, 1964.
136. Moron, M., Sundström, G., and Wachtmeister, C. A., *Acta Chem. Scand.*, 26, 830, 1972.
137. Murray, A. and Williams, D. L., *Organic Syntheses with Isotopes*, Parts 1 and 2, Interscience, New York, 158, 1958.
138. Neale, A. J., Rawlings, T. J., and McCall, E. B., *Tetrahedron*, 21, 1313, 1965.
139. Nefedov, O. M., Levkov, Y. L., and Petrov, A. D., *Dokl. Akad. Nauk S.S.S.R.*, 133, 855, 1960; *Chem. Abstr.*, 54, 24567, 1960.
140. Nilsson, M., *Acta Chem. Scand.*, 20, 423, 1966.
141. Nilsson, M. and Wennerström, O., *Acta Chem. Scand.*, 24, 482, 1970.
142. Noall, M. W., *J. Labeled Comp.*, 5, 192, 1969.
143. Norman, R. O. C. and Taylor, R., *Electrophilic Substitution in Benzenoid Compounds*, Elsevier Publishing Co., Amsterdam, 1965.
144. Normant, J. F., *Synthesis*, 63, 1972.
145. Osborne, D. R., Unilever Research, Sharnbrook, personal communication.
146. Otto, P. P. H. L., Stein, F., and Juppe, G., *J. Labeled Comp.*, 4, 205, 1968.

147. **Pausacker, K. H.,** *Aust., J. Chem.,* 10, 49, 1957.
148. **Pett, M. R. and Tatlow, J. C.,** *J. Chem. Soc.,* 1941, 1954.
149. **Pichat, L.,** Les préparations de molécules marquées au carbon-14, *Bull. Inf. Sci. Tech.,* Commisariat a l'Energie Atomique, No. 118, 24, 1967.
150. **Puchkov, V. A.,** *Khim. Nauka Prom. St'.,* 4, 549, 1959.
151. **Putter, R.,** Methoden zur Herstellung und Umwandlung Aromatischer Diazoniumsalze, in *Methoden der Organischen Chemie,* Vol. X/3, Müller, E., Ed., G. Thieme, Stuttgart, 1965, 7.
152. **Raig, P. and Ammon, R.,** *Arzneimittelforsch.,* 22, 1399, 1972.
153. **Raudnitz, H.,** *Ber.,* 60, 738, 1927.
154. **Relles, H. M.,** *J. Org. Chem.,* 34, 3687, 1969.
155. **Roe, A.,** *Org. React.,* 5, 203, 1949.
156. **Saeki, S., Tsutsui, A., Oguri, K., Yoshimura, H., and Hamana, M.,** *Fukuoka Igaku Zhasshi,* 62, 20, 1971; *Chem. Abstr.,* 74, 146294, 1971. National Research Council of Canada, Technical Translation, Ottawa, 1971.
157. **Safe, S. and Hutzinger, O.,** *J. Chem. Soc.,* Perkin Trans. I, 686, 1972.
158. **Sandoz, A. G.,** Belgian Patent, 665,732, 1965; *Chem. Abstr.,* 64, 19847, 1966.
159. **Saunders, K. H.,** *The Aromatic Diazo-compounds and their Industrial Applications,* Arnold & Co., London, 1949.
160. **Scarborough, H. A. and Waters, W. A.,** *J. Chem. Soc.,* 89, 1927.
161. **Scheidt, F. M.,** German Patent 2,056,441, 1971; *Chem. Abstr.,* 75, 35379, 1971.
162. **Schmidt, H. and Schultz, G.,** *Ann. Chem.,* 207, 311, 1881.
163. **Schwechten, H. W.,** *Ber.,* 65, 1605, 1932.
164. **Semmelhack, M. F., Helquist, P. M., and Jones, L. D.,** *J. Am. Chem. Soc.,* 93, 5908, 1971.
165. **Sengupta, S. K.,** *Ind. J. Chem.,* 4, 235, 1966.
166. **Shaw, D. and Massey, A. G.,** *J. Organometal. Chem.,* 8, 1, 1967.
167. **Shaw, F. R. and Turner, E. E.,** *J. Chem. Soc.,* 285, 1932.
168. **Shu, H.,** *Hua Hsüeh Hsüeh Pao,* 25, 171, 1959; *Chem. Abstr.,* 54, 4489, 1960.
169. **Simon, H. and Floss, H. G.,** *Bestimmung der Isotopenverteilung in Markierten Verbindungen,* Springer Verlag, Berlin, 1967.
170. **Sirchis, J., Ed.,** *Methods of Preparing and Storing Labelled Compounds,* Euratom, Brussels, 1968.
171. **Sommerville, J. L., Ed.,** *The Isotope Index,* Scientific Equipment Co., Indianapolis, 1967.
172. **Stalling, D. L. and Huckings, J. N.,** *J. Assoc. Offic. Anal. Chem.,* 54, 801, 1971.
173. **Starnes, W. H., Jr.,** *J. Am. Chem. Soc.,* 84, 2270, 1962; U.S. Patent No. 3,341,569, 1967; *Chem. Abstr.,* 68, 12670, 1968.
174. **Stroh, R.,** Austausch der Diazoniumgruppe Gegen Chlor (Sandmeyer-Reaktion), in *Methoden der Organischen Chemie,* Vol. V/3, Müller, E., Ed., G. Thieme Stuttgart, 1962, 846.
175. **Stroh, R. and Hahn, W.,** Herstellung von Chlorverbindungen, in *Methoden der Organischen Chemie,* Vol. V/3, Müller, E., Ed., G. Thieme, Stuttgart, 1962, 503.
176. **Sundström, G. and Wachtmeister, C. A.,** *Acta Chem. Scand.,* 26, 3816, 1972.
177. **Tanaka, F. S. and Wien, R. G.,** *J. Labeled Comp.,* 7, 459, 1971.
178. **Tas, A. C. and deVos, R. H.,** *Environ. Sci. Technol.,* 5, 1217, 1971.
179. **Tas, A. C. and Kleinpool, R. J. C.,** *Bull. Environ. Contam. Toxicol.,* 8, 32, 1972.
180. **Tas, A. C. and Kleinpool, R. J. C.,** unpublished results, personal communication.
181. **Taylor, E. C., Kienzle, F., and McKillop, A.,** *J. Am. Chem. Soc.,* 92, 6088, 1970.
182. **Thomas, A. F.,** *Deuterium Labelling in Organic Chemistry,* Appleton-Century-Croft, New York, 1971.
182a. **Tsutsui, M.,** *Ann. N.Y. Acad. Sci.,* 93, 135, 1961.
183. **Ullmann, F.,** *Ann. Chem.,* 332, 38, 1904.
184. **Ullenius, C.,** *Acta Chem. Scand.,* 26, 3383, 1972.
185. **Van Roosmalen, F. L. W.,** *Rec. Trav. Chim.,* 53, 359, 1934.
186. **Wachtmeister, C. A.,** PCB Conference II, Stockholm, December, 1972.
187. **Wallnöfer, P. R., Engelhardt, G., Safe, S., and Hutzinger, O.,** *Chemosphere,* 2, 69, 1973.
188. **Webb, R. G. and McCall, A. C.,** *J. Assoc. Offic. Anal. Chem.,* 55, 746, 1972 and private communication.
189. **Weber, A. and Söllscher, C.,** *Ber.,* 16, 882, 1883.
190. **Weingarten, H.,** *J. Org. Chem.,* 25, 1066, 1960.
191. **Weingarten, H.,** *J. Org. Chem.,* 26, 730, 1961.
192. **Weingarten, H.,** *J. Org. Chem.,* 26, 4347, 1961.
193. **Weingarten, H.,** *J. Org. Chem.,* 27, 2024, 1962.
194. **Wenzel, M. and Schulze, P. E.,** *Tritium-Markierung,* W. DeGruyter, Berlin, 1962.
195. **Weygand, F. and Simon, H.,** Herstellung Isotopenhaltiger Organischer Verbindungen, in *Methoden der Organischen Chemie (Houben-Weyl),* Vol. IV/2, Müller, E., Ed., G. Thieme, Stuttgart, 1955, 539.
196. **Williams, G. H.,** *Homolytic Aromatic Substitution,* Pergamon Press, New York, 1960.
197. **Willis, D. E. and Addison, R. F.,** *J. Fish. Res. Bd. Can.,* 29, 592, 1972.
198. **Wynberg, H. and Wolf, A. P.,** *J. Am. Chem. Soc.,* 85, 3308, 1963.
199. **Zerweck, W. and Schütz, K.,** U.S. Patent 2,280,504, 1942; *Chem. Abstr.,* 36, 5658, 1942.
200. **Zerweck, W. and Schütz, K.,** German Patent 870, 106, 1953; *Chem. Abstr.,* 50, 16863, 1956.

Chapter 4

# OTHER BIPHENYL SYNTHESES

**Newer Synthesis of the Biphenyl Ring System**

In this section, newer methods for the synthesis of substituted biphenyls which have not been used for the preparation of chlorobiphenyls will be described. Some are of obvious potential utility for chlorobiphenyl synthesis, others may be useful for the preparation of intermediates or metabolites.

Many of the syntheses described are applicable for the preparation of terphenyls and polyphenyls.[5,155] However, these compounds will not be mentioned here. Discussion of biphenyl syntheses generally can be found in References 118, 119, and 155. Reviews dealing with specific methods are listed in the appropriate sections in this and the previous chapter.

*Biphenyls Via Aryl Grignard Reagents or Aryl Lithium Compounds and Cyclic Ketones*

Phenyl lithium compounds condense with cyclohexanones to give tertiary alcohols which can be dehydrated and dehydrogenated by heating with sulfur and zinc to give biphenyls.[1]

The analogous preparation of biphenyls from Grignard reagents has been briefly discussed.[9,82]

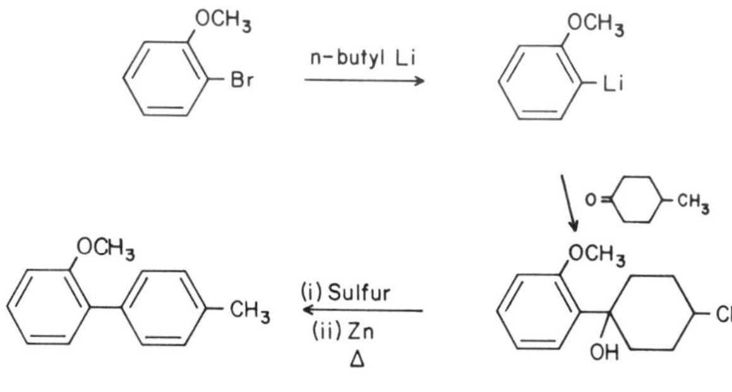


*Biphenyls by Coupling of Aromatic Compounds with Thallium (III) Acetate*

When benzene or toluene is heated with thallic triacetate in acetic acid, poor yields of biphenyls are obtained.[59] Yields are improved in the presence of methanesulfonic acid and palladium (II) acetate. It is interesting to note that a study of the isomer distribution in dimethylbiphenyl (from toluene) showed very little substitution in positions 2 and 2′ but usually well over 50% of the 3,3′ plus 3,4′ derivative.

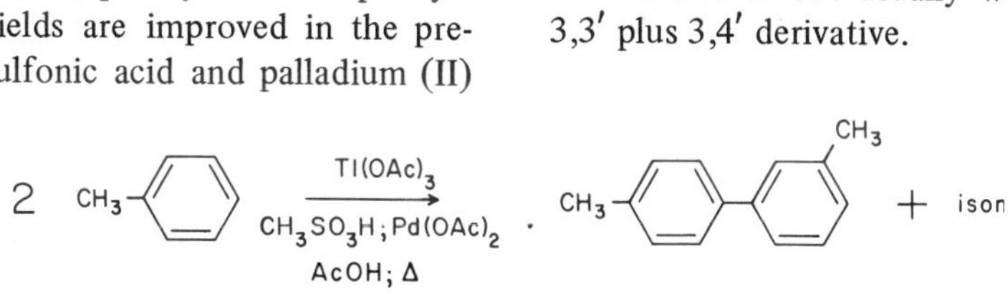


*Biphenyls by Oxidation of Aromatic Compounds with Lead Tetraacetate in Trifluoroacetic Acid*

Oxidation of aromatic compounds with lead tetraacetate in trifluoroacetic acid gives a number of derivatives including biphenyls, phenols, and aryl-lead compounds.[106] In certain cases (e.g., with methoxybenzene) reasonable yields of the corresponding biphenyl are obtained whereas in many cases other reaction products predominate (e.g., benzene gives 80% phenol).

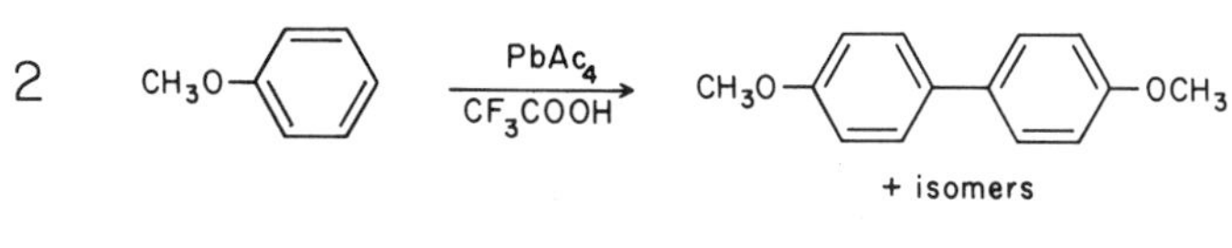

*Biphenyls by Oxidative Decomposition of N-phenyl-N' (tri-n-butylstannyl) hydrazines*

Unsymmetrical biphenyls can be prepared in yields of up to 90% by oxidative decomposition of *N*-phenyl-*N'*-(tri-*n*-butylstannyl) hydrazines.[76] The oxidizing agent used was chloranil and the preparation of methyl- and methoxybiphenyls were reported. The stannylhydrazines can be conveniently prepared in high yields from diethylaminotri-*n*-butylstannate and a phenylhydrazine.

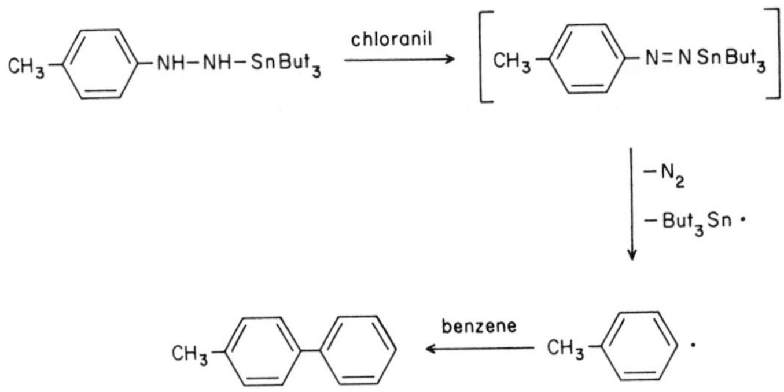

*Biphenyls from Acyclic Cross-Conjugated Dienamines and Aromatic Acid Chlorides*

Aromatic acid chlorides and cross-conjugated dienamines condense to give biphenyls dihydro-γ-pyrones, or γ-pyrones depending on the reaction conditions.[61] When the mixture of the reactants is heated in benzene under reflux, HCl is slowly evolved and biphenyls are formed almost exclusively. A number of morpholino-biphenyls with methyl-, methoxy-, and nitro-substituents were prepared using this procedure.

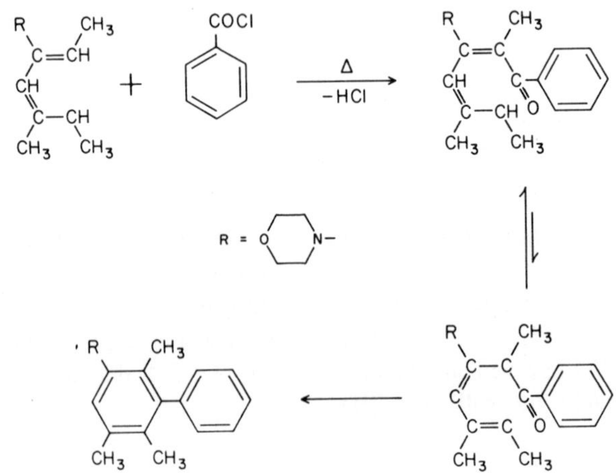

*Biphenyls by Reaction of Acetylenes with α-methylene Carbonyl Compounds*

The addition of acetylene derivatives to α-methylene carbonyl compounds was recently shown to yield biphenyls.[22] Only biphenylcarboxylic acids were prepared by this procedure.

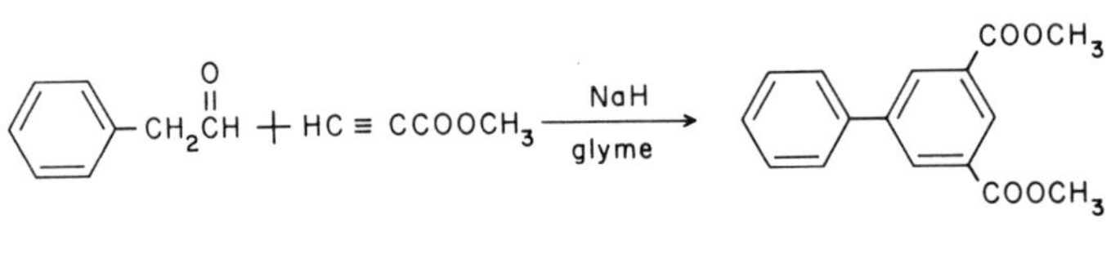

*Biphenyls via Intermediate Iron Complexes*

A simple stepwise process for the synthesis of biphenyls and related hydrocarbons was recently reported by Helling and Braitsch.[58] Bisphenyliron (II) complexes undergo addition with phenyllithium reagents and the adduct, on oxidation with potassium permanganate, gives high yields of biphenyl. Only methyl-biphenyl derivatives were prepared.

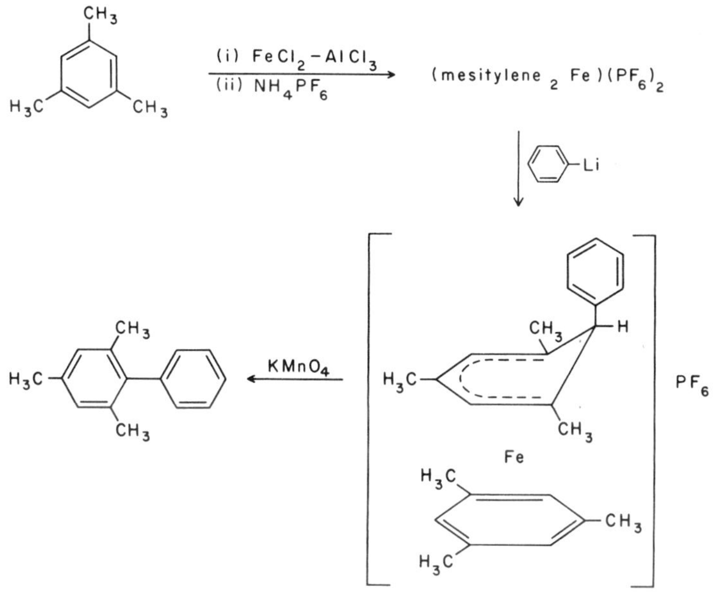


*Biphenyls from Aryl Halides and Bis (1,5-cyclooctadiene) Nickel (0)*

Semmelhack and co-workers recently reported[131] that a variety of aryl halides react directly with bis (1,5-cyclooctadiene) nickel (0) at moderate temperatures in dimethylformamide to produce biaryls. Good yields were obtained when the halide was bromine, but even chlorobenzene gave biphenyl (14%). Symmetrical methyl-, methoxy-, cyano-, formyl- and acetylbiphenyls can be prepared by this method but attempts to prepare nitro-, hydroxy-, or carboxy- biphenyls failed.

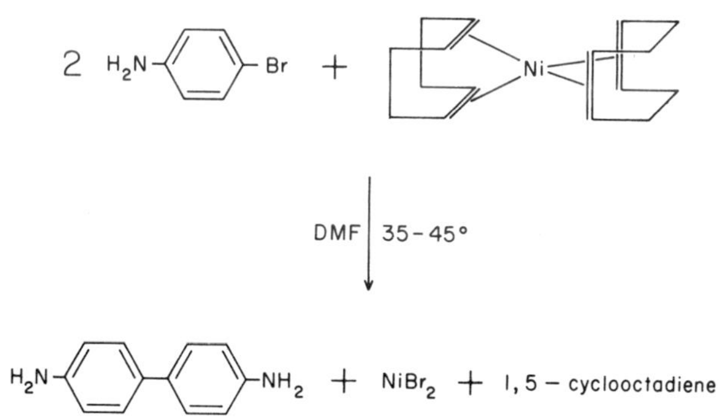


*Biphenyls by Photolysis of N-phenylsulfonyl-dimethyl Sulfoximine and Aromatic Bromo and Iodo Compounds*

A number of biphenyls carrying different substituents (methyl, hydroxy, amino, formyl, nitro, and carboxy) are accessible by photolysis of the appropriately substituted aromatic iodo compound in benzene. A UV lamp with a maximum output at λ 2537 Å was used, irradiation time was about 20 hr, and yields generally better than 50% were obtained.[151]

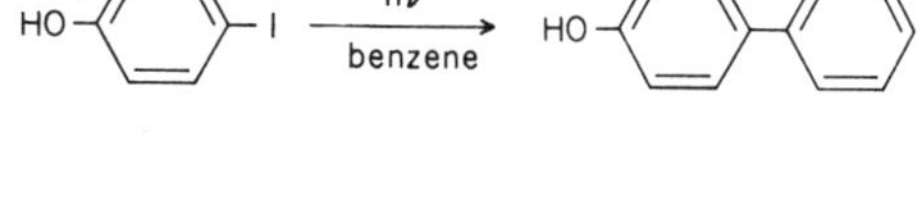

Irradiation of aromatic bromo compounds in benzene gives somewhat lower yields and these starting materials are apparently not useful for the synthesis of amino-and nitrobiphenyls.[95] Photolysis of pentafluoroiodobenzene in 1,2,4,5-tetrafluorobenzene gave 65% of the corresponding nonafluorobiphenyl.[20]

Biphenyl and methyl biphenyls were prepared by photolysis of *N*-phenylsulfonyldimethyl sulfoximines in benzene.[3]

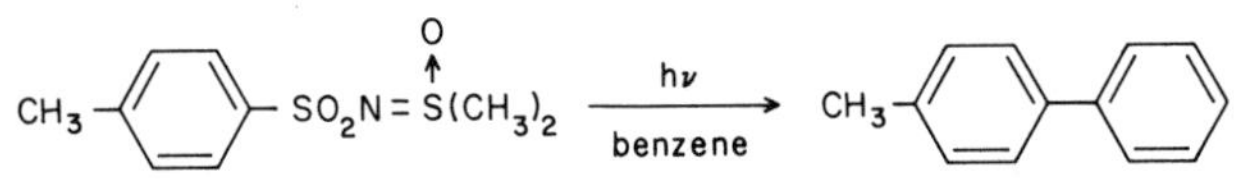


*Biphenyls from Aromatic Sulfur Compounds and Raney-Nickel*

A variety of aromatic sulfur compounds such as disulfides, thioethers and thiols were shown to be transformed by Raney nickel into the corresponding biaryl derivatives at high temperatures.[56] No substituted biphenyls were reported in this paper.

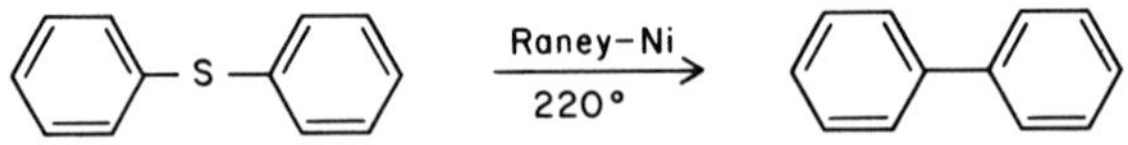


*Biphenyls by Nucleophilic Aromatic Substitution*

By treating aminobenzenes with reactive halobenzenes biphenyls were recently obtained.[43] The reaction proceeds via charge-transfer (CT) complexes which can be isolated if conditions are kept sufficiently mild. These CT complexes lose HCl when heated to give biphenyls via a nucleophilic substitution reaction.

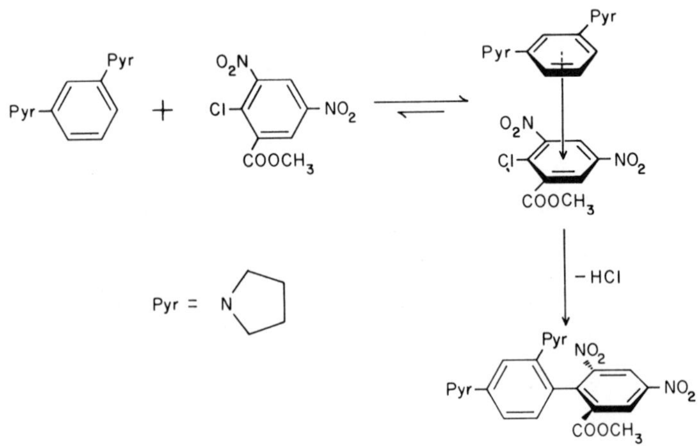


*Biphenyls from Sulfonamides*

Waldau and Pütter have shown[148] that isomer-free hydroxybiphenylamines can be obtained by treating certain sulfonamides with alkali. For the reaction to occur, the following requirements must be met: (a) in ring A, one position *ortho* or *para* to the sulfonamide group must be unsubstituted; (b) the hydroxy group on ring A must enable electrophilic entry of B to occur through its electron-donating effect (bond formation will occur *o*- or *p*- to the hydroxy group); and (c) ring B must be activated by electron-withdrawing substituents such as $NO_2$ or C=O in *o*- or *p*- position to the sulfonamide linkage.

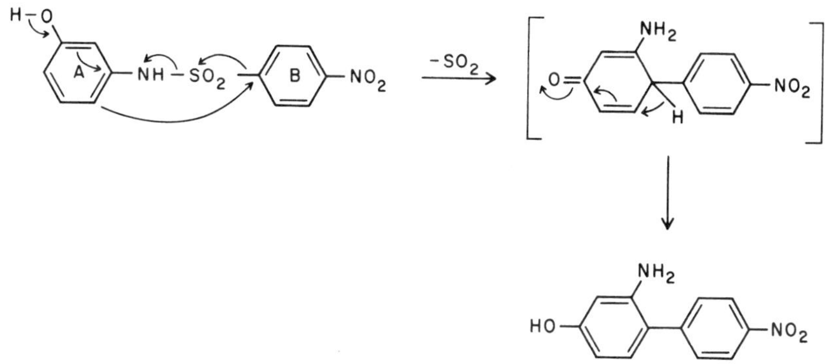

*Biphenyls by Rearrangement of N,O-Diarylhydroxylamines*

Recently, it was found that *N,O*-diarylhydroxylamines rearrange to form aminobiphenylols (aminohydroxybiphenyls)[133,134] which is a reaction analogous to the benzidine rearrangement. However, in contrast to the benzidine rearrangement, no acid catalysis is necessary and the reaction also occurs readily with rings carrying strong electron-withdrawing groups. The *N,O*-diarylhydroxylamines which are prepared from *N*-hydroxycarbamates and substituted fluorobenzenes are not isolated but undergo spontaneous rearrangement.

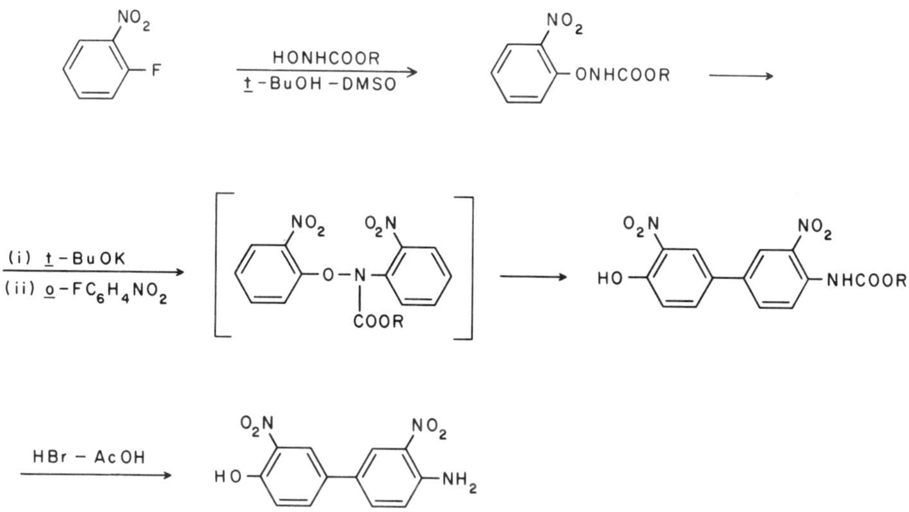

Aminobiphenylols (aminohydroxybiphenyls) have also been obtained from the phenylation (with diphenyliodonium salts) of the anion of *N*-acetyl-N-phenylhydroxylamine[37] and as a by-product in the reaction of 2-methylsulfonylnitrobenzene with *N*-(2-methylsulfonylphenyl) hydroxylamine in the presence of sodium hydroxide.[132] In both reactions, *N,O*-diarylhydroxylamine intermediates have been postulated.

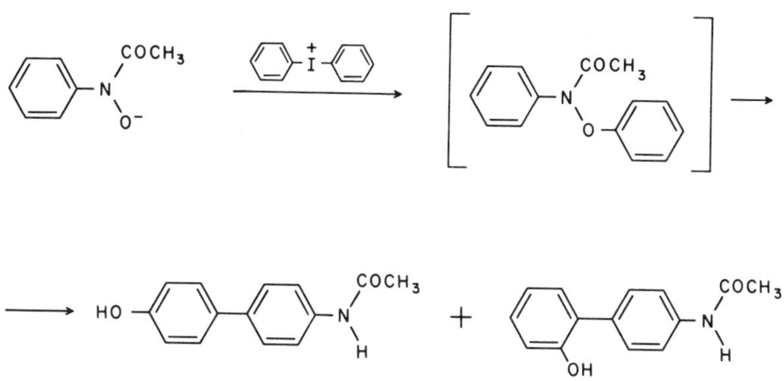

major product

*Biphenyls by Photochemical Synthesis from Phenols*

Unsymmetrical dihydroxybiphenyls were obtained when mixtures of a halogenophenol and a *p*-substituted phenol in aqueous potassium hydroxide were irradiated with a low pressure mercury lamp (λ max 2537 Å).[108,109] A series of alkyl and nitro-dihydroxybiphenyls were prepared; reaction times of 60 to 100 hr were used and yields of 40 to 80% were obtained. *Ortho, meta,* and *para* halophenols can be used in this reaction and iodine as well as bromine and chlorine have successfully been used as the halogen substituent. Similarly, irradiation of bromophenols gave dihydroxybiphenyls among other products.[79]

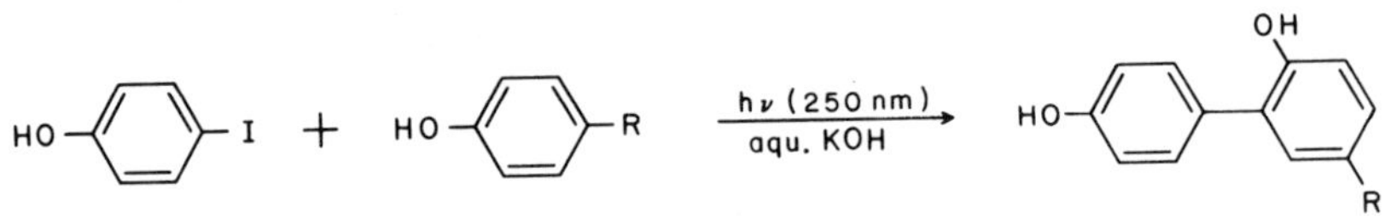


*Biphenyls from o-Bromobenzoates and Phloroglucinol*

2-Bromo-4-methylbenzoic acid reacts with phloroglucinol in the presence of slightly alkaline aqueous copper sulfate to give 5-methyl-2′,4′,6′-trihydroxybiphenyl-2-carboxylic acid in 50% yield.[116]

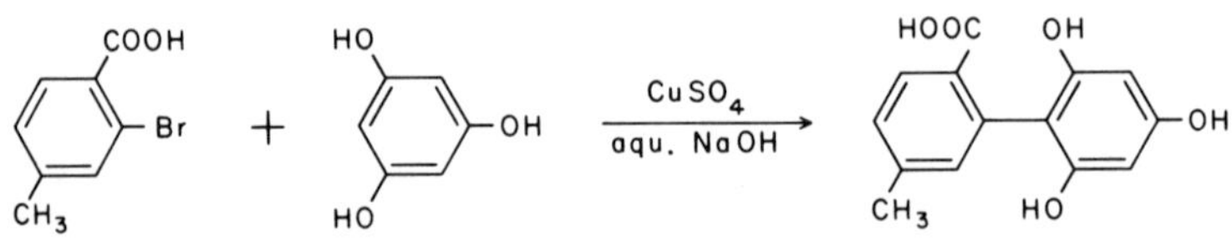


*Biphenyls by Oxidative Coupling*

Oxidative coupling of phenols and related compounds to give biaryl type derivatives is an accepted synthetic route in natural product chemistry.[80,104]

Reagents which have recently been used for the oxidative coupling to form biphenyl derivatives include isoamyl nitrite,[78] thallium(III)trifluoroacetate[144] and silver carbonate.[11]

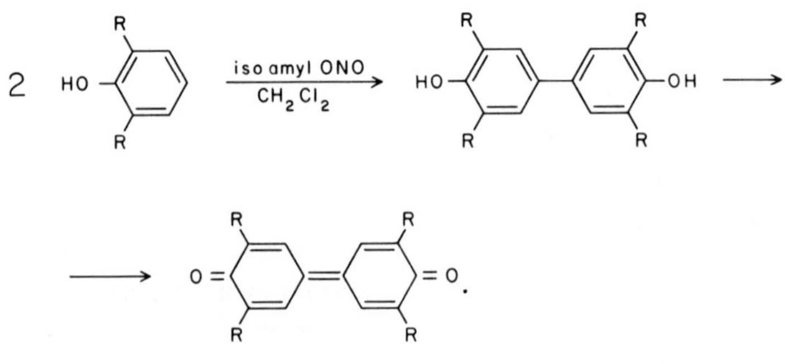


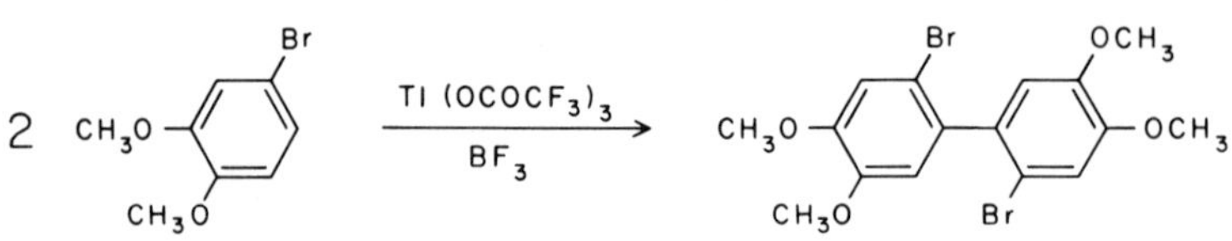


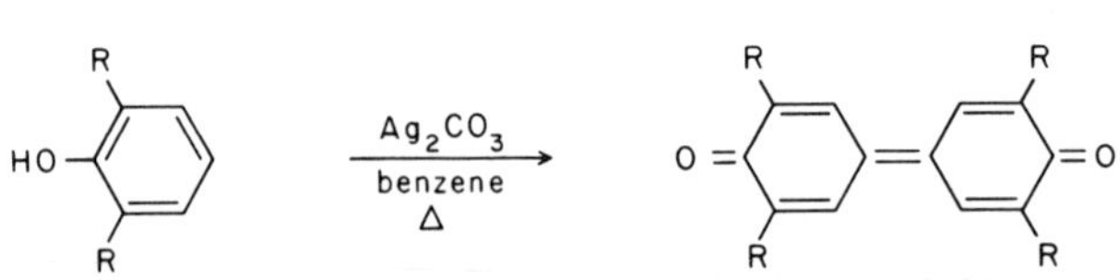

The mechanism of phenolic oxidative coupling reactions has been discussed.[98]

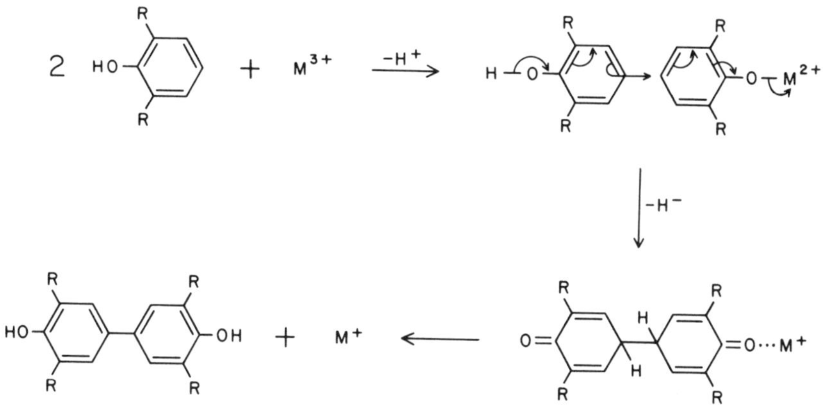

*Biphenyls by the Reduction of Diazonium Salts*

Diazonium salts, prepared *in situ* from the corresponding amines can be converted, most often by the reduction with Cu(I) salts, to biphenyls.[7,8,33,113] Diphenic acid for example has been prepared by this method in yields as high as 90%.[7]

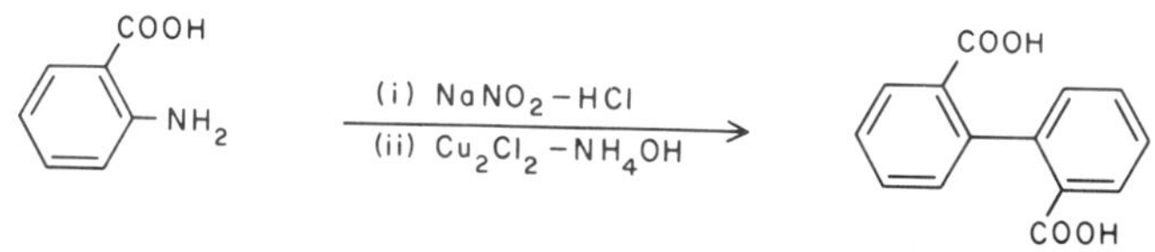


*Biphenyls by the Condensation of Phenols or Diazonium Salts with Quinones*

A large number of biphenyls were recently prepared by the acid catalyzed reaction of acylquinones with phenols and related compounds.[85]

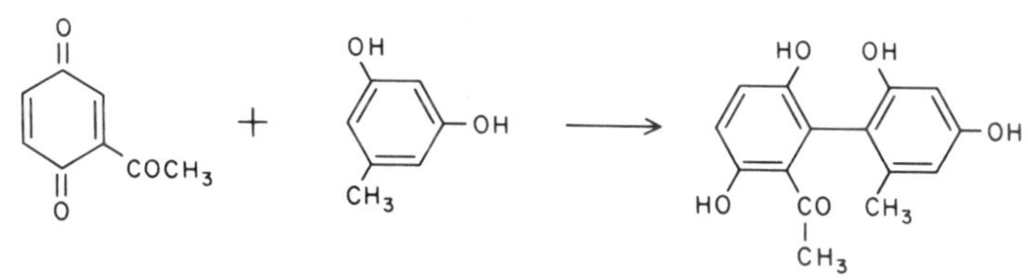

It has been known for many years that diazotized amines and quinones react to give phenylquinones[86] and, therefore, provide access to 2,5-dihydroxybiphenyls.

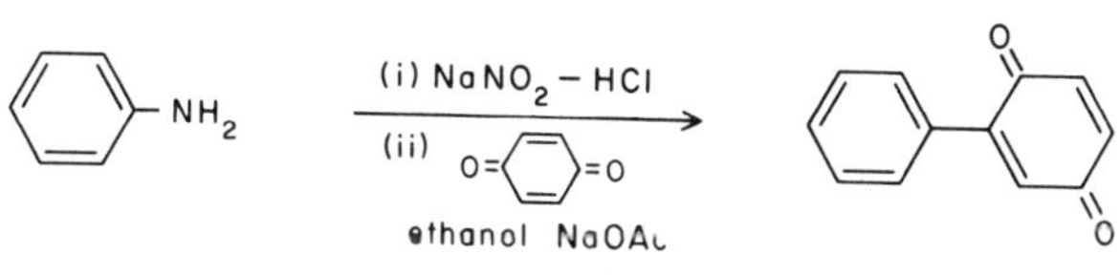


**Synthesis of Chlorobiphenylamines (Aminochlorobiphenyls)**

Chlorobiphenylamines are useful intermediates in the synthesis of chlorobiphenyls chlorobiphenylols (chlorohydroxybiphenyls), and chlorobiphenyls labeled with deuterium, tritium, and chlorine-36. In this section, methods which have been used for the preparation of chlorobiphenylamines are listed with appropriate examples. A list of compounds which have been described in the literature is given in the following section.

Methods for the introduction of amino group into aromatic compounds generally are discussed in References 50 and 103.

*1. Chlorobiphenylamines by Chlorination of Biphenylamines and Chlorobiphenylamines*

This method, which is briefly discussed in Chapter 3 (see for further information), usually gives well-defined polychlorobiphenylamine derivatives. Because of the increase in electron densities in positions ortho and para to the amino group, chlorination takes place on the same ring and products formed can easily be predicted. Depending on the conditions, chlorine may enter the second, unsubstituted ring and this usually takes place at the 4-position.

Chlorination is generally carried out on the *N*-acetyl derivative and the free amine regenerated by acid hydrolysis.

*2. Chlorobiphenylamines by Reduction of Chloronitrobiphenyls*

A wide variety of reducing agents have been used in the reduction of chloronitrobiphenyls to aminochlorobiphenyls (*cf.* 50, 103); examples are zinc and acetic acid, zinc, or tin and hydrochloric acid, and catalytic reduction.

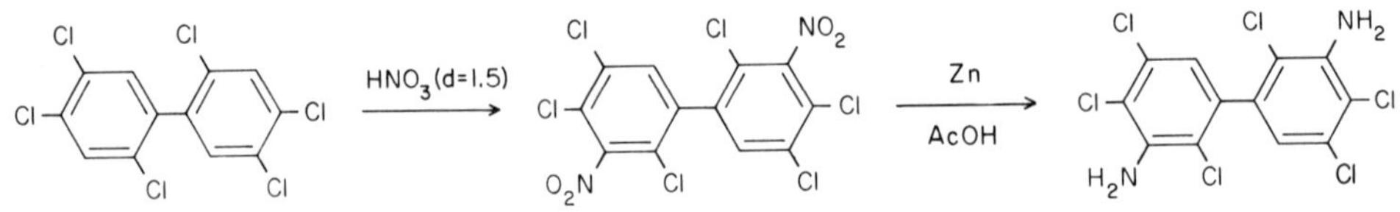


Chloronitrobiphenyls are easily accessible by nitration (*cf.* 130) but are also prepared by Ullmann coupling of halonitrobenzenes.[45,46] For certain synthetic routes it may be desirable to reduce only one of two nitro groups of a chloronitrobiphenyl (*cf.* 73, 87, 91, 94).

*3. Benzidine Rearrangement of Chlorohydroazobenzenes and Related Reactions*

The benzidine rearrangement[12,135] is a convenient method for the the preparation of several chloro-4,4′-biphenyldiamines. Different amounts of the corresponding 2,4′-diamino and 2,2′-diamino products are usually formed during this reaction[31] and in addition, diphenylamines may be observed in certain instances.

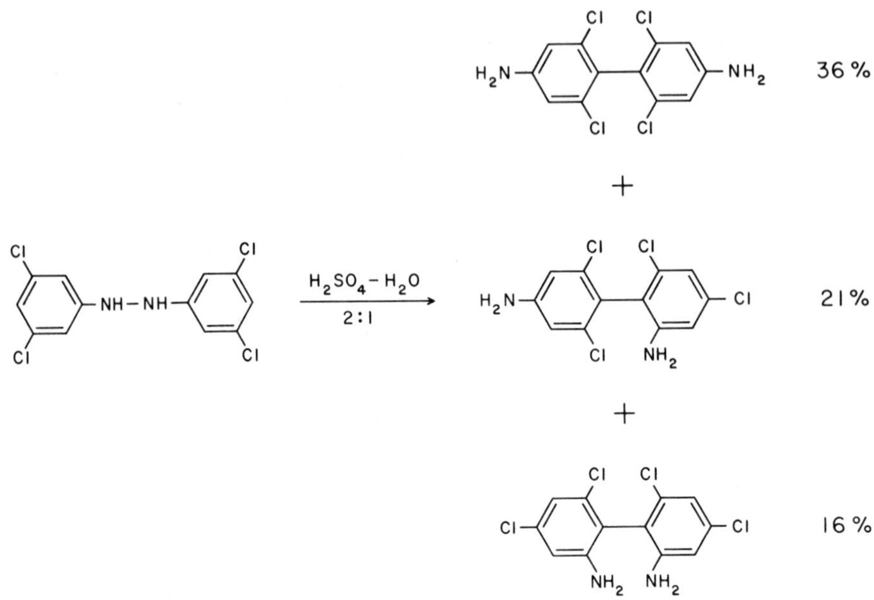

The starting materials for the benzidine rearrangement are aromatic hydrazo compounds, and aqueous hydrochloric or sulfuric acid is commonly used to effect the rearrangement. The hydrazo compounds are usually prepared by the reduction of the corresponding azo compound or by related methods.[44] In earlier years, the hydrazo compound was not always isolated before being rearranged, but the azo compound was reduced with stannous chloride in hydrochloric acid. Under these conditions (sometimes referred to as the Jacobson-method), the hydrazo compound rearranged as it was formed.

The azobenzene derivatives are commonly prepared by the reduction of nitrobenzene derivatives with zinc and alkali, but several other methods are available.[129] A convenient method which has been used for the preparation of unsymmetrical [e.g., 10] and symmetrical [e.g., 38] chloroazobenzenes is the coupling of nitrosobenzene with aniline derivatives.

Chlorobiphenylamines have also been obtained, as by-products, by the UV-radiation induced rearrangement of chloroazobenzenes.[10]

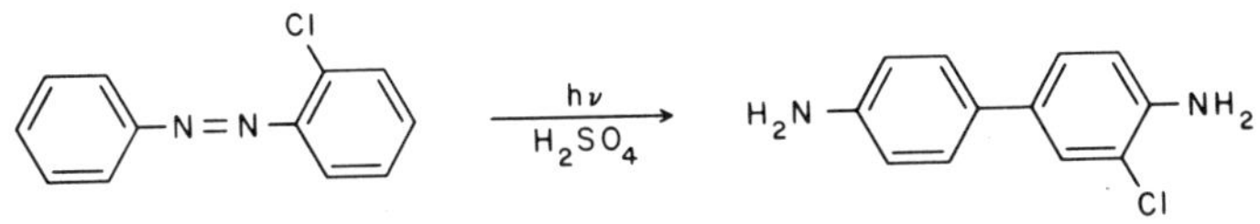


Chlorobiphenylamines are obtained as the main products from chloroazoxybenzenes by reduction and rearrangement with Al-Hg and hydrochloric acid,[153] iron powder and acetic acid[154] and zinc/acetic acid.[84]

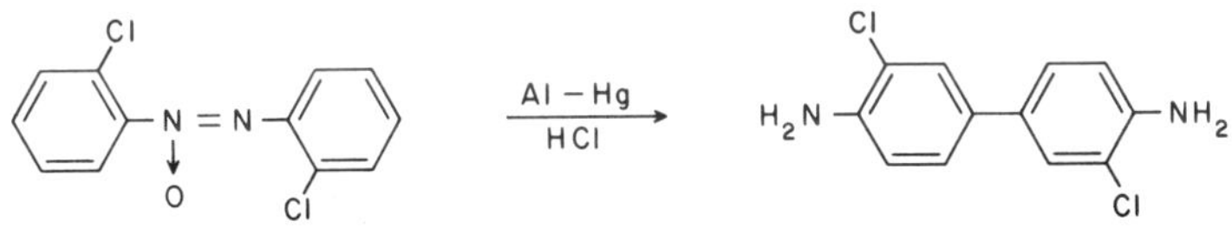


*4. Chlorobiphenylamines from Biphenylpolyamines ($-NH_2 \rightarrow Cl$)*

This procedure usually involves the replacement of an amino group by chlorine via the Sandmeyer reaction while other amino groups in the molecule are protected.

Chlorobiphenyldiamines have been monoacetylated by using only one equivalent of acetylating reagent as, for example, in the monoacetylation of 2,2′-dichlorobenzidine in aqueous ethanol.[30] A similar monoacetylation of dimethylbenzidine[143] makes use of the monoacetylbenzidine's lower solubility in ether; as the reaction with acetic anhydride proceeds, the monoacetyl derivative precipitates out.

*5. Chlorobiphenylamines from Phenylhydroxylamine and Chlorobenzene*

In this new substitution reaction, phenylhydroxylamine condenses with aromatic compounds such as chlorobenzene to give 2- and 4-aminobiphenyls as the main product.[110]

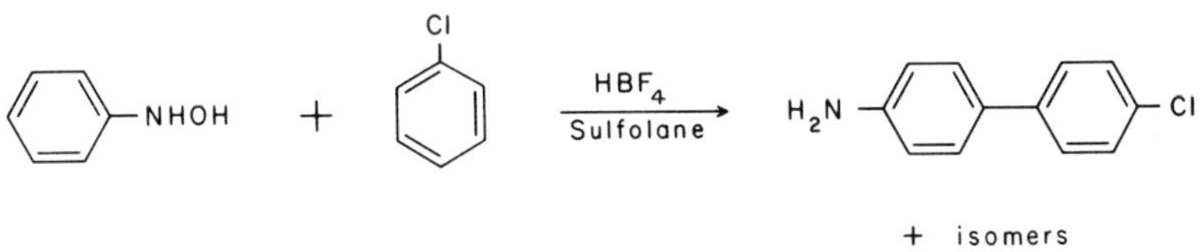


*6. Chlorobiphenylamines by Condensation of Aminochlorobenzenediazonium Compounds with Copper*

The copper powder mediated condensation of diazotized 4-amino-2,5-dichloroacetanilide yielded *N,N*-diacetyl-2,2′,5,5′-tetrachlorobenzidine.[41] On hydrolysis, the corresponding tetrachlorobenzidine was obtained which was difficult to prepare by other condensation reactions.

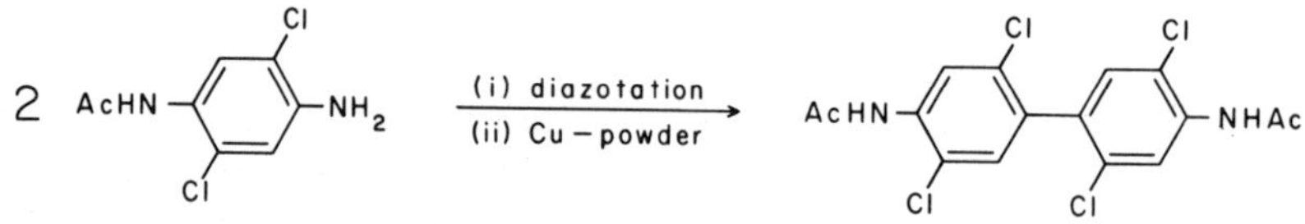


*7. Chlorobiphenylamines by Hofmann Degradation of Carboxamides*

4′-Chloro-2-biphenylamine was prepared from 4′-chloro-2-biphenylcarboxylic acid via the acid chloride and amide which was treated with bromine and sodium hydroxide in the usual manner.[70]

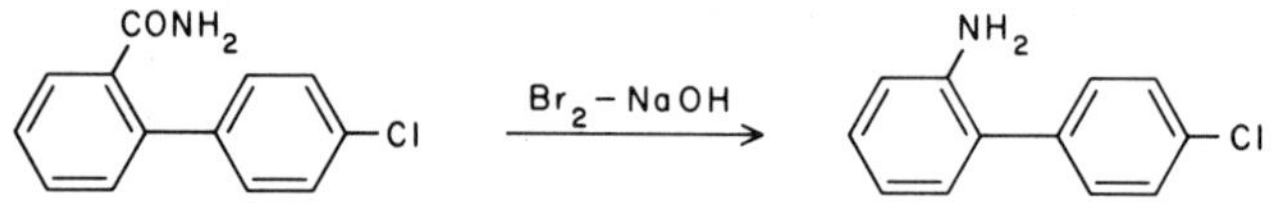


*8. Chlorobiphenylamines by Reduction of Azo Compounds*

3,3′-Dichloro-2,2′-biphenylamine was prepared by reduction of diazo dyes which were prepared by the coupling of diazotized 2,6-dichloroaniline and β-naphthol.[88]

*9. Chlorobiphenylamines from Polychlorobiphenyls and Ammonia*

Polychlorobiphenyldiamines, usually of uncertain structure, are obtained when highly chlorinated biphenyls are treated with ammonium hydroxide at 280°[141] or ammonia and metal halides at ∼200°.[137]

**List of Chlorobiphenylamines (Aminochlorobiphenyls) Described in the Literature**

In this section, chlorobiphenyl mono- and diamines (mono- and diaminochlorobiphenyls) which have been described in the literature are listed together with their melting points and method of preparation. Generally, the most convenient methods of preparation are listed in the Tables and References and are chosen from easily accessible journals whenever possible. A complete bibliography until 1929 can be found in *Beilsteins Handbuch der Organischen Chemie.*[13]

**Synthesis of Chlorobiphenylols (Chlorohydroxybiphenyls, Chlorophenylphenols)**

Chlorohydroxybiphenyls are found as metabolites of chlorobiphenyls in a number of organisms (see Chapter 7); the chemistry of these compounds is, therefore, of interest.

In this section, methods which have been used for the preparation of chlorobiphenylols are listed with appropriate examples. A list of compounds which have been described in the literature is given in the following section.

Many biphenyl syntheses discussed earlier in this chapter yield hydroxy- (or alkoxy-) biphenyl derivatives; most of these may be adaptable for the preparation of chlorobiphenylols.

Removal of a methyl group (for instance with boron tribromide; see Reference 99) as the last step in a synthetic route has not been listed as separate reaction.

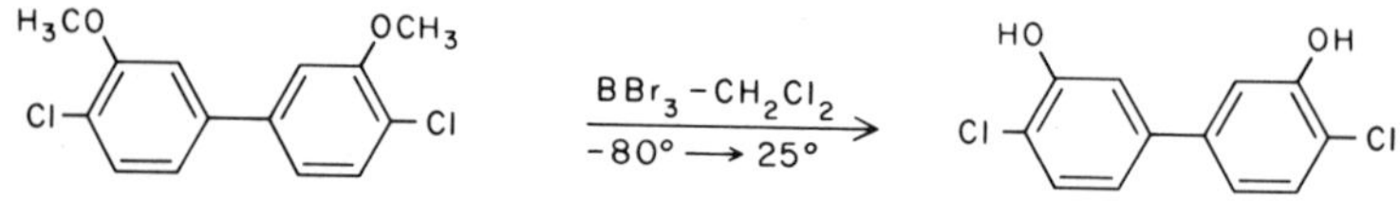


The preparation of the etherial sulfate conjugate of 3-chloro-2-biphenylol has been described.[54]

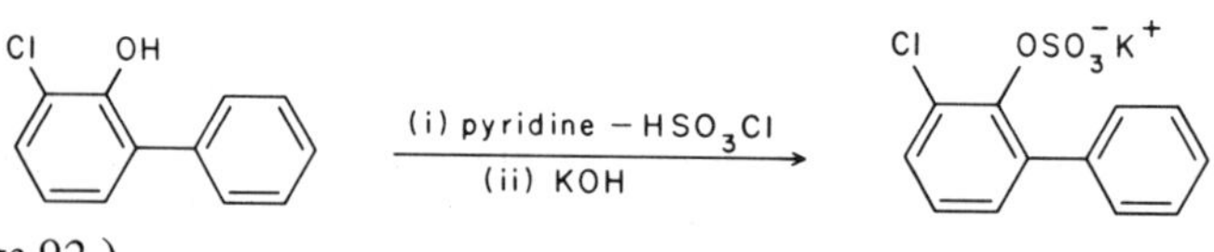


(Text continued on page 92.)

## Monoamino-Monochloro-Derivatives

| Compound: Name | Compound: Formula | Melting points | Method of preparation | References |
|---|---|---|---|---|
| 3-Chloro-2-biphenylamine | 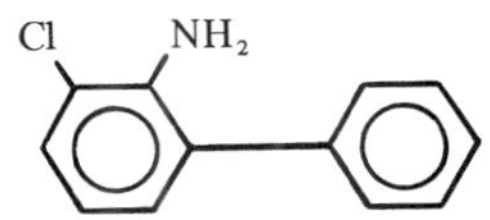 | 15 | 1 | 38 |
| 4-Chloro-2-biphenylamine | 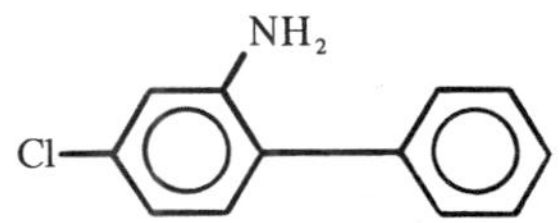 | oil | 2 | 66 |
| | | $bp_{12}$ mm 185–6° | 2 | 67 |
| | | 52 | 2 | 114 |
| 5-Chloro-2-biphenylamine | 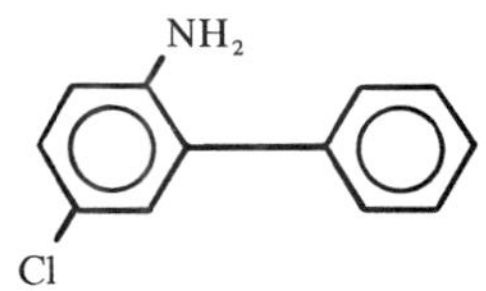 | 54 | 1 | 38 |
| | | 51 | 1 | 17 |
| | | | 1 | 32 |
| | | 47–50 | 1 | 139 |
| 2′-Chloro-2-biphenylamine | 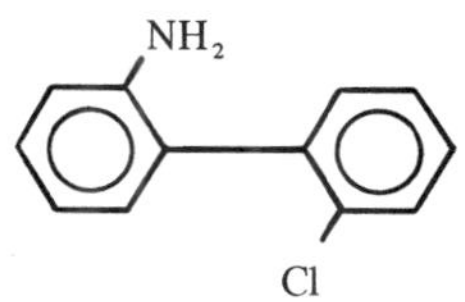 | 56–57 | 2 | 94 |
| | | 50–51 | 5 | 110 |
| | | 54 | 4 | 121 |
| 4′-Chloro-2-biphenylamine | 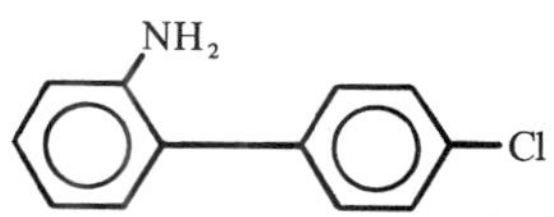 | 71 | 1 | 38 |
| | | 47 | 2 | 52 |
| | | 42–44 | 4 | 24 |
| | | 43–43.5 | 5 | 110 |
| | | 47–48 | 7 | 70 |
| 4-Chloro-3-biphenylamine | 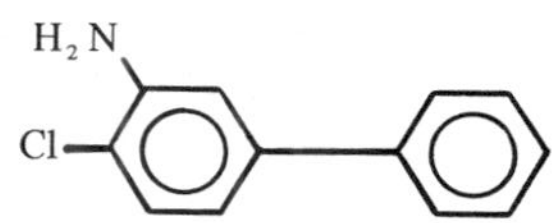 | oil<br>HCl salt<br>247 dec. | 1 | 21 |

| | | | | |
|---|---|---|---|---|
| 2′-Chloro-3-biphenylamine | 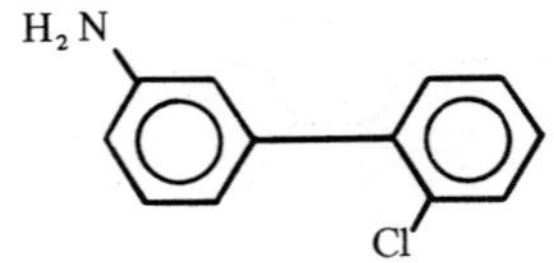 | 47–48 | 2 | 81 |
| 3′-Chloro-3-biphenylamine | 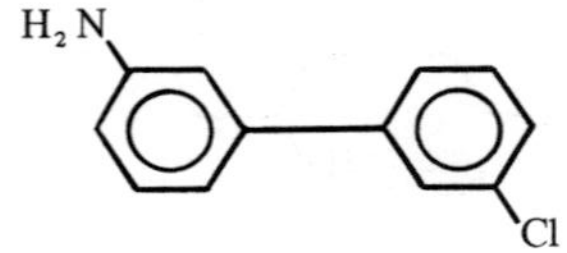 | HCl salt<br>227 dec. | 2 | 152 |
| 4′-Chloro-3-biphenylamine | 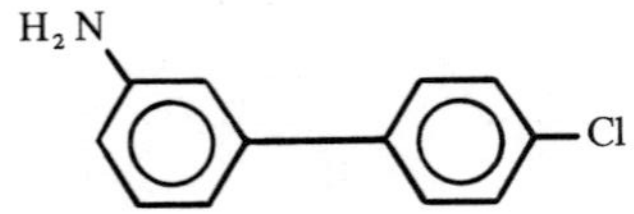 | 82 | 2 | 21 |
| 2-Chloro-4-biphenylamine | 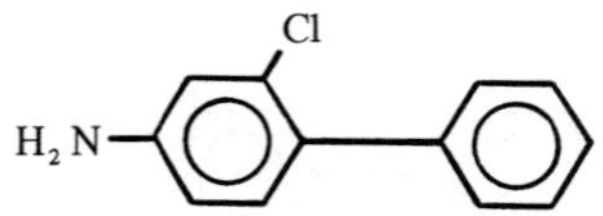 | oil<br>$bp_{16}$ mm 205–6°<br>oil | 2<br><br>2 | 105<br><br>26 |
| 3-Chloro-4-biphenylamine | 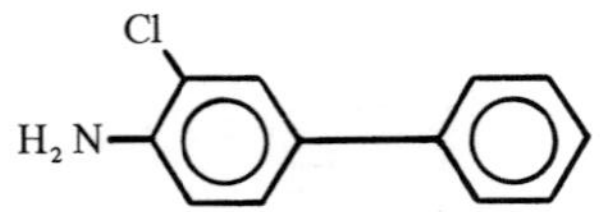 | 71 | 1 | 125 |
| 2′-chloro-4-biphenylamine<br><br>acetyl derivative | 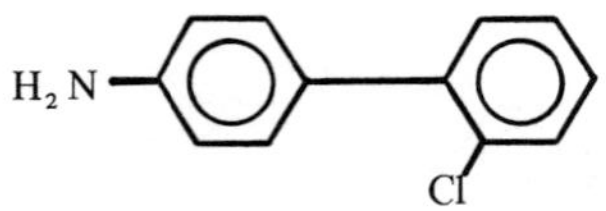 | $bp_7$ mm 183–5°<br>42–43 | 2<br>2<br>5 | 105, 53<br>26<br>110 |
| 3′-Chloro-4-biphenylamine | 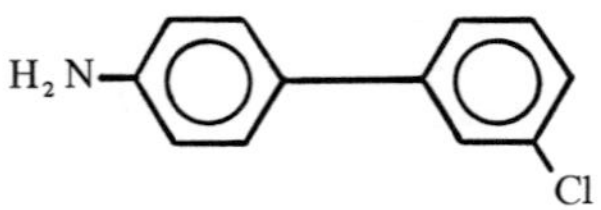 | 47–48 | 2 | 26 |

| Name | Formula | Melting points | Method of preparation | References |
|---|---|---|---|---|
| 4′-Chloro-4-biphenylamine | 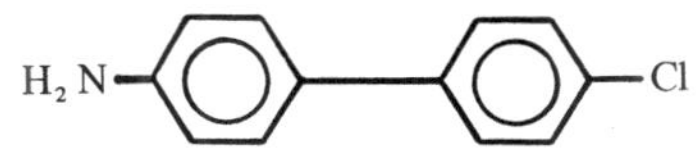 | 133<br>132.5<br>134 | 5<br>2<br>4 | 110<br>26<br>49 |

## Monoamino-Dichloro-Derivatives

| Compound | | | | |
|---|---|---|---|---|
| Name | Formula | Melting points | Method of preparation | References |
| 3,5-Dichloro-2-biphenylamine | 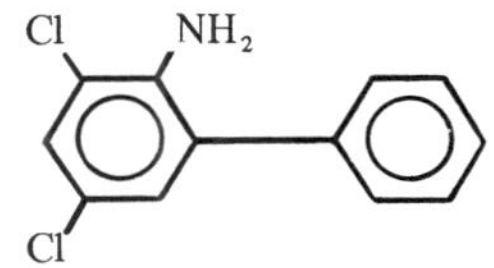 | 51 | 1 | 126 |
| 4,4′-Dichloro-2-biphenylamine | 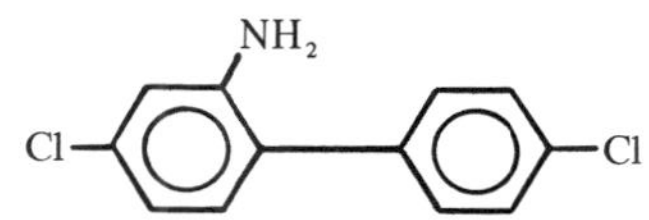 | 91<br>94–95 | 2<br>2 | 77<br>72 |
| 3′,5′-Dichloro-2-biphenylamine | 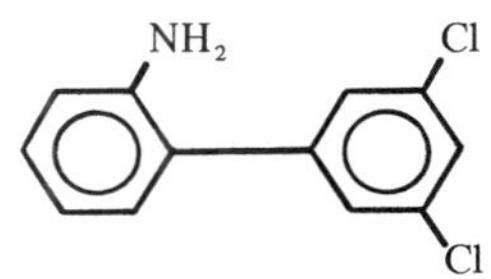 | 74 | 2 | 63 |
| 2′,4-Dichloro-3-biphenylamine | 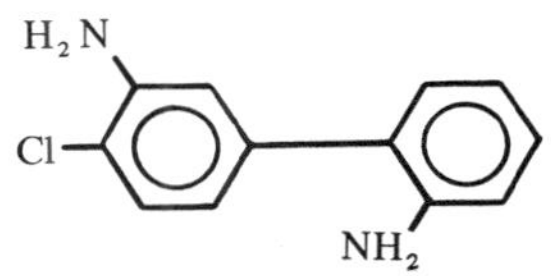 | 44 | 2 | 18 |

| | | | | |
|---|---|---|---|---|
| 3′,5′-Dichloro-3-biphenylamine | 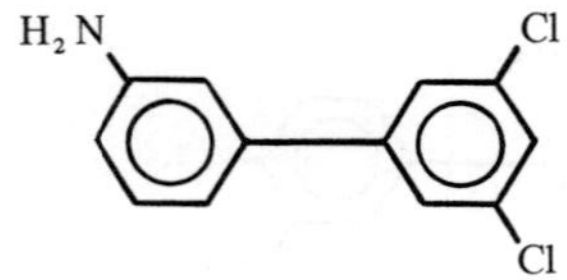 | oil | 2 | 62 |
| 2,2′-Dichloro-4-biphenylamine | 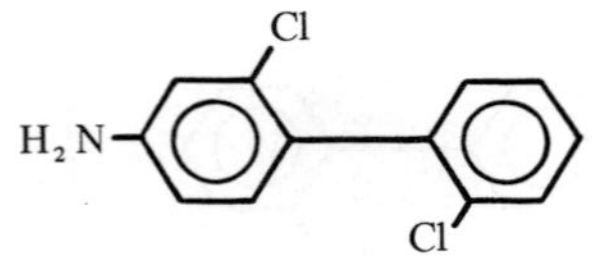 | 73–74 | 2 | 91 |
| 2,4′-Dichloro-4-biphenylamine | 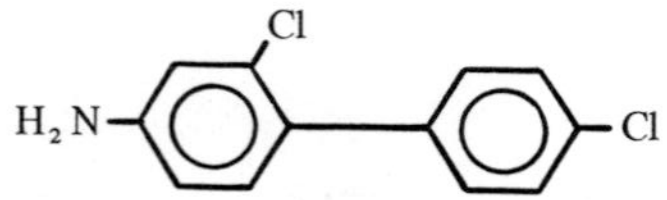 | 83 | 2 | 48 |
| 3,4′-Dichloro-4-biphenylamine | 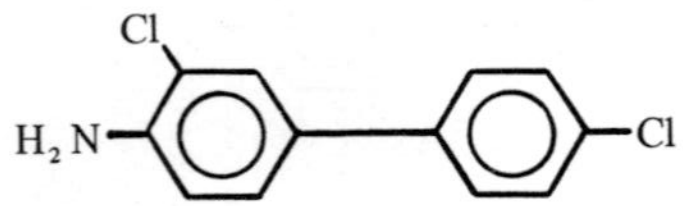 | 100 | 1 | 16 |
| 3′,5′-Dichloro-4-biphenylamine | 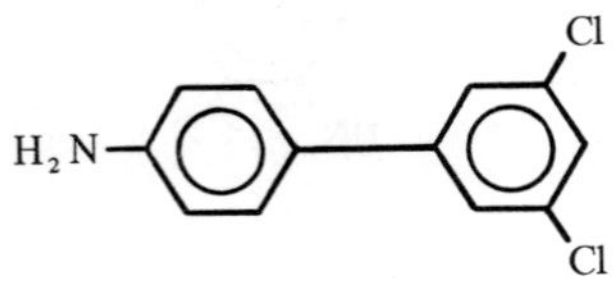 | 124 | 2 | 63 |

Monoamino-Trichloro-Derivatives

| Compound: Name | Compound: Formula | Melting points | Method of preparation | References |
|---|---|---|---|---|
| 2′,4,4′-Trichloro-3-biphenylamine | 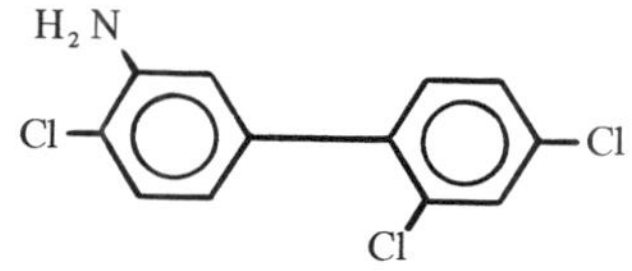 | 105 | 2 | 21 |
| 3,4′,5-Trichloro-4-biphenylamine | 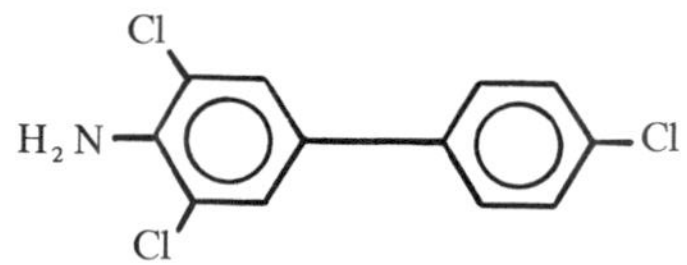 | 128 | 1 | 125 |

Monoamino-Tetrachloro-Derivatives

| Compound: Name | Compound: Formula | Melting points | Method of preparation | References |
|---|---|---|---|---|
| 3′,4,4′,5-Tetrachloro-2-biphenylamine | 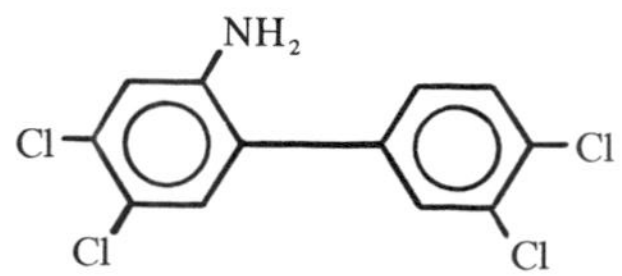 | 79–82 | 2 | 60 |

Diamino-Monochloro-Derivatives

| Compound | | Melting points | Method of preparation | References |
|---|---|---|---|---|
| Name | Formula | | | |
| 2′-Chloro-3,4-biphenyldiamine | 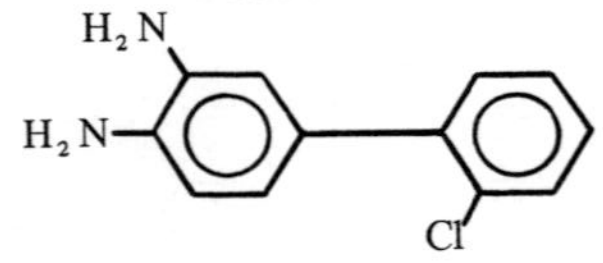 | HCl salt 300–305(dec.) | 2 | 92 |
| 5-Chloro-2,4′-biphenyldiamine | 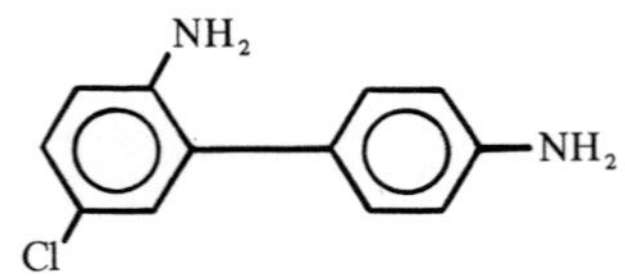 | oil<br>HCl salt 255 | 1<br>3<br>3 | 18<br>83<br>10 |
| 2-Chloro-4,4′-biphenyldiamine | 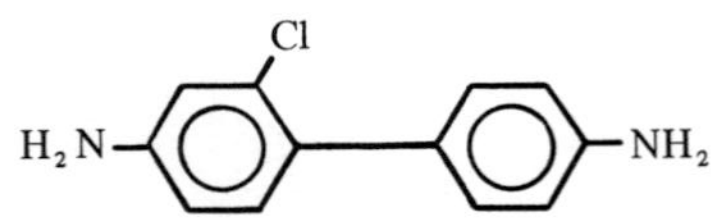 | 113<br>101.5–102 | 3<br>3 | 25<br>10 |
| 3-Chloro-4,4′-biphenyldiamine | 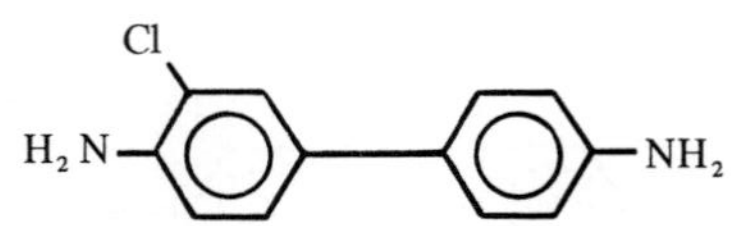 | 74.5–75<br>75 | 3<br>3 | 10<br>15 |

Diamino-Dichloro-Derivatives

| Compound | | Melting points | Method of preparation | References |
|---|---|---|---|---|
| Name | Formula | | | |
| 3,3′-Dichloro-2,2′-b:phenyldiamine | 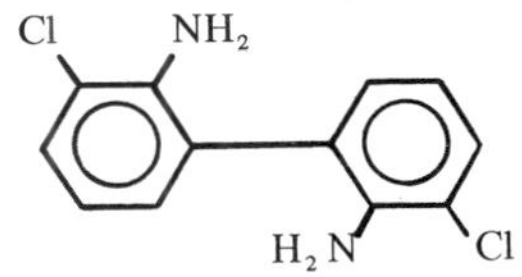 | 123–4 | 7 | 88 |
| 4,4′-Dichloro-2,2′-biphenyldiamine | 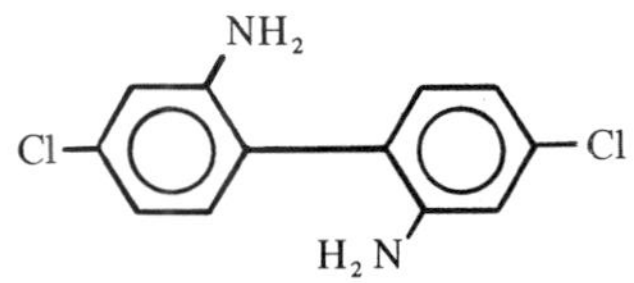 | 87 | 2 | 64 |
| 4,4′-Dichloro-2,3′-biphenyldiamine | 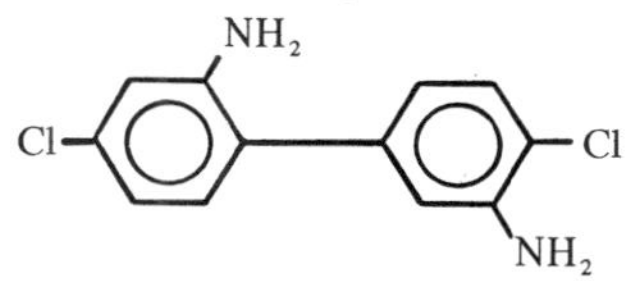 | 83 | 2 | 64 |
| 2′4-Dichloro-2,4′-biphenyldiamine | 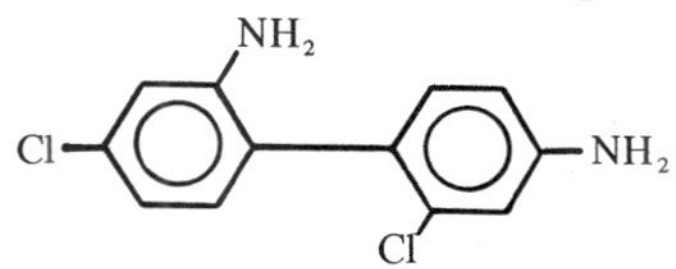 | 162.5–163.5 | 3 | 147 |
| 3′,5-Dichloro-2,4′biphenyldiamine | 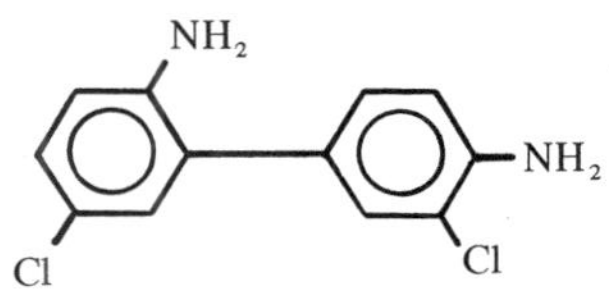 | oil<br>HCl salt<br>225 dec. | 1 | 18 |
| 2′,6-Dichloro-2,4′-biphenyldiamine | 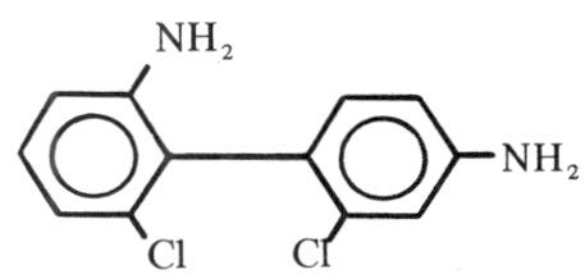 | 110–110.5 | 3 | 147 |

| | | | | |
|---|---|---|---|---|
| 4,4′-Dichloro-3,3′-biphenyldiamine | 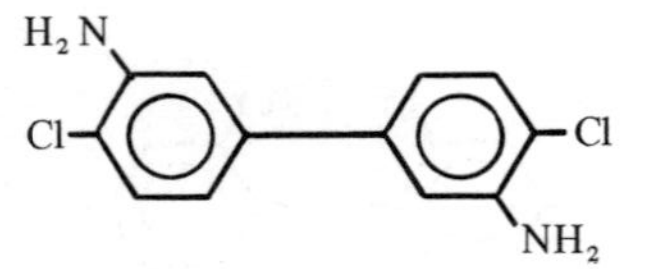 | 133.5 | 2 | 65 |
| | | 133.5 | 2 | 64 |
| 4,6′-Dichloro-3,3′-biphenyldiamine | 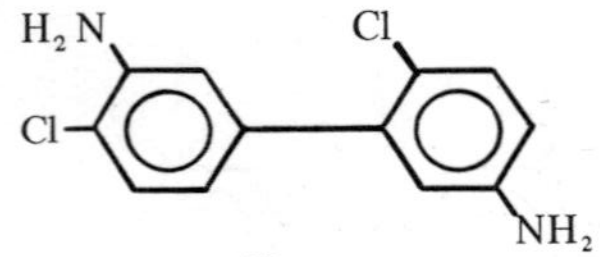 | 105 | 2 | 18 |
| 2,5-Dichloro-4,4′-biphenyldiamine | 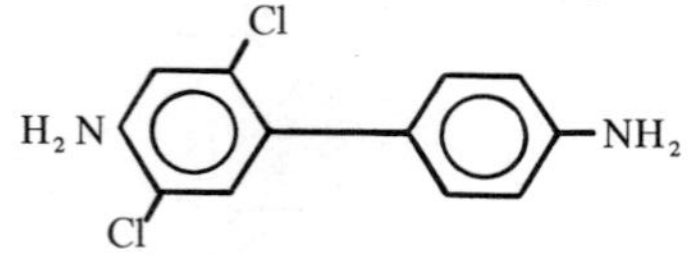 | 95 | 3 | 38 |
| 2,2′-Dichloro-4,4′-biphenyldiamine | 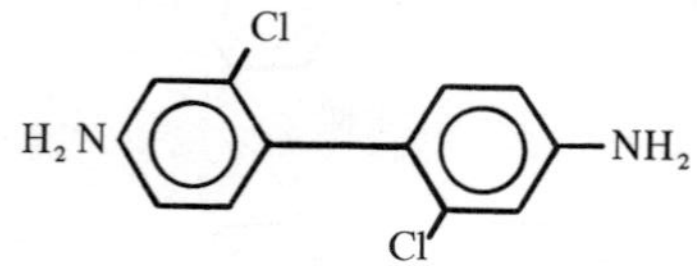 | 165–166 | 3 | 19 |
| | | 166.8 | 3 | 128 |
| | | | 2 | 91 |
| | | 167.5–168 | 3 | 147 |
| 3,3′-Dichloro-4,4′-biphenyldiamine | 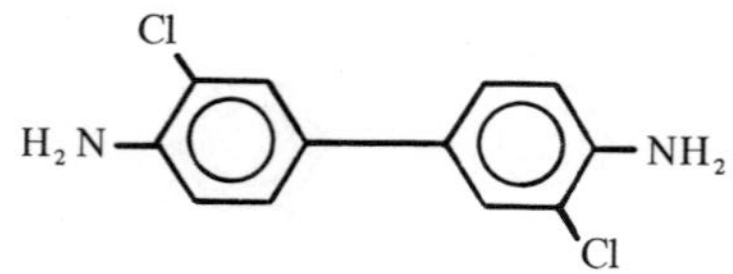 | 133 | 1 | 146 |
| | | 132–133 | 3 | 39 |
| | | 132–133 | 3 | 19 |

Diamino-Trichloro-Derivatives

| Compound | | | | |
|---|---|---|---|---|
| Name | Formula | Melting points | Method of preparation | References |
| 3′4,5-Trichloro-2,4′-biphenyldiamine | 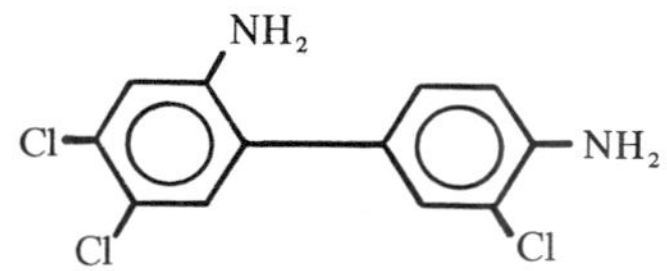 | – | – | – |
| diacetyl derivative | | 213–214 | 2 | 60 |

Diamino-Tetrachloro-Derivatives

| Compound | | | | |
|---|---|---|---|---|
| Name | Formula | Melting points | Method of preparation | References |
| 4,4′,5,5′-Tetrachloro-2,2′-biphenyldiamine | 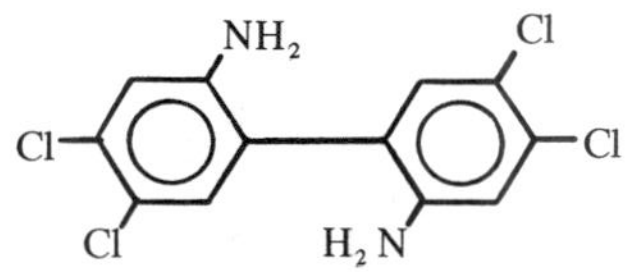 | 167 | 2 | 146 |
| 4,4′,6,6′-Tetrachlorc-2,2′-biphenyldiamine | 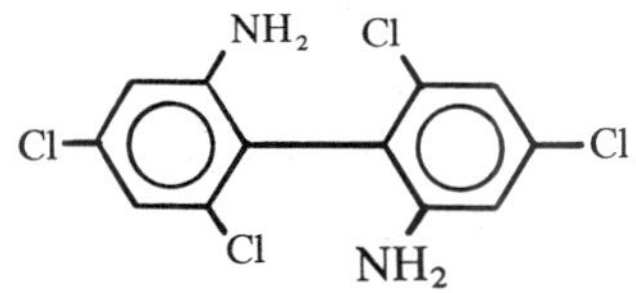 | 120–121 | 3 | 31 |

| | | | | |
|---|---|---|---|---|
| 2′,4,6,6′-Tetrachloro-2,4′-biphenyldiamine | 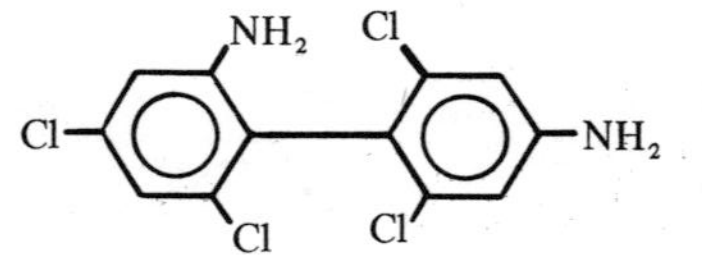 | 141–141.5 | 3 | 31 |
| 2,2′,5,5′-Tetrachloro-4,4′-biphenyldiamine | 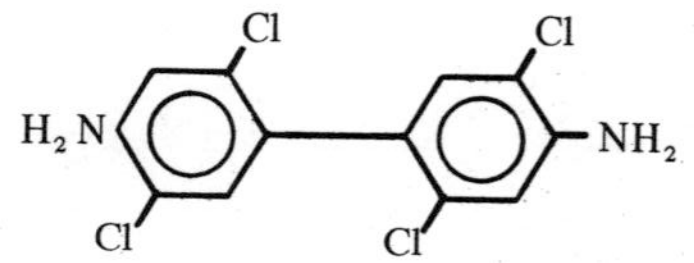 | 137.5 | 3<br>6 | 40<br>41 |
| 2,2′,6,6′-Tetrachloro-4,4′-biphenyldiamine | 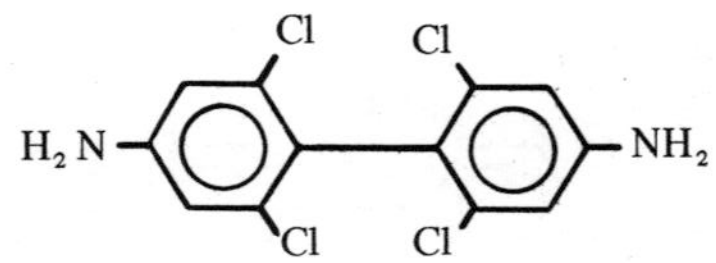 | 129<br>212.5–213.5 | 3<br>3 | 146<br>31 |
| 3,3′,5,5′-Tetrachloro-4,4′-biphenyldiamine | 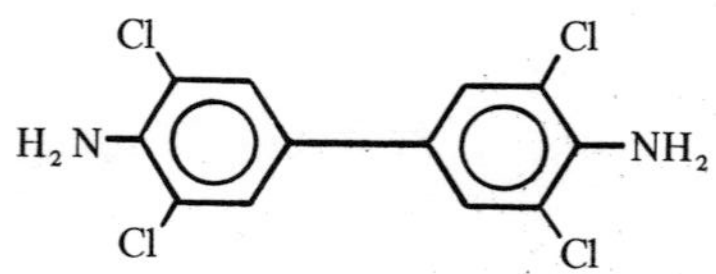 | 226 | 1 | 146 |

Diamino-Hexachloro-Derivatives

| Compound: Name | Compound: Formula | Melting points | Method of preparation | References |
|---|---|---|---|---|
| 2,2′,4,4′,5,5′-Hexachloro-3,3′-biphenyldiamine | 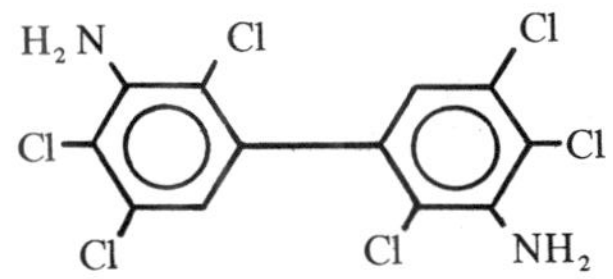 | 191–192 | 2 | 120 |
| 2,2′,4,4′,6,6′-Hexachloro-3,3′-biphenyldiamine | 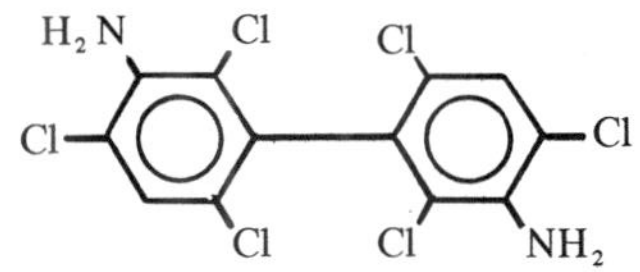 | 167.5–168.5 | 2 | 150 |
| 2,2′,3,3′,5,5′-Hexachloro-4,4′-biphenyldiamine | 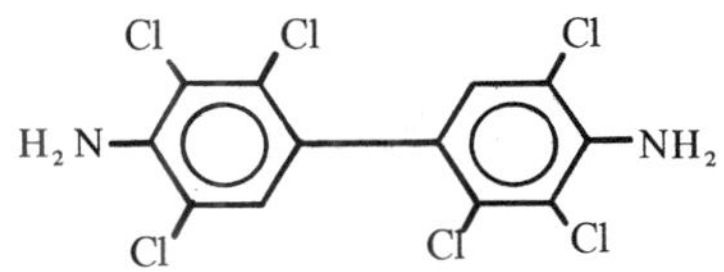 | 186–187 | 1 | 120 |

Diamino-Octachloro-Derivatives

| Compound: Name | Compound: Formula | Melting points | Method of preparation | References |
|---|---|---|---|---|
| 2,2′,3,3′,5,5′,6,6′-Octachloro-4,4′-biphenyldiamine | 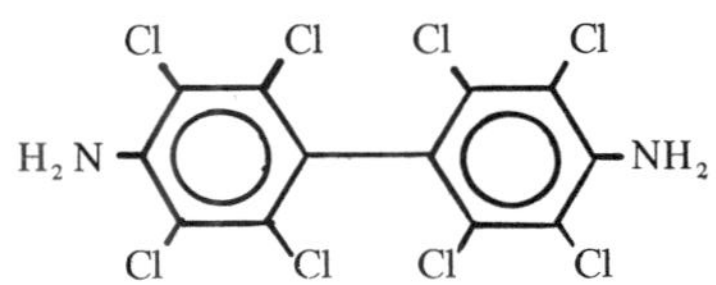 | 289 | 1 | 146 |

*1. Chlorobiphenylols from Chlorobiphenylamines Via Diazonium Salts ($NH_2 \rightarrow OH$)*

Chlorobiphenylols have been prepared from the corresponding chlorobiphenylamines[38,77,107,111] by diazotization and replacement of the diazonium salt by the hydroxyl group[68,122] usually by heating with aqueous sulfuric acid. In the authors' laboratories only poor yields of chlorobiphenylols were usually obtained by this method.

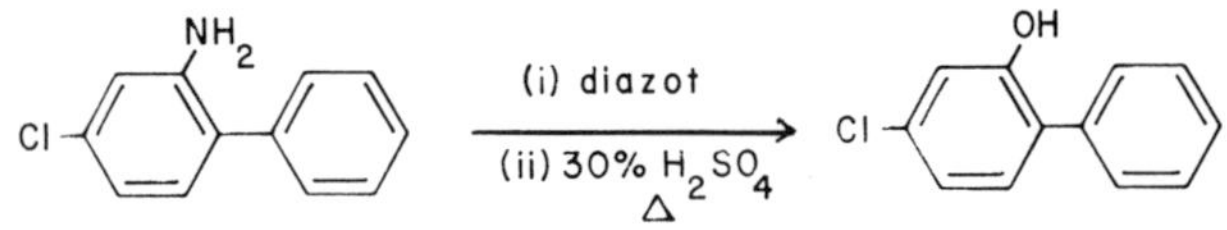

*2. Chlorobiphenylols by Chlorination of Hydroxybiphenyls*

The hydroxy group has both a directing (*o*-, *p*-) and activating influence in electrophilic aromatic substitution. Predicable mono- and dichlorobiphenyl derivatives (per hydroxy group) can usually be obtained on chlorination mostly without the use of catalysts.[23,29,35,42,90,117,123,124]

The availability of electrons from the hydroxy group can be canceled by acetylation; 4-acetoxybiphenyl, for instance, yields 4-acetoxy-4′-chlorobiphenyl with iodine as catalyst.

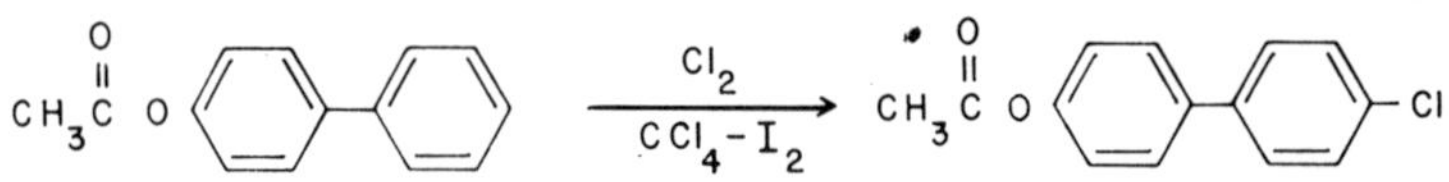

When 4-hydroxybiphenyl was treated with the BMC-reagent in the usual manner, nonachlorohydroxybiphenyl was formed along with several by-products, some of which gave mass spectra corresponding to hepta- and octachlorohydroxybiphenyls.[71] Penta- and heptachlorohydroxybiphenyls were obtained on prolonged chlorination of 4-hydroxybiphenyl in acetic acid.[36]

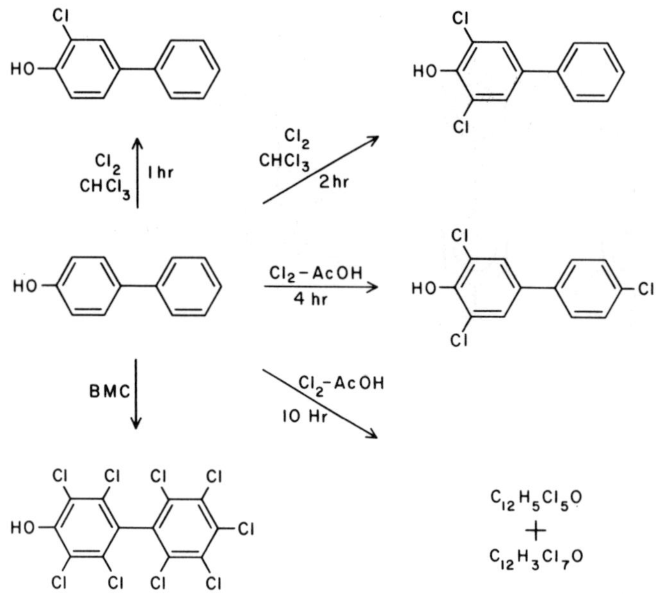

*3. Chlorobiphenylols by Arylation Reaction (Decomposition of Diazonium Salts in Aromatic Substrate)*

This type of reaction does not have the importance for the preparation of chlorobiphenylols as it does for the chlorobiphenyls (Chapter 3). The simplest version, diazotation of chloroanilines and reaction with phenol,[34] gives low yields.

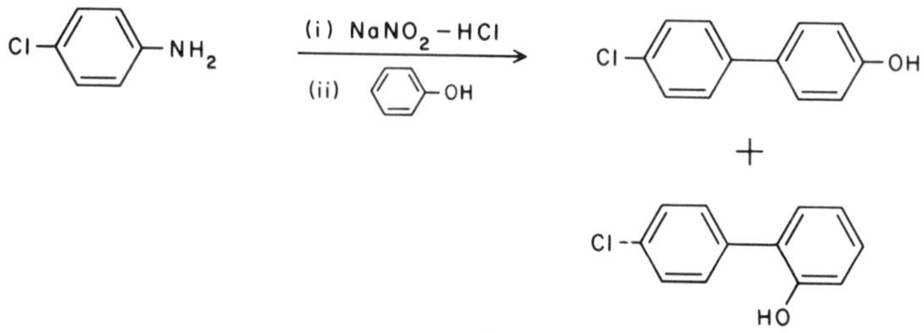

A patent described the preparation of 4,5-dichloro-2-methoxybiphenyl from 4,5-dichloro-2-methoxyaniline by diazotation, treatment with dimethylamine and reaction of the triazene so formed with benzene and HCl gas.[156]

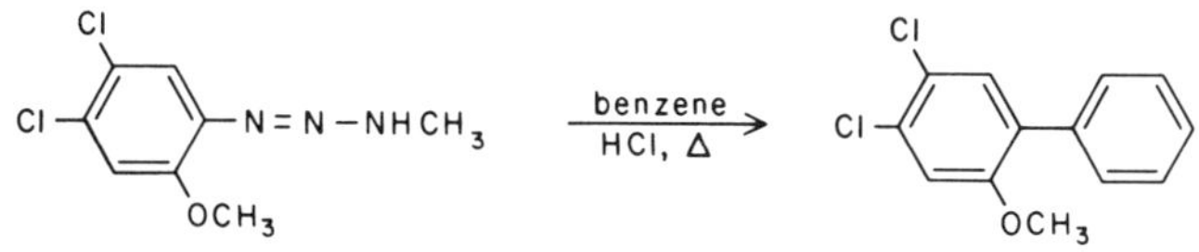

A number of chlorobiphenylols have recently been prepared[72] utilizing the simplified Gomberg procedure (one-pot reaction with amyl nitrite) of Shu[136] and Cadogan.[27] The decomposition of

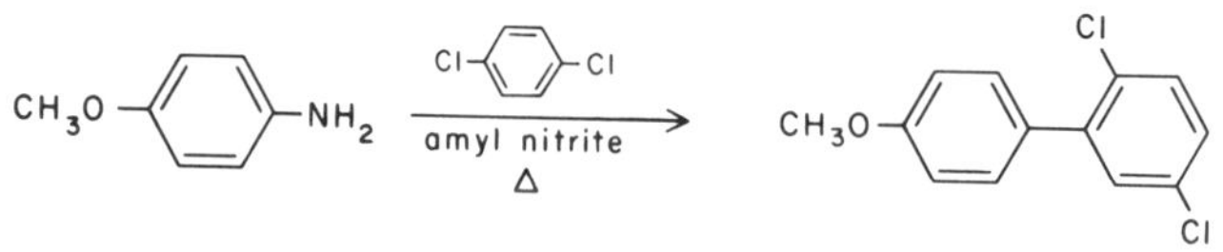

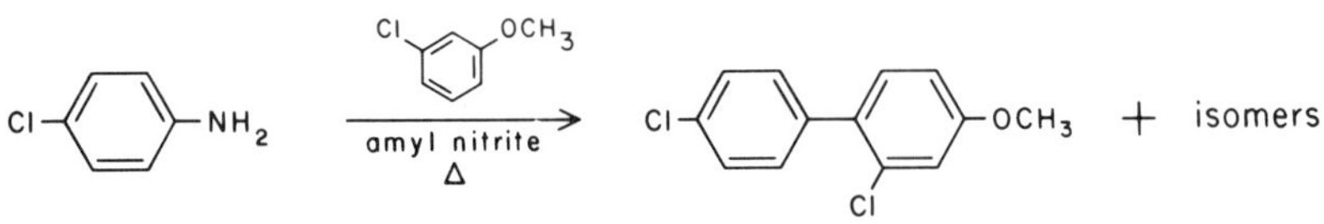

aryl diazonium fluoroborates in aryl substrate in the presence of 1 ≡ of pyridine[2] is also applicable for the preparation of these compounds.[72] In all instances, methoxy-derivatives were used in the coupling reaction.

*4. Chlorobiphenylols by Decarboxylation of Chlorohydroxybiphenylcarboxylic Acids*

4′-Chloro-3-biphenylol was prepared by decarboxylation of 4′-chloro3-hydroxybiphenyl-2,4-dicarboxylic acid.[112] The dicarboxylic acid was prepared by condensation of α-formyl-4-chloroacetophenone with diethyl acetone-1,3-dicarboxylate and hydrolysis of the diester.

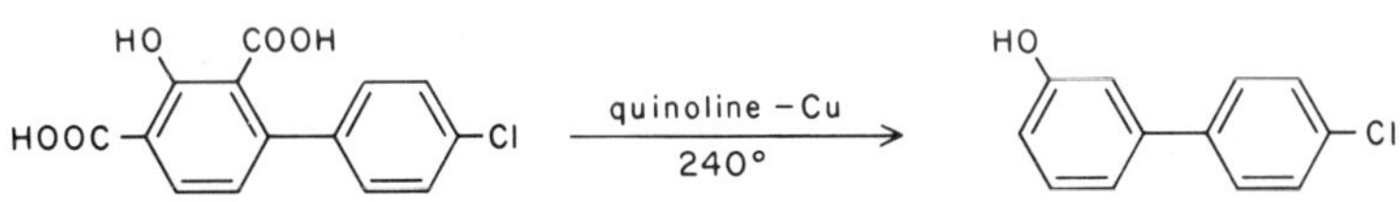

*5. Chlorobiphenylols by the Action of Alkali or Sodium Methoxide on Chlorobiphenyls*

Decachlorobiphenyl when treated with aqueous alkali in an autoclave at high temperatures yields octachloro-4,4′-biphenylol.[138,140]

The chief product on heating 2,5-dichlorobiphenyl with sodium methylate for prolonged times was 2-chloro-5-biphenylol.[38]

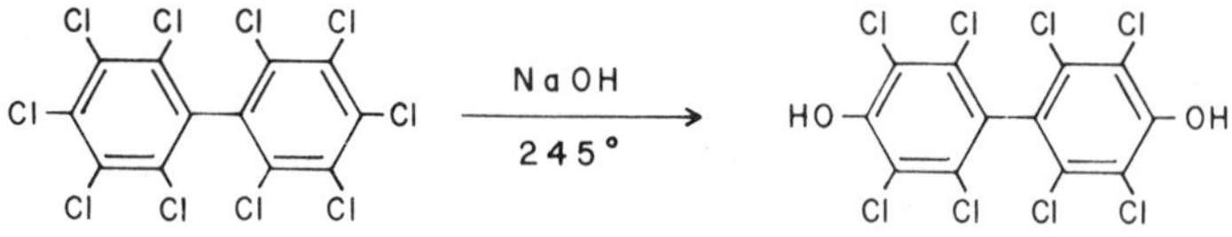

*6. Chlorobiphenylols by Isomerization Reaction*

*o*-Hydroxy substituted chlorobiphenylols can be isomerized with $AlCl_3$ to give the corresponding *m*-hydroxy derivative. 4-Chloro-3-biphenylol has been prepared from 3-chloro-2-biphenylol in this way.[57]

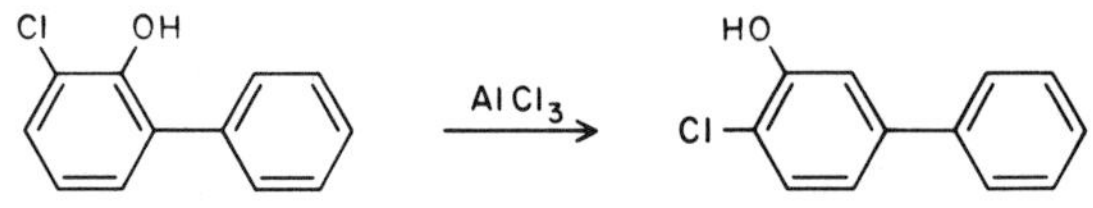

*7. Chlorobiphenylols from Aminobiphenylols Via Diazonium Salts ($NH_2 \rightarrow Cl$)*

Aminobiphenylols can be converted to the corresponding chlorobiphenylols[51,72,93,149] by the usual Sandmeyer procedure.

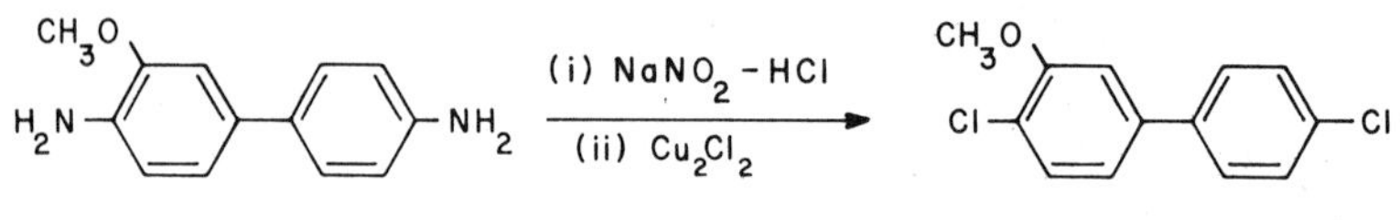

2,4′-Dichloro-4-biphenylol was recently prepared by this method;[72] the intermediate 4′-nitro-2-amino-4-biphenylol was obtained by rearrangement of the corresponding sulfonamide.

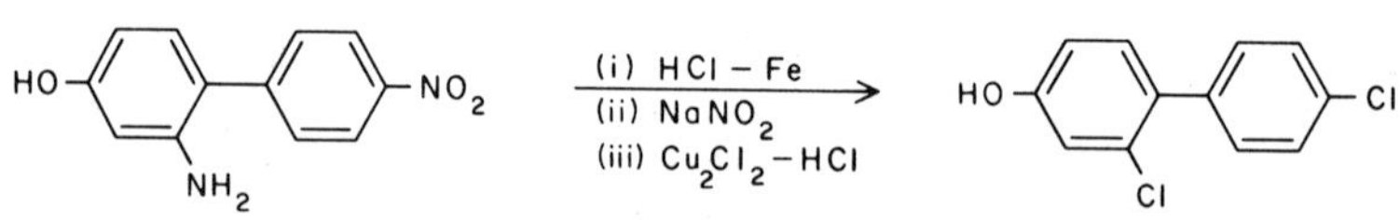

*8. Chlorobiphenylols by Photochemical or Thermal Reaction of Chlorobenzene-2-diazo-1-oxides in Benzene*

Irradiation of tetrachloro-2-diazo-1-oxide in benzene leads to the formation (37%) of 3,4,5,6-tetrachloro-2-biphenylol.[69]

The same diazooxide heated in chlorobenzene to 130° gives 3,4,4′,5,6-pentachloro-2-biphenylol in 14% yield.

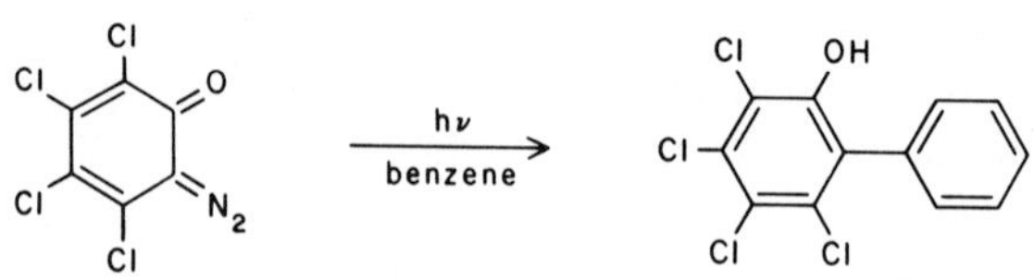

*9. Chlorobiphenylols by Condensation of Chlorobenzene-2-diazo-1-oxides and Subsequent Reduction*

Tetrachloro-2-diazo-1-oxide when treated with bromine in chloroform solution condenses to give the diazo-dioxo compound shown in the equation. On reduction with Raney Nickel a heptachlorobiphenyldiol was obtained whose most likely structure is shown below.[115]

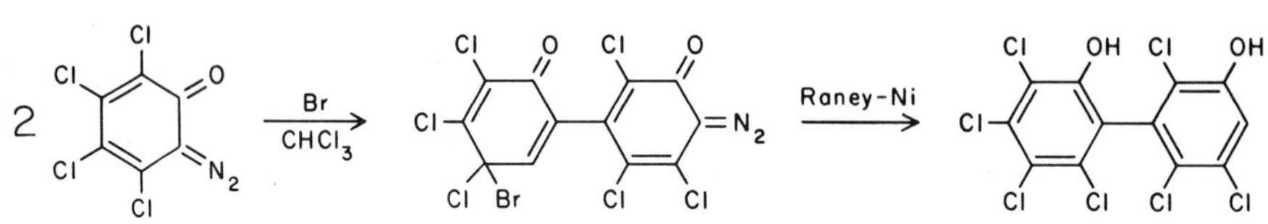

*10. Chlorobiphenylols by Coupling Diazonium Salts with Quinones and Reduction*

Diazotized *o*-chloroaniline was condensed with benzoquinone to give 2-(*o*-chlorophenyl)-1,4-benzoquinone. This compound on reduction gave 2′-chloro-2,5-biphenyldiol.[102]

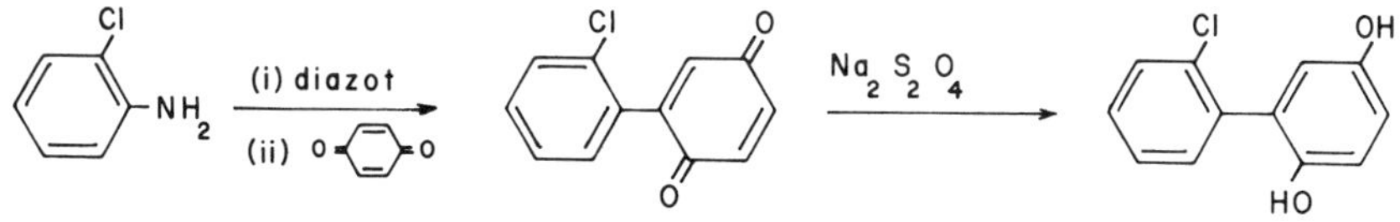

A number of dichloro-2,5-biphenyldiols have been prepared by a similar method.[47]

*11. Chlorobiphenylols by Condensation of Chlorophenols with Benzoyl Peroxide*

3,3′,5,5′-tetrachloro-4,4′-biphenyldiol was prepared by refluxing 2,6-dichlorophenol and benzoyl peroxide in carbon tetrachloride.[75]

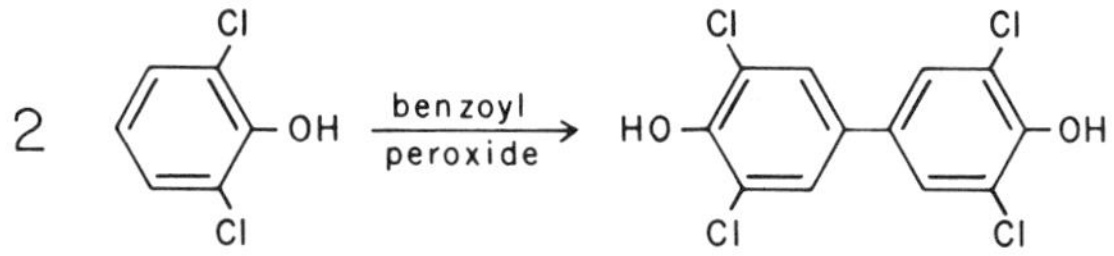

*12. Chlorobiphenylols by Electrochemical Oxidation of Chlorobiphenyls*

Oxidation of 4,4′-dichlorobiphenyl at very high anodic potentials in moderately dried actonitrite gave a dichlorobiphenyl quinone and an (intermediate) hydroxy derivative.[142] The reaction was formulated to proceed as shown below.

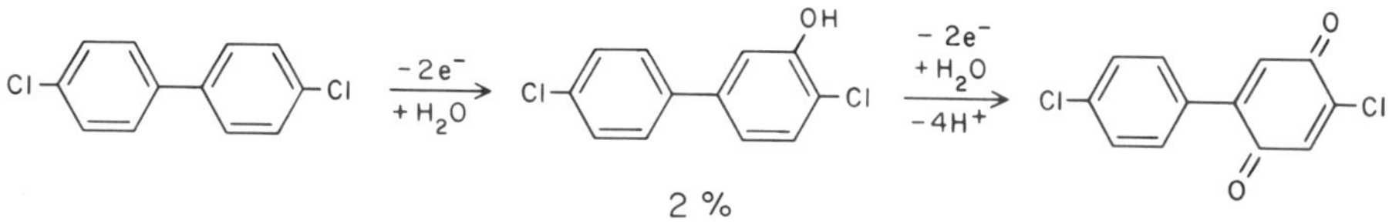

*13. Chlorobiphenylols by Hydroxylation of Chlorobiphenyls*

The hydroxylation of aromatic substrates by chemical[4] and photochemical[96] methods is being actively investigated. Fair yields of hydroxylated compounds are obtained when the corresponding aromatic substrate is treated with peroxytrifluoroacetic acid and boron trifluoride.[55] Poor to moderate yields of dichlorobiphenylols were recently obtained from 4,4′-dichlorobiphenyl[72] by this method. The structures of the products, however, were not further investigated.

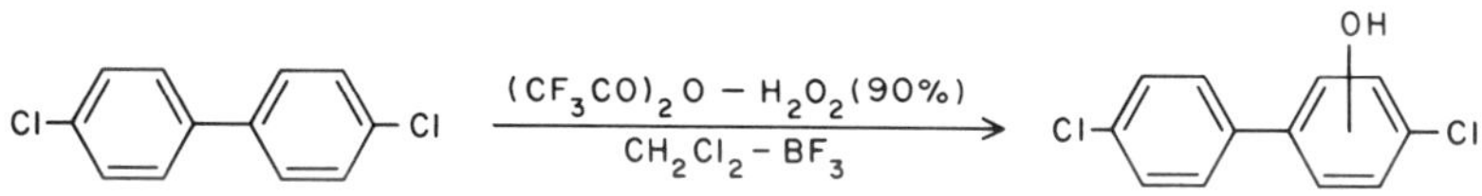

No significant amounts of hydroxylated products were obtained when 4,4′-dichlorobiphenyl was heated with lead tetraacetate in trifluoroacetate (*cf.* 106) or treated with a number of the common (Fenton, Udenfriend) hydroxylating systems (*cf.* 4).

*14. Chlorobiphenylols from Chlorobiphenyls Via Organothallium Compounds*

A method for the preparation of phenols from the corresponding benzene derivatives has recently been described.[145] It involves thallation of the aromatic compound and the aryl thallium ditrifluoroacetate so formed in trifluoroacetic acid is added to one equivalent of lead tetraacetate in trifluoroacetic acid. This mixture is stirred at room temperature for *ca.* 15 min after which time one equivalent of triphenylphosphine is added and the phenol recovered. A 56% yield of 4-chlorophenol, for instance, was obtained from chlorobenzene by this method.

Poor to moderate yields of dichlorobiphenylols were recently obtained from 4,4′-dichlorobiphenyls by this method but the materials were not further investigated.[72]

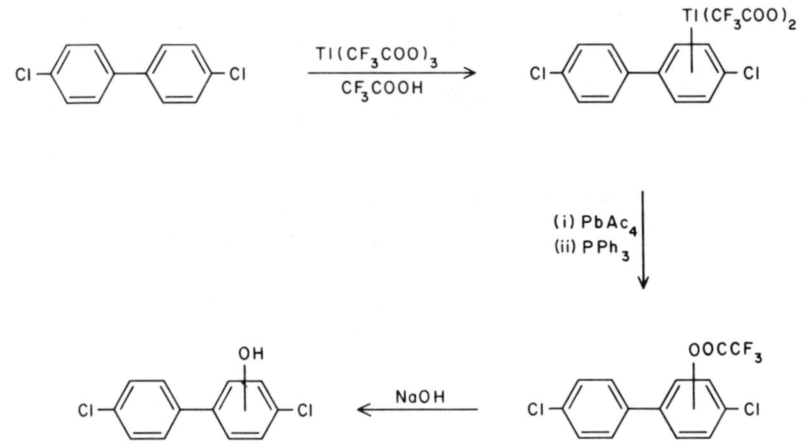

*15. Chlorobiphenylols by the Ullmann Reaction*

Two chloromethoxybiphenyls have recently been prepared in relatively poor yield by a mixed Ullmann reaction[72] (see also Chapter 3). The separation of three products formed was easily accomplished by TLC due to their different polarities.

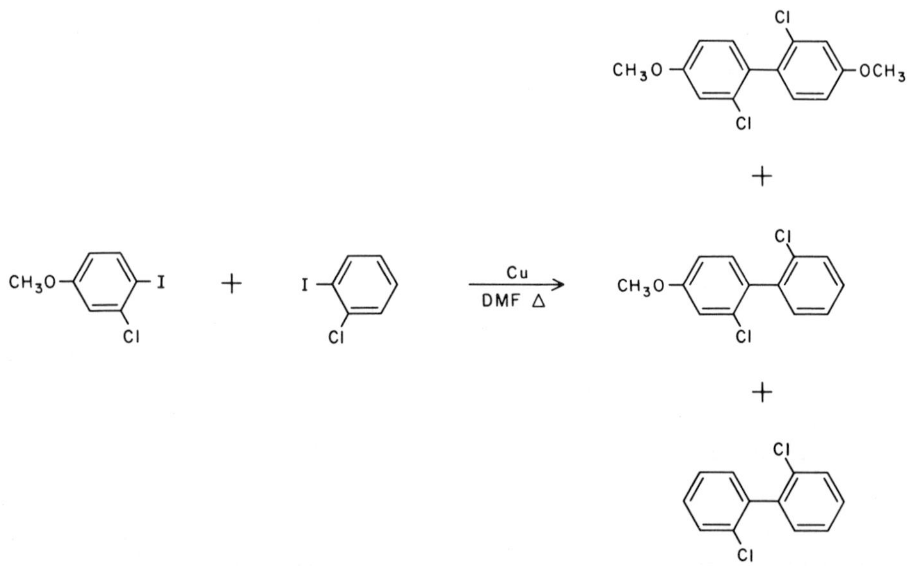

*16. Chlorobiphenylols by the Decomposition of Aryl-sulfonylchlorides in Aromatic Substrate*

Thermal decomposition of acyloxy- or alkyloxyphenylsulfonyl chloride in chlorobenzenes, in the presence of cuprous chloride as catalyst, gives acyloxy- or alkyloxychlorobiphenyls[97] which can be deacetylated or demethylated[101] to give chlorobiphenylols.

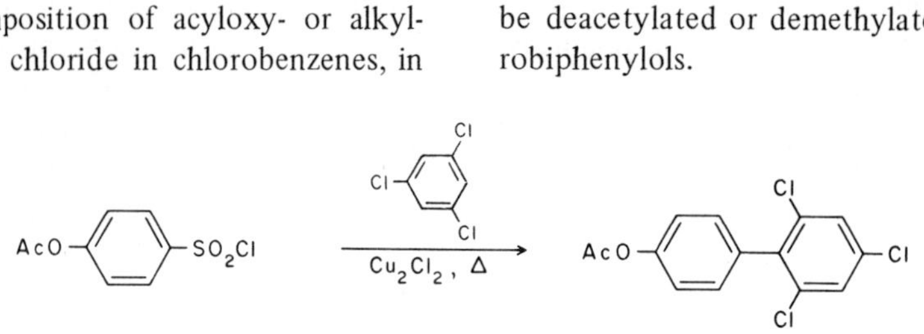

**List of Chlorobiphenylols (Chlorohydroxybiphenyls) Described in the Literature**

In this section chlorobiphenyl-mono- and diols (mono- and dihydroxychlorobiphenyls) which have been described in the literature are listed together with their melting points and method of preparation. Emphasis is on practical procedures and accessible references although all reactions which are of potential use are included. Derivatives such as methyl ethers are listed only where the parent compound has not been described. Removal of a methyl group[99] as the last step in a synthetic route has not been listed as a separate reaction.

For complete literature to 1949, *Beilsteins Handbuch der Organischen Chemie* may be consulted.[14]

Monochloro-Monohydroxy-Derivatives

| Compound | | Melting points | Method of preparation | References |
|---|---|---|---|---|
| Name | Formula | | | |
| 3-Chloro-2-biphenylol | 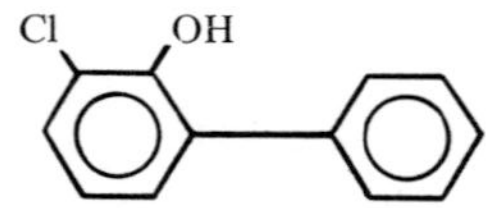 | 73–74 | 7<br>3 | 149<br>34 |
| 4-Chloro-2-biphenylol | 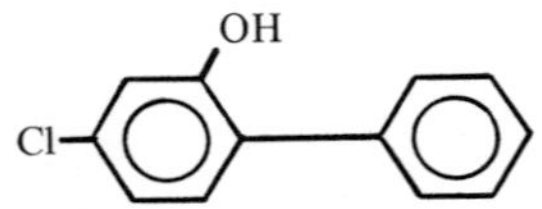 | 38.5–39 | 1 | 107 |
| 5-Chloro-2-biphenylol | 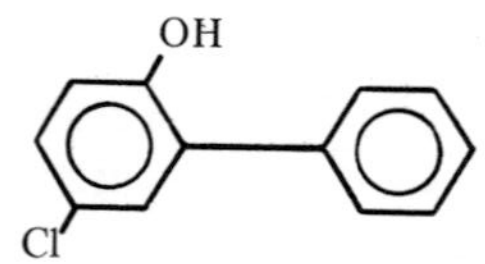 | 46<br>oil | 1<br>7 | 38<br>149 |
| 2′-Chloro-2-biphenylol<br>methyl ether | 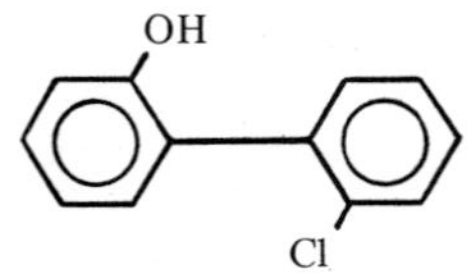 | –<br>53–54 | –<br>7 | –<br>93 |
| 4′-Chloro-2-biphenylol | 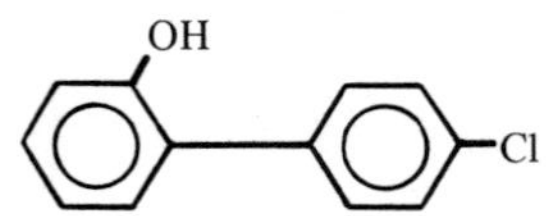 | 53 | 3 | 34 |
| 4-Chloro-3-biphenylol | 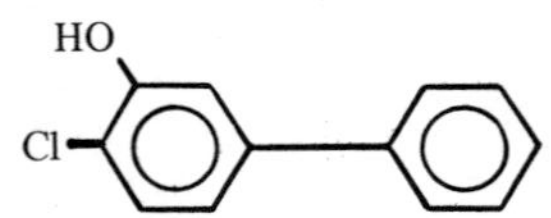 | 50 | 6 | 57 |

| | | | | |
|---|---|---|---|---|
| 6-Chloro-3-biphenylol | 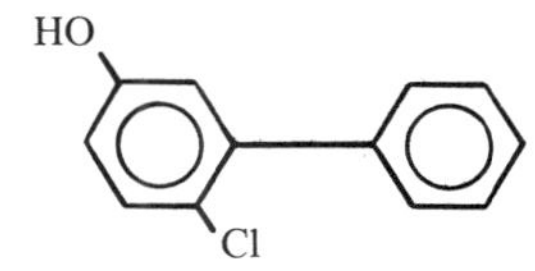 | 63 | 5<br>3 | 38<br>89 |
| 4′-Chloro-3-biphenylol | 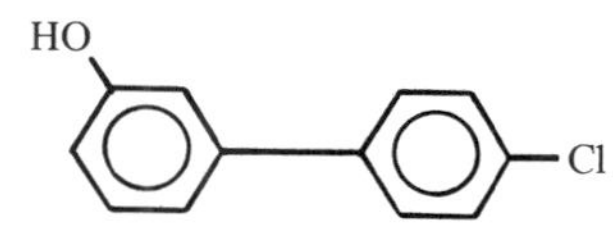 | 74–75.5<br>86 | 7<br>4 | 111<br>112 |
| 2-Chloro-4-biphenylol | 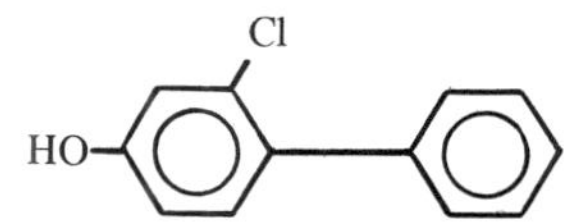 | 80–80.5 | 3 | 72 |
| 3-Chloro-4-biphenylol | 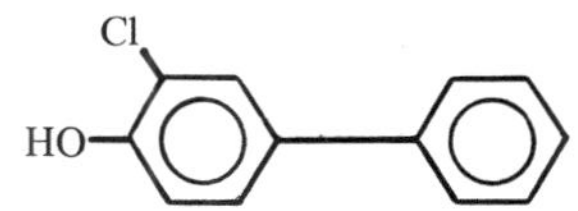 | 76–77.5<br>78.5–79.5 | 2<br>2 | 35<br>72 |
| 2′-Chloro-4-biphenylol | 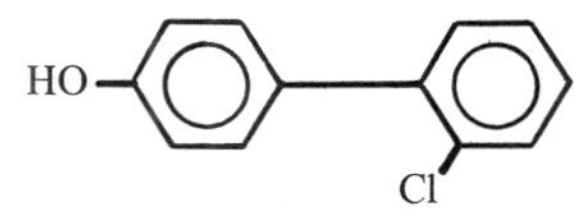 | 90.5–91 | 1 | 100 |
| 4′-Chloro-4-biphenylol | 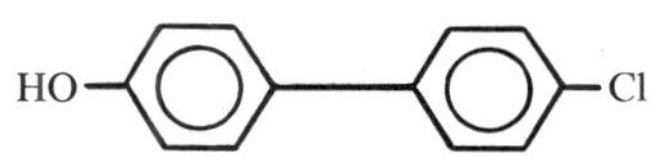 | 146–147<br>145–146<br><br>147–147.5 | 1<br>2<br>3<br>2 | 6<br>123<br>34<br>72 |

Dichloro-Monohydroxy-Derivatives

| Compound | | Melting points | Method of preparation | References |
|---|---|---|---|---|
| Name | Formula | | | |
| 3,5-Dichloro-2-biphenylol | 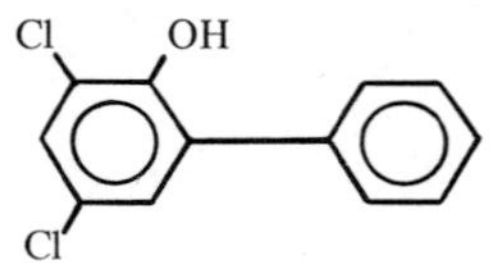 | 46–48<br>54–56 | 2<br>2 | 90<br>72 |
| 4,5-Dichloro-2-biphenylol methyl ether | 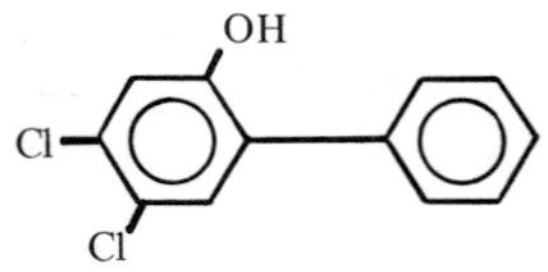 | –<br>85–86 | –<br>3 | –<br>156 |
| 4,4′-Dichloro-2-biphenylol | 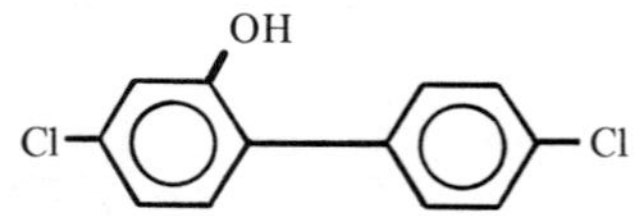 | 78 | 1 | 77 |
| 2′,5′-Dichloro-2-biphenylol | 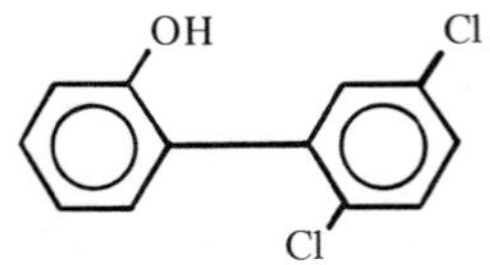 | 97–98 | 3 | 72 |
| 4,4′-Dichloro-3-biphenylol | 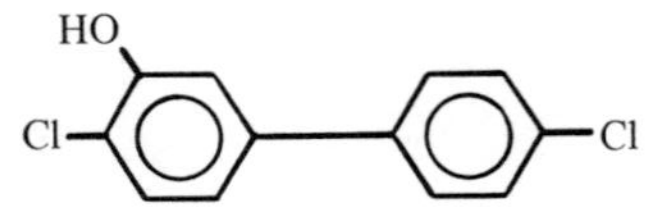 | 74–75 | 7 | 72 |
| 2′,5′-Dichloro-3-biphenylol | 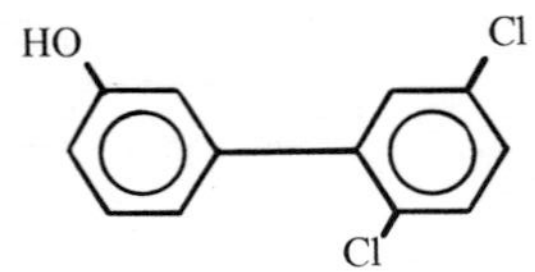 | 117.5–118.5 | 3 | 72 |

| | | | | |
|---|---|---|---|---|
| 3,5-Dichloro-4-biphenylol | 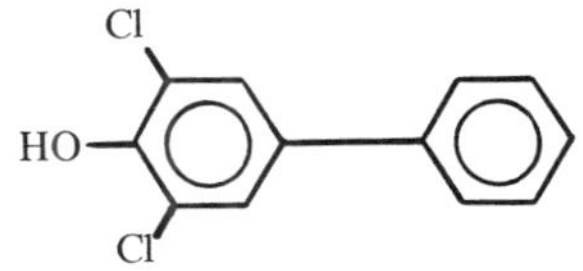 | 80.5–82<br>84–86 | 2<br>2 | 35<br>72 |
| 2,2′-Dichloro-4-biphenylol | 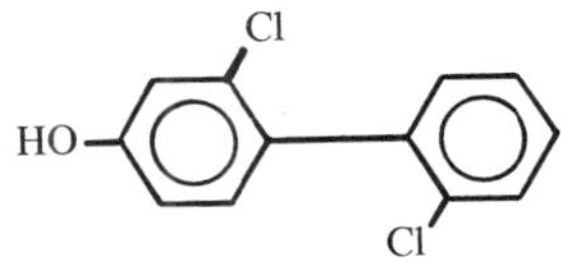 | * | 15 | 72 |
| 2,4′-Dichloro-4-biphenylol | 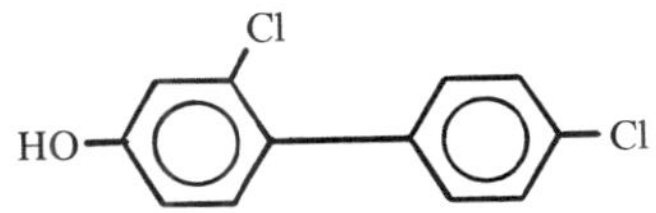 | 116–118 | 3<br>7 | 72<br>72 |
| 3,4′-Dichloro-4-biphenylol | 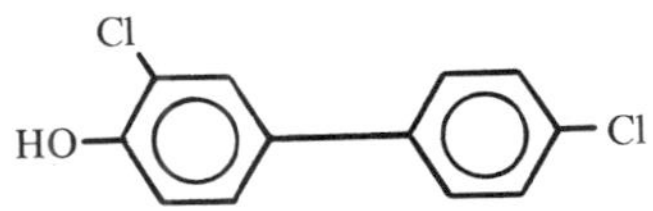 | 71–72<br>78–79 | 2<br>2 | 124<br>72 |
| 2′,5′-Dichloro-4-biphenylol | 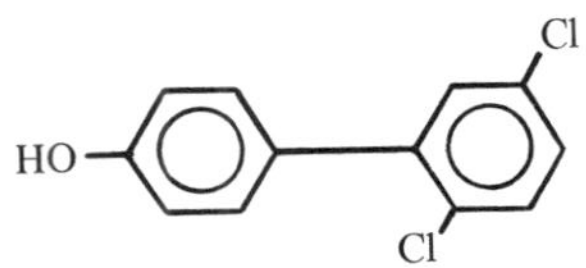 | 95–97 | 3 | 72 |

*Characterized by mass spectrum only.

## Trichloro-Monohydroxy-Derivatives

| Compound: Name | Compound: Formula | Melting points | Method of preparation | References |
|---|---|---|---|---|
| 2′,4′,6′-Trichloro-2-biphenylol | 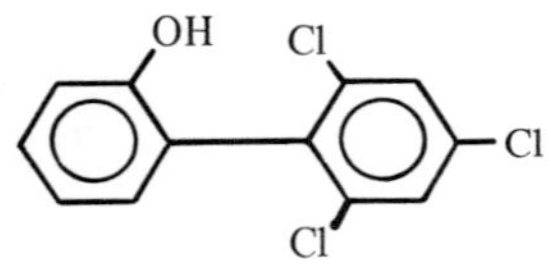 | 107–108.5 | 16 | 97, 101 |
| 2′,4′,6′-Trichloro-3-biphenylol | 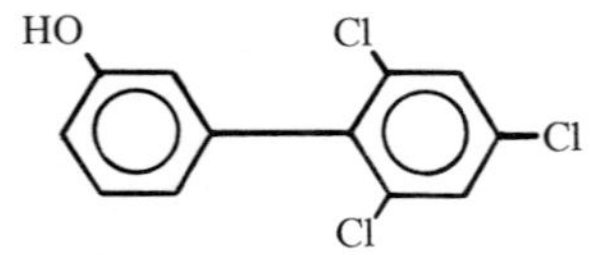 | 94.5–96 | 16 | 97, 101 |
| 3,4′,5-Trichloro-4-biphenylol | 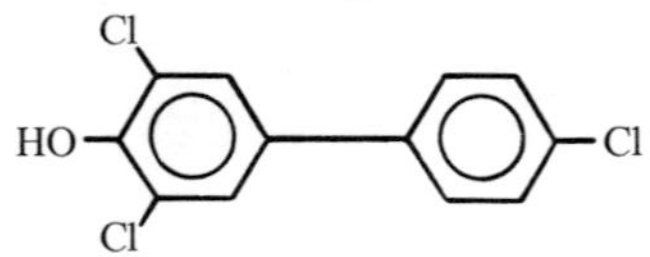 | 133.5–137<br>144<br>147–150 | 2<br>2<br>2 | 35<br>124<br>72 |
| 2,2′,5′-Trichloro-4-biphenylol | 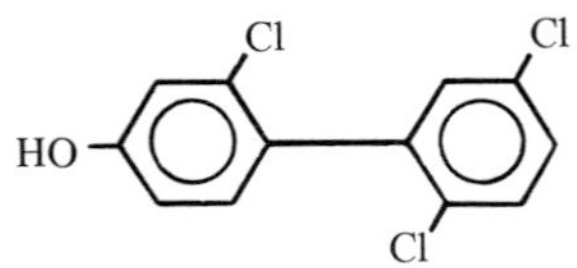 | 98.5–100 | 3 | 72 |
| 2′,4′,6′-Trichloro-4-biphenylol | 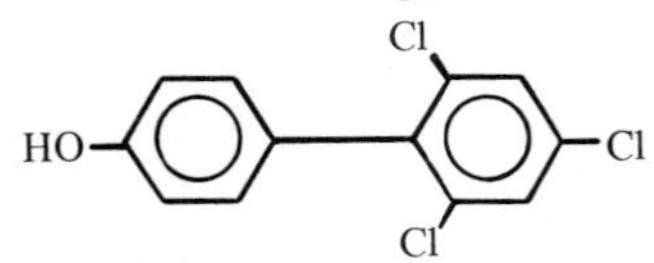 | 130–131 | 16 | 97, 101 |

Tetrachloro-Monohydroxy-Derivatives

| Compound: Name | Compound: Formula | Melting points | Method of preparation | References |
|---|---|---|---|---|
| 3,4,5,6-Tetrachloro-2-biphenylol | 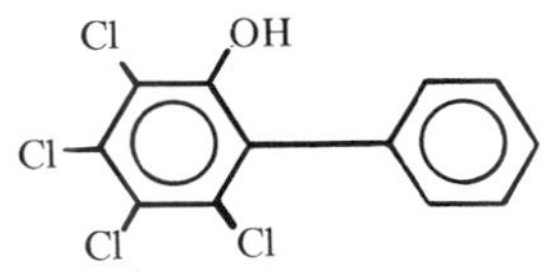 | 81–83 | 8 | 69 |
| 2′,3′,5′,6′-Tetrachloro-3-biphenylol | 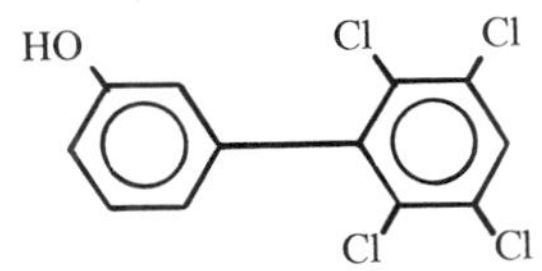 | 104.5–105.5 | 16 | 97, 101 |
| 2′,3′,5′,6′-Tetrachloro-4-biphenylol | 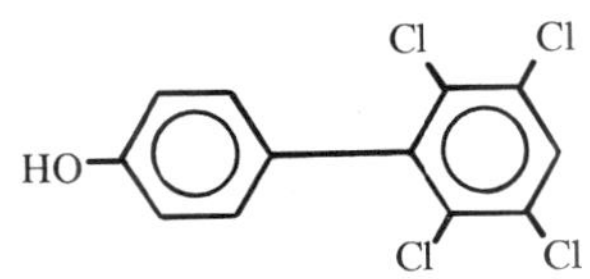 | 138.5–139.5 | 16 | 97, 101 |

Pentachloro-Monohydroxy-Derivatives

| Compound: Name | Compound: Formula | Melting points | Method of preparation | References |
|---|---|---|---|---|
| 3,4,4′5,6-Pentachloro-2-biphenylol | 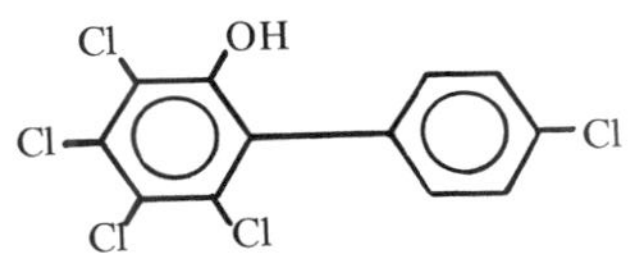 | 124–126 | 8 | 69 |

Monochloro-Dihydroxy-Derivatives

| Compound: Name | Compound: Formula | Melting points | Method of preparation | References |
|---|---|---|---|---|
| 2′-Chloro-2,5-biphenyldiol | 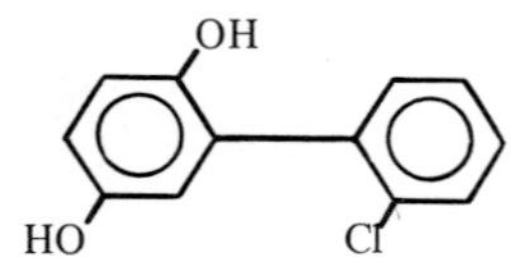 | 98.5 | 10 | 102 |
| 4′-Chloro-2,5-biphenyldiol dimethyl ether | 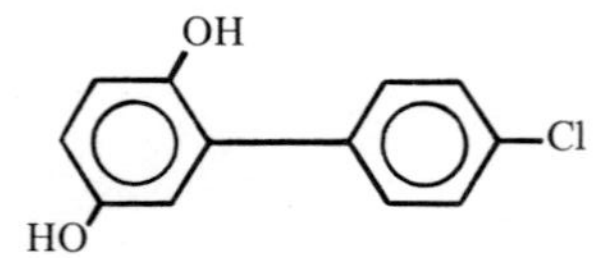 | –<br>63–64 | –<br>– | –<br>74 |
| 4-Chloro-3,3′-biphenyldiol dimethyl ether | 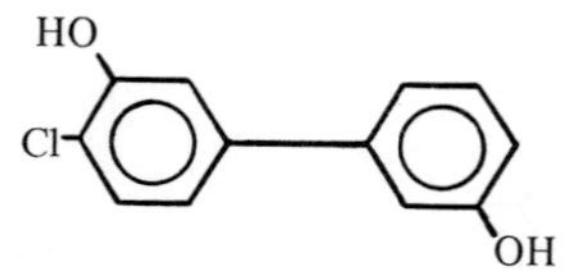 | –<br>74 | –<br>7 | –<br>51 |
| 3-Chloro-4,4′-biphenyldiol | 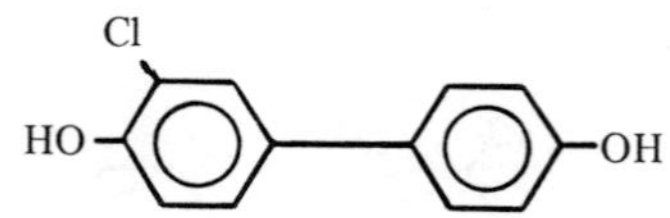 | 215<br>143–144 | 2<br>2 | 29<br>72 |

Dichloro-Dihydroxy-Derivatives

| Compound | | Melting points | Method of preparation | References |
|---|---|---|---|---|
| Name | Formula | | | |
| 2′,3′-Dichloro-2,5-biphenyldiol | 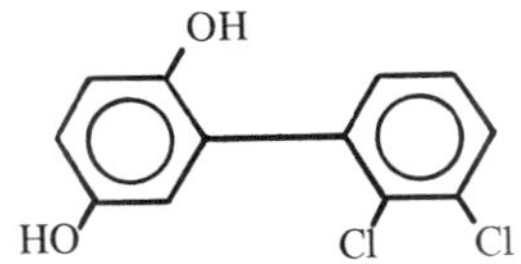 | 170–172 | 10 | 47 |
| 2′,4′-Dichloro-2,5-biphenyldiol | 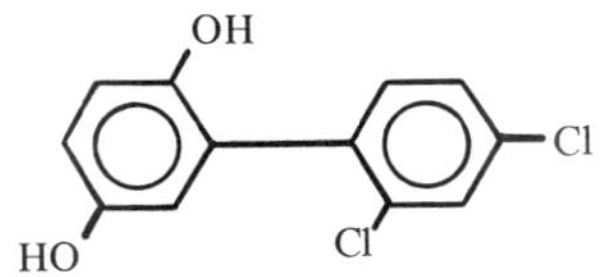 | 182–183 | 10 | 47 |
| 2′,5′-Dichloro-2,5-biphenyldiol | 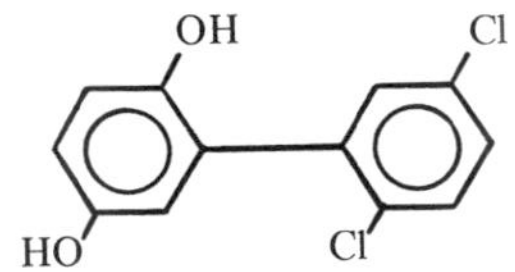 | 173–174 | 10 | 47 |
| 2′,6′-Dichloro-2,5-biphenyldiol | 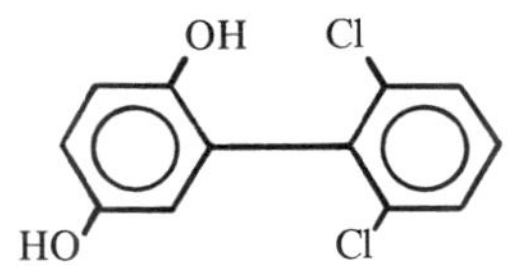 | 134–135 | 10 | 47 |
| 3,3′-Dichloro-2,2′-biphenyldiol | 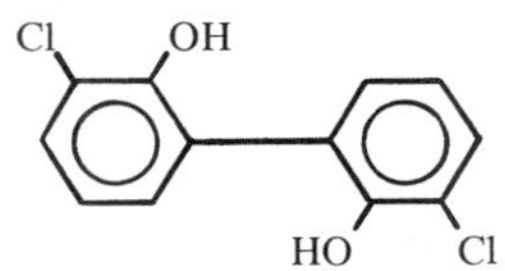 | 129 | 2 | 117 |
| 5,5′-Dichloro-2,2′-biphenyldiol | 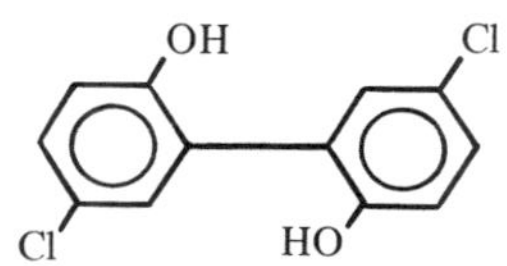 | 170 | 2 | 117 |

| Name | Formula | Melting points | Method of preparation | References |
|---|---|---|---|---|
| 4,4′-Dichloro-3,3′-biphenyldiol dimethyl ether | 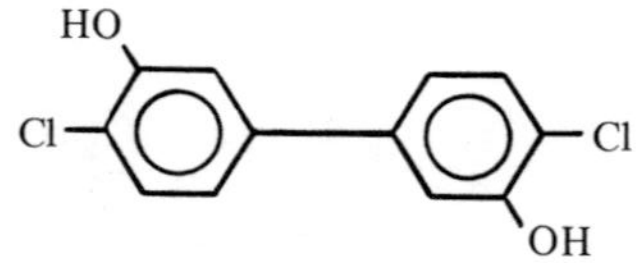 | 124.5–125.5<br>130 | 7<br>7 | 72<br>51 |
| 2,2′-Dichloro-4,4′-biphenyldiol | 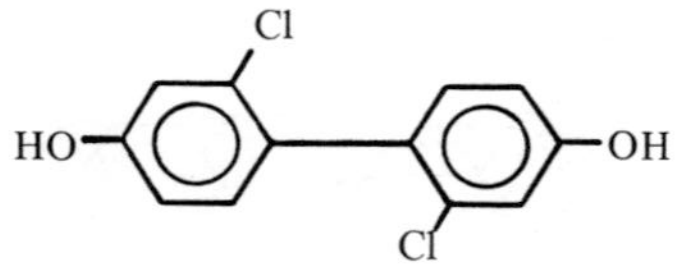 | 148–150 | 1<br>15 | 72<br>72 |
| 3,3′-Dichloro-4,4′-biphenyldiol | 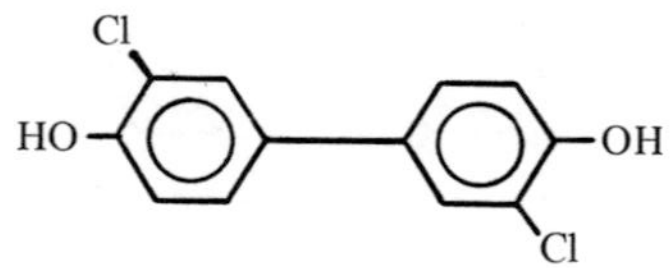 | 124<br><br>130–132 | 2<br>1<br>2 | 29<br>28<br>72 |

Trichloro-Dihydroxy-Derivatives

| Compound | | | | |
|---|---|---|---|---|
| Name | Formula | Melting points | Method of preparation | References |
| 3,3′,5-Trichloro-4-4′-biphenyldiol | 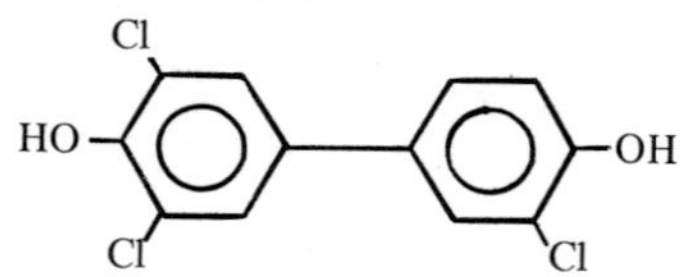 | 193–195 | 2 | 72 |

## Tetrachloro-Dihydroxy-Derivatives

| Compound: Name | Compound: Formula | Melting points | Method of preparation | References |
|---|---|---|---|---|
| 3,3′,5,5′-Tetrachloro-2,2′-biphenyldiol | 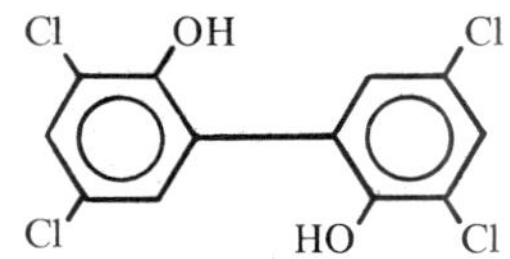 | 178 | 2 | 42 |
| 3,3′,5,5′-Tetrachloro-4,4′-biphenyldiol | 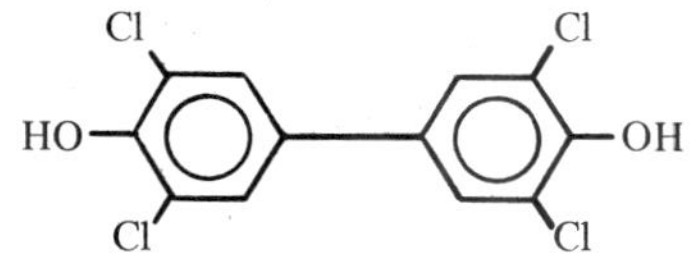 | 235<br>234<br>233 | 2<br>11<br>2 | 127<br>75<br>29 |

## Heptachloro-Dihydroxy-Derivatives

| Compound: Name | Compound: Formula | Melting points | Method of preparation | References |
|---|---|---|---|---|
| 2′,3,4,5,5′,6,6′-Heptachloro-2,3′-biphenyldiol | 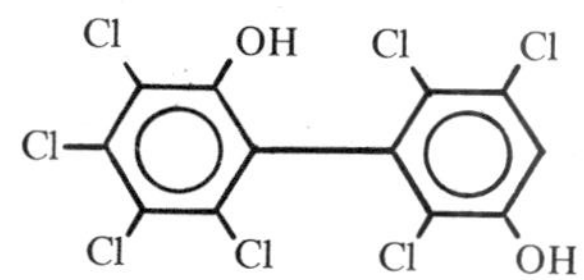 | 114–116 | 9 | 115 |

Octachloro-Dihydroxy-Derivatives

| Compound | | Melting points | Method of preparation | References |
|---|---|---|---|---|
| Name | Formula | | | |
| 2,2′,3,3′,5,5′,6,6′-Octachloro-4,4′-biphenyldiol | 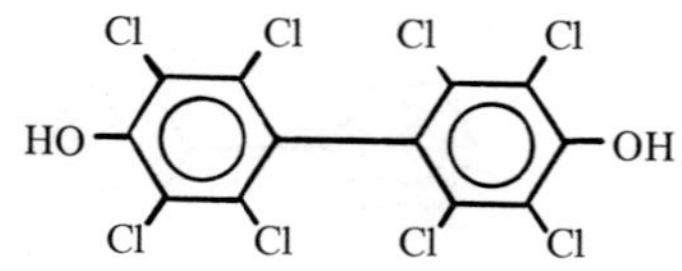 | 235–238 | 2<br>5<br>5 | 23<br>140<br>138 |

# REFERENCES


1. **Abramovitch, R. A. and Koleoso, O. A.,** *J. Chem. Soc.,* (B), 779, 1969.
2. **Abramovitch, R. A. and Saha, J. G.,** *Tetrahedron,* 21, 3297, 1965.
3. **Abramovitch, R. A. and Takaya, T.,** *Chem. Commun.,* 1369, 1969.
4. **Albert, K. S.,** Aromatic Hydroxylation as an Analytical Reaction, Ph.D. Thesis, The University of Wisconsin, 1972.
5. **Ames, G. R.,** *Chem. Rev.,* 58, 895, 1958.
6. **Angeletti, A. and Gatti, D.,** *Gazz. Chim. Ital.,* 58, 630, 1929.
7. **Atkinson, E. R., Lawler, H. J., Heath, J. C., Kimball, E. H., and Read, E. R.,** *J. Am. Chem. Soc.,* 63, 730, 1941.
8. **Atkinson, E. R., Morgan, C. R., Warren, H. H., and Manning, T. J.,** *J. Am. Chem. Soc.,* 67, 1513, 1945.
9. **Bachmann, W. E. and Hoffman, R. A.,** *Org. React.,* 2, 224, 1944.
10. **Badger, G. M., Drewer, R. J., and Lewis, G. E.,** *Aust. J. Chem.,* 17, 1036, 1964.
11. **Balogh, V., Fetzion, M., and Golfier, M.,** *J. Org. Chem.,* 36, 1339, 1971.
12. **Banthrope, D. V.,** in *Topics in Carbocyclic Chemistry,* Lloyd, D., Ed., Logos Press, London, 1969, 1.
13. *Beilsteins Handbuch der Organischen Chemie,* Vol. 12 (monoamino), Springer, Berlin, Hauptwerk, 1317; Ergänzungsband I, 546, Ergänzungsband, II, 747; Vol. 13 (diamino): Hauptwerk, 210, Ergänzungsband, I, 57; Ergänzungsband, II, 87.
14. *Beilsteins Handbuch der Organischen Chemie,* Vol. 6 (monohydroxy), Springer, Berlin, Hauptwerk, 672; Ergänzungsband I, 323; Ergänzungsband II, 623; Ergänzungsband III, 3281; (dihydroxy), Hauptwerk, 989, Ergänzungsband I, 484; Ergänzungsband II, 960; Ergänzungsband III, 5371.
15. **Belcher, R., Nutten, A. J., and Stephen, W. I.,** *J. Chem. Soc.,* 2336, 1958.
16. **Bell, F. and Gibson, J. A.,** *J. Chem. Soc.,* 24, 1955.
17. **Bell, F. and Gibson, J. A.,** *J. Chem. Soc.,* 3560, 1955.
18. **Bellavita, V.,** *Gazz. Chim. Ital.,* 65, 632, 1935.
19. **Bilbo, A. J. and Wyman, G. M.,** *J. Am. Chem. Soc.,* 75, 5312, 1953.
20. **Birchall, J. M., Haszeldine, R. N., and Speight, J. G.,** *J. Chem. Soc.,* (C), 2187, 1970.
21. **Blakey W. and Scarborough, H. A.,** *J. Chem. Soc.,* 3000, 1927.
22. **Boekelheide, V. and Nottke, J. E.,** *J. Org. Chem.,* 34, 4134, 1969.
23. **Bourdon, J. and Calvin, M.,** *U.S.A.E.C.,* UCRL-3466, 1956; *Chem. Abstr.,* 8045, 1957.
24. **Bradsher, C. K. and Wissow, L. J.,** *J. Am. Chem. Soc.,* 68, 404, 1946.
25. **Burns, J., McCombie, H., and Scarborough, H. A.,** *J. Chem. Soc.,* 2928, 1928.
26. **Byron, D. J., Gray, G. W., and Wilson, R. C.,** *J. Chem. Soc.,* (C), 840, 1966.
27. **Cadogan, J. I. G.,** *J. Chem. Soc.,* 4257, 1962.
28. **Cain, J. C.,** *J. Chem. Soc.,* 688, 1903.
29. **Cain, J. C.,** *J. Chem. Soc.,* 7, 1904.
30. **Cain, J. C. and May, P.,** *J. Chem. Soc.,* 720, 1910.
31. **Carlin, R. B. and Forshey, W. O., Jr.,** *J. Am. Chem. Soc.,* 72, 793, 1950.
32. **Chaix, M. and de Röchebovet, F.,** *Bull. Soc. Chim.,* 2, 273, 1935.
33. **Chardonnens, L. and Würmli, A.,** *Helv. Chim. Acta,* 29, 922, 1946.
34. **Colbert, J. C. and Lacy, R. M.,** *J. Am. Chem. Soc.,* 68, 270, 1946.
35. **Colbert, J. C., Meigs, W., and Mackin, B.,** *J. Am. Chem. Soc.,* 56, 202, 1934.
36. **Colbert, J. C., Meigs, W., and Stuerke, A. H.,** *J. Am. Chem. Soc.,* 56, 2128, 1934.
37. **Cox, J. R. and Dunn, M. F.,** *Tetrahedron Lett.,* 985, 1963.
38. **de Crauw, R.,** *Rec. Trav. Chim.,* 59, 753, 1931.
39. **Dey, B. B., Govindachari, T. R., and Rajagopalan, S. C.,** *J. Sci. Ind. Res. (India),* 4, 645, 1946; *Chem. Abstr.,* 40, 6347, 1946.
40. **Dey, B. B., Govindachari, T. R., and Rajagopalan, S. C.,** *J. Sci. Ind. Res. (India),* 5B, 75, 1946; *Chem. Abstr.,* 41, 4046, 1947.
41. **Dey, B. B., Govindachari, T. R., Rajagopalan, S. C., and Venkatakrishna Udupa,** *J. Sci. Ind. Res. (India),* 6B, 103, 1947; *Chem. Abstr.,* 43, 3729, 1949.
42. **Diels, O. and Biebergeil, A.,** *Ber.,* 35, 302, 1902.
43. **Effenberger, F., Nagel, K., and Agster, W.,** *Angew. Chem. Internat. Ed.,* 10, 566, 1971.
44. **Enders, E.,** Methoden zur Herstellung und Umwandlung von Aromatischen Hydrazoverbindungen sowie Tri- und Tetraarylhydrazinen, in *Methoden der Organischen Chemie,* Vol. X/2, (Houben Weyl) Müller, E., Ed., Thieme, Stuttgart, 1967, 697.
45. **Fanta, P. E.,** *Chem. Rev.,* 38, 139, 1946.
46. **Fanta, P. E.,** *Chem. Rev.,* 64, 613, 1964.
47. **Farbenindustrien, A. G.,** French Patent 816 719, 1937; *Chem. Abstr.,* 32, 2145, 1938.
48. **Finzi, C. and Bellavita, V.,** *Gazz. Chim. Ital.,* 64, 335, 1934.
49. **Gelmo, P.,** *Ber.,* 39, 4175, 1906.

50. **Gibson, M. S.,** The introduction of the amino group, in *The Chemistry of the Amino Group,* Patai, S., Ed., Interscience, New York, 1968, 37.
51. **Gottlieb-Billroth, H.,** *J. Am. Chem. Soc.,* 49, 482, 1927.
52. **Harris, M.,** U.S. Patent 2,079,450, 1937; *Chem. Abstr.,* 31, 4346, 1937.
53. **Harris, M.,** U.S. Patent 2,126,009; *Chem. Abstr.,* 32, 7478, 1938.
54. **Harris, G. E. and Christiansen, W. G.,** *J. Am. Pharm. Assoc.,* 24, 553, 1935.
55. **Hart, H.,** *Accounts Chem. Res.,* 4, 337, 1971.
56. **Hauptmann, H., Walter, W. F., and Marino, C.,** *J. Am. Chem. Soc.,* 80, 5832, 1958.
57. **Hay, A. S.,** U.S. Patent 3,363,008, 1968; *Chem. Abstr.,* 69, 2702, 1968.
58. **Helling, J. F. and Braitsch, D. M.,** *J. Am. Chem. Soc.,* 92, 7207, 1970.
59. **Henry, P. M.,** *J. Org. Chem.,* 35, 3083, 1970.
60. **Hickmott, P. W. and Hudson, A. T.,** *J. Chem. Soc.,* (C), 762, 1971.
61. **Hickmott, P. W. and Yoxall, C. T.,** *J. Chem. Soc.,* (C), 1829, 1971.
62. **Hinkel, L. E. and Dippy, J. F. J.,** *J. Chem. Soc.,* 1387, 1930.
63. **Hinkel, L. E. and Hey, D. H.,** *J. Chem. Soc.,* 2786, 1928.
64. **Hodgson, H. H. and Holt, P. F.,** *J. Chem. Soc.,* 37, 1937.
65. **Hodgson, H. H. and Holt, P. F.,** *J. Chem. Soc.,* 1413, 1934.
66. **Hoegerle, K. and L'Ecuyer, P.,** *Can. J. Chem.,* 37, 2068, 1959.
67. **Hollingsworth, B. L. and Petrow, U.,** *J. Chem. Soc.,* 3771, 1961.
68. **Holzach, K.,** *Die Aromatischen Diazoverbindungen,* Ferdinand Enke, Stuttgart, 1947.
69. **Huisgen, R., Binsch, G., and König, H.,** *Ber.,* 97, 2884, 1964.
70. **Huntress, E. H. and Seikel, M. K.,** *J. Am. Chem. Soc.,* 61, 816, 1939.
71. **Hutzinger, O.,** unpublished results.
72. **Hutzinger, O., Safe, S., and Zitko, V.,** in preparation.
73. **Idoux, J. P.,** *J. Chem. Soc.,* (C), 435, 1970.
74. **Farbenindustrien, I. G.,** German Patent 566,521, 1931; Beilstein, Vol. 6, Ergänzungswerk III, 5372.
75. **Inoue, H., Simamura, O., and Takamizawa, K.,** *Bull. Chem. Soc. Jap.,* 35, 1958, 1962.
76. **Ishikawa, H. and Mukaiyama, T.,** *Bull. Chem. Soc. Jap.,* 45, 967, 1972.
77. **Jenkins, R. L.,** U.S. Patent 2,084,033, 1937; *Chem. Abstr.,* 31, 5386, 1937.
78. **Jerussi, R. A.,** *J. Org. Chem.,* 35, 2105, 1970.
79. **Joschek, H. I. and Miller, S. I.,** *J. Am. Chem. Soc.,* 88, 3269, 1966.
80. **Kametani, T. and Fukumoto, K.,** *Synthesis,* 657, 1972.
81. **Kaslow, C. E. and Summers, R. M.,** *J. Org. Chem.,* 20, 1738, 1955.
82. **Kharasch, M. S. and Reinmuth, O.,** *Grignard Reactions of Nonmetallic Substances,* Prentice-Hall, New York, 1954.
83. **Kobayashi, M.,** *J. Chem. Soc. Jap.,* 74, 968, 1953.
84. **Krasovitskii, B. M. and Pereyaslova, D. G.,** *Zhur. Vsesoyuz. Khim. Obshchestra im. D. I. Mendeleeva,* 6, 446, 1961; *Chem. Abstr.,* 56, 362, 1962.
85. **Kuser, P., Inderbitzin, M., Brauchli, J., and Engster, C. H.,** *Helv. Chim. Acta,* 54, 980, 1971.
86. **Kvalnes, D. E.,** *J. Am. Chem. Soc.,* 56, 2478, 1934.
87. **LeFevre, R. J. W. and Turner, E. E.,** *J. Chem. Soc.,* 254, 1928.
88. **Legon'kova, L. M. and Stepanov, B. I.,** *Tr. Mosk. Khim. Tekhnol.,* Inst. No. 48, 120, 1965; *Chem. Abstr.,* 65, 20251, 1966.
89. **Leigh, T., Thorp, J. M., and Waring, W. S.,** British Patent 1,121,722, 1968; *Chem. Abstr.,* 70, 57471, 1969.
90. **Leupold, I. and Musso, H.,** *Ann. Chem.,* 746, 134, 1971.
91. **Litvinenko, L. M., Grekov, A. P., Verkhovod, N. N., and Dzyuba, V. P.,** *Zhur. Obshch. Khim.,* 26, 2524, 1956; *J. Gen. Chem. U.S.S.R.* (English translation), 2817; *Chem. Abstr.,* 51, 5009, 1957.
92. **Mascarelli, L. and Gatti, D.,** *Gazz. Chim. Ital.,* 63, 654, 1933.
93. **Mascarelli, L. and Pirona, M.,** *Gazz. Chim. Ital.,* 68, 117, 1938.
94. **Mascarelli, L., Gatti, D., and Pirona, M.,** *Gazz. Chim. Ital.,* 61, 782, 1931.
95. **Matsuura, T. and Omura, K.,** *Bull. Chem. Soc. Jap.,* 39, 944, 1966.
96. **Matsuura, T. and Omura, K.,** *Synthesis,* in press.
97. **McCall, E. and Cummings, W.,** British Patent 1,014,554, 1965; *Chem. Abstr.,* 64, 9632, 1966.
98. **McDonald, P. D. and Hamilton, G. A.,** Mechanism of phenolic oxidative coupling reactions, in *Oxidation in Organic Chemistry,* Part B., Trahanovsky, W. S., Ed., Academic Press, New York, 1973, 97.
99. **McOmie, J. F. W. and West, D. E.,** *Organic Syntheses,* 49, 50, 1969.
100. Monsanto Chemicals Ltd., U.S. Patent 2,084,027, 1935; *Beilsteins Handbuch der Organischen Chemie,* Vol. 6, Ergänzungswerk III, 3332.
101. Monsanto Chemicals Ltd., French Patent 1,436,883, 1966; *Chem. Abstr.,* 67, 43552, 1967.
102. **Moualim, R. J. and Venkataraman, K.,** *J. Sci. Ind. Res. (India),* 3, 447, 1945; *Chem. Abstr.,* 39, 4605, 1945.
103. **Müller, E., Ed.,** Stickstoffverbindungen II, Amine (Herstellung), in *Methoden der Organischen Chemie,* Vol. II/1, Thieme, Stuttgart, 1957.
104. **Musso, H.,** *Angew. Chem. Intern. Ed.,* 2, 723, 1963.

105. Niwa, H., *Tohoku Yakka Daigaku Kiyo*, 4, 31, 1957; *Chem. Abstr.*, 53, 7235, 1958.
106. Norman, R. O. C., Thomas, C. B., and Willson, J. S., *J. Chem. Soc.*, (B), 518, 1971.
107. Oki, M., Hosoya, H., and Iwamura, H., *Bull. Chem. Soc. Jap.*, 34, 1391, 1961.
108. Omura, K. and Matsuura, T., *Chem. Commun.*, 1394, 1969.
109. Omura, K. and Matsuura, T., *Synthesis*, 28, 1971.
110. Parish, J. H. and Whiting, M. C., *J. Chem. Soc.*, 4713, 1964.
111. Polaczkowa, W. and Kwiatkowska, S., *Rocz. Chem.*, 44, 375, 1970; *Chem. Abstr.*, 72, 131888, 1970.
112. Prelog, V., Würsch, J., and Königsbacher, K., *Helv. Chim. Acta*, 34, 258, 1951.
113. Pütter, R., Methoden zur Herstellung und Umwandlung Aromatischer Diazoniumsalze, in *Methoden der Organischen Chemie*, Vol. X/3, Müller, E., Ed., G. Thieme, Stuttgart, 1965, 7.
114. Raiford, L. C. and McNulty, G. M., *J. Am. Chem. Soc.*, 56, 680, 1934.
115. Ried, W. and Torok, E., *Ann. Chem.*, 687, 232, 1965.
116. Rigby, W., *J. Chem. Soc.*, (C), 765, 1971.
117. Robertson, P. W. and Briscoe, H. V. A., *J. Chem. Soc.*, 1964, 1912.
118. Rodd, E. H., Ed., *The Chemistry of Carbon Compounds*, Vol. III, Elsevier, Amsterdam, 1954, 296.
119. Rodd, E. H., Ed., *The Chemistry of Carbon Compounds*, Vol. III, Elsevier, Amsterdam, 1956, 1028.
120. Safe, S. and Hutzinger, O., *J. Chem. Soc.*, Perkin Trans. I, 686, 1972.
121. Sandin, R. B. and Hay, A. S., *J. Am. Chem. Soc.*, 74, 274, 1952.
122. Saunders, K. H., *The Aromatic Diazo-Compounds and Their Industrial Applications*, E. Arnold & Co., London, 1949.
123. Savoy, C. M. S. and Abernethy, J. L., *J. Am. Chem. Soc.*, 64, 2219, 1942.
124. Savoy, C. M. S. and Abernethy, J. L., *J. Am. Chem. Soc.*, 65, 1464, 1943.
125. Scarborough, H. A. and Waters, W. A., *J. Chem. Soc.*, 557, 1926.
126. Scarborough, H. A. and Waters, W. A., *J. Chem. Soc.*, 89, 1927.
127. Schraufstätter, E., Meiser, W., and Gönnert, R., *Z. Naturforsch.*, 16b, 95, 1961.
128. Schultz, G., *Ber.*, 17, 463, 1884.
129. Schündehütte, K. H., Methoden zur Herstellung und Umwandlung von Diarylazoverbindungen, in *Methoden der organischen Chemie*, Vol. X/3 (Houben-Weyl), Müller, E., Ed., Thieme, Stuttgart, 1965, 219.
130. Seidenfaden, W. and Pawellek, D., Methoden zur Herstellung und Umwandlung von aromatischen Nitroverbindungen, in *Methoden der organischen Chemie*, Vol. X/1 (Houben-Weyl), Müller, E., Ed., Thieme, Stuttgart, 1971, 471.
131. Semmelhack, M. F., Helquist, P. M., and Jones, L. D., *J. Am. Chem. Soc.*, 93, 5908, 1971.
132. Shaw, K. B. and Miller, R. K., *Can. J. Chem.*, 48, 1394, 1970.
133. Sheradsky, T. and Salemnik, G., *Tetrahedron Lett.*, 645, 1971.
134. Sheradsky, T. and Salemnik, G., *Isr. J. Chem.*, 10, 857, 1972.
135. Shine, H. J., The Benzidine Rearrangements, in *Mechanisms of Molecular Migrations*, Vol. 2, Thyagarajan, B. S., Ed., Interscience Publ., New York, 1969, 191.
136. Shu, H., *Hua Hsüeh Hsüeh Pao*, 25, 171, 1959; *Chem. Abstr.*, 54, 4489, 1960.
137. Smith, E., U.S. Patent 3,480,673, 1969; *Chem. Abstr.*, 72, 31423, 1970.
138. Smith, F. D., U.S. Patent 2,449,088, 1948; *Chem. Abstr.*, 43, 813, 1949.
139. Smith, P. A. S., Clegg, J. M., and Hall, J. H., *J. Org. Chem.*, 23, 524, 1958.
140. Societe d'Electro-Chimie, Belgian Patent 613,066, 1962; *Chem. Abstr.*, 57, 16492, 1962.
141. Societe d'Electro-Chimie, British Patent 1,054,406, 1967; *Chem. Abstr.*, 66, 46191, 1967.
142. Stuart, J. D., Keenan, R. R., Fenn, R. J., Jensen, R. G., and Pudelkiewicz, W. J., Paper No. 24, American Chemical Society, National Meeting, New York, 1972, Division of Water, Air, and Waste Chemistry; and private communication.
143. Taber, R. L. and Daub, G. H., *Synthesis*, 261, 1971.
144. Taylor, E. C. and McKillop, A., *Accounts Chem. Res.*, 3, 338, 1970.
145. Taylor, E. C., Altland, H. W., Danforth, R. H., McGillivray, O., and McKillop, A., *J. Am. Chem. Soc.*, 92, 3520, 1970.
146. Van Roosmalen, F. L. W., *Rec. Trav. Chim.*, 53, 359, 1934.
147. Večeřa, M. and Petránek, J., *Collect. Czech. Chem. Commun.*, 25, 2005, 1960.
148. Waldau, E. and Pütter, R., *Angew. Chem. Int. Ed.*, 11, 826, 1972.
149. Weissberger, A. and Salminen, I. F., *J. Am. Chem. Soc.*, 67, 58, 1945.
150. White, J. and Adams, R., *J. Am. Chem. Soc.*, 54, 2104, 1932.
151. Wolf, W. and Kharasch, N., *J. Org. Chem.*, 30, 2493, 1965.
152. Woods, G. F., Reed, F. T., Arthur, T. E., and Ezekiel, H., *J. Am. Chem. Soc.*, 73, 3854, 1951.
153. Yoshimura, M., Japan Patent 4276, 1954; *Chem. Abstr.*, 50, 4223, 1956.
154. Yoshimura, M., Japan Patent 7316, 1954; *Chem. Abstr.*, 50, 9448, 1956.
155. Zander, M., Biaryle, in *Methoden der Organischen Chemie* (Houben-Weyl), G. Thieme, Stuttgart, in press.
156. Zerweck, W. and Schütz, K., U.S. Patent 2,280,504, 1942; *Chem. Abstr.*, 36, 5658, 1942.

Chapter 5

# CHEMICAL REACTIONS OF CHLOROBIPHENYLS

One of the main reasons for the usefulness of PCB as industrial compounds is their low reactivity. For instance, PCB's are stable to conditions of hydrolysis and oxidation encountered in industrial use.

In this chapter, the most important reactions known to occur with chlorobiphenyls will be discussed.

## Oxidation

In 1938, an attempt was made to elucidate the composition of a PCB mixture by oxidation with nitric acid and analysis of the chlorobenzoic acids formed[55] (see also Figure 12 in Chapter 2). Oxidations were carried out with boiling $HNO_3$ (d = 1.4) for 100 hr. Potassium permanganate, chromic acid, and nitric acid (d = 1.2) apparently did not oxidize chlorobiphenyl mixtures with averages of five or seven chlorine atoms per molecule.

Mono-, di-, and trichlorobiphenyls can be oxidized to the corresponding chlorobenzoic acids with chromic anhydride and acetic acid.[1,50]

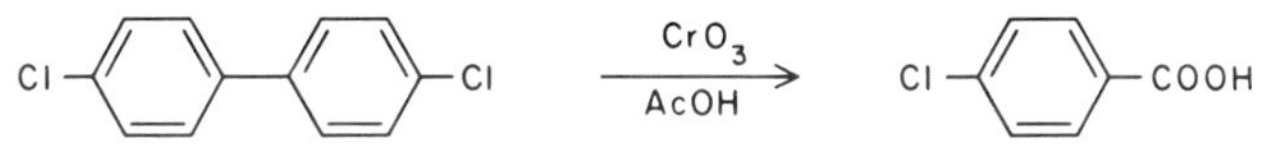


The reaction of chlorobiphenyls with hydroxylating reagents is described in Chapter 4.

## Reduction

The reductive dechlorination of PCB to give a single compound, biphenyl, has been described.[3] This reaction, which has been recommended as an analytical method, is carried out on a precolumn during gas chromatography.

Replacement of chlorine in chlorobiphenyls by hydrogen via chemical reaction has been described for decachlorobiphenyl[4] using lithium aluminum hydride, butyl lithium and water, or the Grignard reagent and water. Two chlorine atoms in the 4 and 4′ positions were replaced when 2 mol of reagents were used. With 1 mol reagent, a mixture of starting material octa- and nonachlorobiphenyl was obtained.

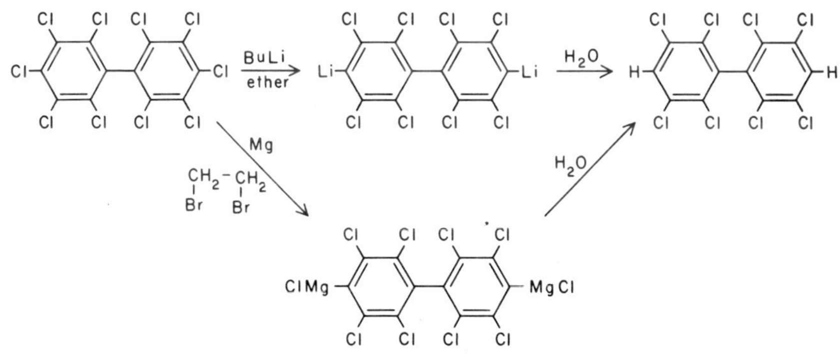


Other reagents which have been used for the reductive dechlorination of other compounds are $Na_2Te$,[29] Raney nickel,[7] sodium in liquid ammonia,[34] zinc dust[31] and sodium hydrazide, and hydrazine.[23]

As far as the authors are aware, reduction of the aromatic ring in chlorobiphenyls has not been reported. However, alkali metals and ammonia have been shown to reduce the parent compound biphenyl to phenylcyclohexane[46] and phenyl-2,5-cyclohexadiene.[39]

## Chlorination

Chlorination of biphenyl, chlorobiphenyls (Chapter 5), aminobiphenyls (Chapter 4) and hydroxybiphenyls (Chapter 4) are discussed from a synthetic point of view in other parts of this book.

All chlorobiphenyls undergo complete chlorination to decachlorobiphenyl when treated with antimony pentachloride[3,19] antimony pentachloride-iodine or reagent BMC ($SO_2Cl_2$, $AlCl_3$, $S_2Cl_2$)[20] or trichlorosulfur tetrachloroaluminate ($SCl_3AlCl_4$).[18] This reaction is the basis of an analytical method.

**Nitration**

Nitration of chlorobiphenyls is one of the few reactions which gives well defined derivatives in good yield even of highly chlorinated members of the series. For example, 2,2′,4,4′,6,6′-hexachlorobiphenyl[53] and 2,2′,4,4′,5,5′-hexachlorobiphenyl[38] were easily converted to the 2,2′,4,4′,-6,6′-hexachloro-3,3′-dinitrobiphenyl and 2,2′,4,4′,-5,5′-hexachloro-3,3′-dinitrobiphenyl respectively on treatment with nitric acid (d = 1.5).

Mono-, di-, tri-, or tetranitroderivatives are obtained, depending on the conditions of nitration from chlorobiphenyls with fewer chlorine atoms per molecule.[1] From 4,4′-dichlorobiphenyl, for instance, 4,4′-dichloro-2-nitrobiphenyl, 4,4′-dichloro-2,3′-dinitrobiphenyl, and 4,4′-dichloro-2,-3′,5′-trinitrobiphenyl were obtained[44,48] and 2,2′-dichlorobiphenyl yielded either a 5,5′-di- or a 3,3′,5,5′-tetranitro derivative as the main product.[10]

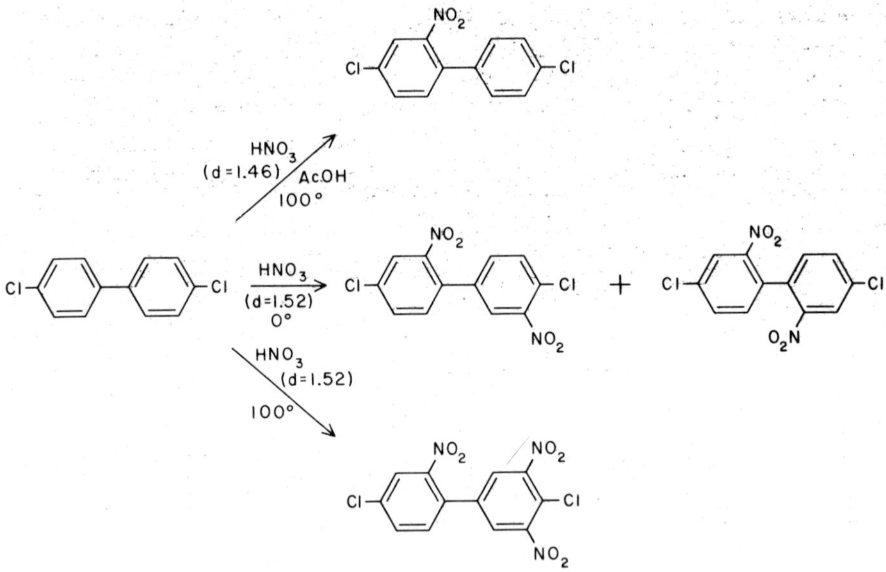


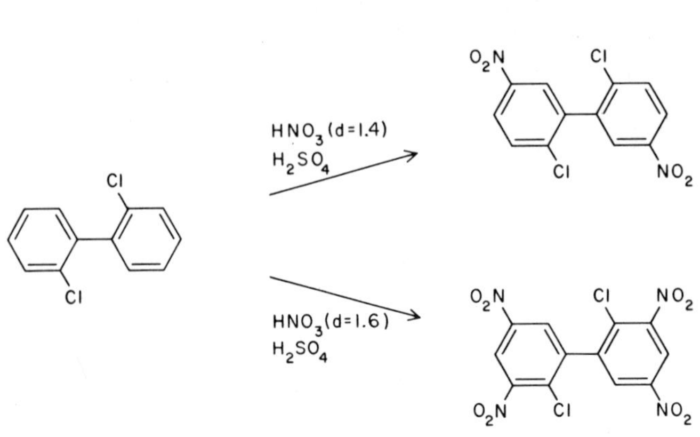


Nitration of biphenyl, nitrobiphenyls,[9,42,43] hydroxybiphenyls,[5,11] and aminobiphenyls[8,27] have also been described. The separation of the nitration products of biphenyl by thin-layer and gas chromatography have recently been reported.[12]

Nitrobiphenyls and chloronitrobiphenyls are a convenient source of aminobiphenyls which in turn are valuable intermediates (Chapter 4) in the synthesis of chlorobiphenyls and their metabolites.

Nitration was proposed as an analytical reaction to differentiate between PCB and aromatic pesticides such as DDT.[22] This method was later found to be unsuitable partly because the biphenyls with lower chlorine content were nitrated along with the DDT-type compounds.[35]

**Nucleophilic Displacement of Chlorine**

Nucleophilic displacement in chlorobiphenyls has mainly been studied with decachlorobiphenyl. Substitution in this compound (and related highly chlorinated diphenyls) occurs preferentially at the 4,4′-positions which can be rationalized by considering a para-quinonoid structure as the most likely intermediate.

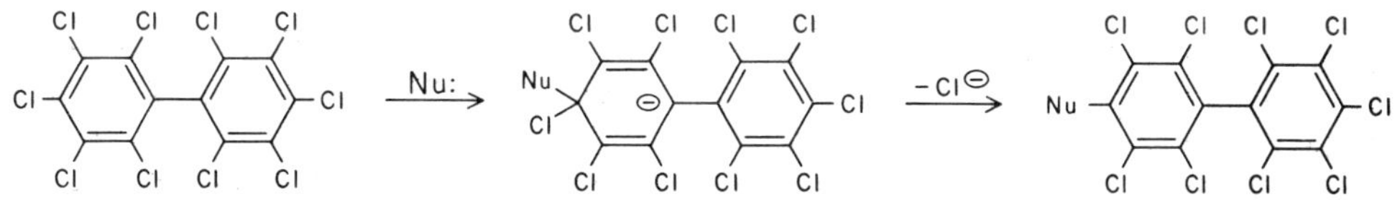

*Reaction with Alkoxide*

Nucleophilic displacement of chlorine by alkoxy groups employing sodium alkoxides or alkali and alcohol at high temperatures and pressures is a patented process[49] (see also Chapter 4) for the preparation of alkoxy and dialkoxypolychlorobiphenyls.

The analogous reaction of sodium methoxide with hexachlorobenzene was shown to proceed at much lower temperatures in the presence of pyridine,[36] a reaction which has been developed as confirmatory test in the gas chromatographic analysis of hexachlorobenzene.[21] Similarly, decachlorobiphenyl was shown to give the 4-methoxy- or 4,4′-dimethoxy-chlorobiphenyl when heated with 1 or 2 mol respectively of sodium methoxide in pyridine.[4]

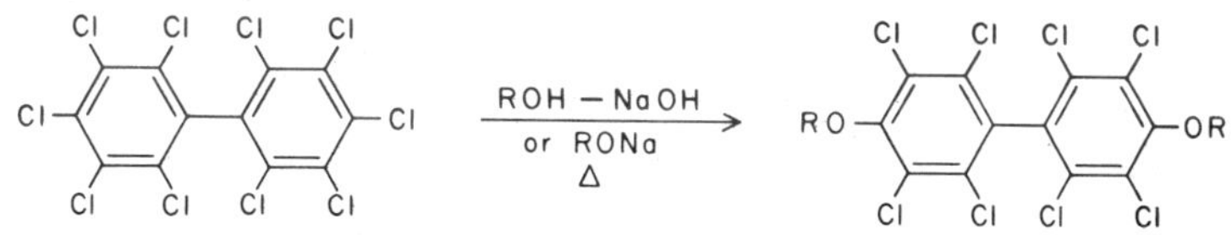

*Reaction with Amines*

Mono and diamino derivatives of polychlorobiphenyls were obtained when a mixture of octa- and nonachlorobiphenyls was treated with ammonia in the presence of metal halides at 175° to 350°.[47] Similarly, 4-methylaminobiphenyl was obtained in 87% yield when 4-chlorobiphenyl and aqueous methylamine were heated at 220° in the presence of cuprous chloride.[24]

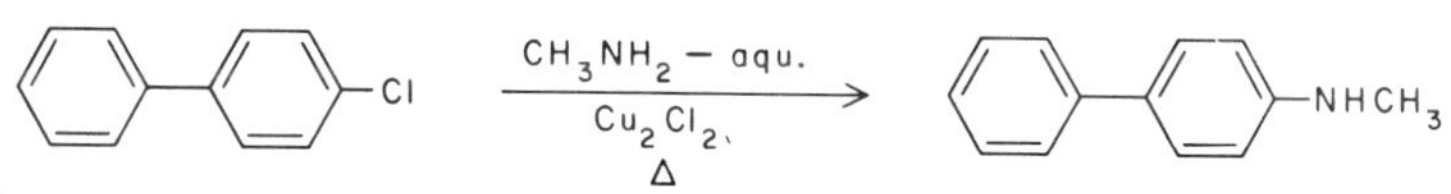

Decachlorobiphenyl, when treated with 1 mol or excess piperidine, yields the 4- or 4,4′-di-piperidino derivative, respectively.[47]

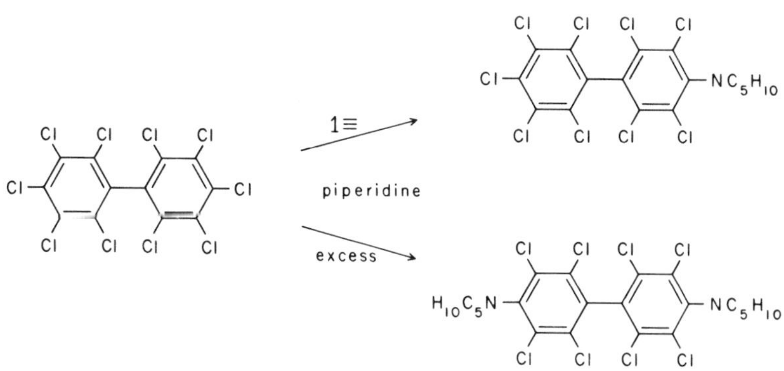

*Reaction with Thiolate Anion*

The reaction of a chlorobiphenyl with a thiolate anion has not been described. However, hexachlorobenzene was shown to react with $CH_3S^-$[26] to give products formed by the replacement of two ($C_6Cl_4(SCH_3)_2$) and four ($C_6Cl_2(SCH_3)_4$) chlorine atoms and similar reactions might be anticipated to occur with the more highly chlorinated biphenyls.

**Organometallic Derivatives**

Decachlorobiphenyl gives a Grignard reagent by the entrainment technique using dibromoethane[4] and lithium derivatives with butyl lithium.[4,13] Hydrolysis of these derivatives give octachlorobiphenyl when 2 mol of the reagents were used and a mixture of starting material, octa- and nonachlorobiphenyls when 2 mol of the reagents were used. The octa and nonabiphenyls obtained were shown to be mixtures of isomers.[13] The Grignard reagent of hexachlorobenzene forms readily when the reaction was initiated by ethyl magnesium bromide.[28]

**Isomerization Reactions**

When treated with hydrogen chloride and aluminum chloride at 160,° the three monochlorobiphenyls give the same, presumably thermodynamic mixture (3% *o*; 64% *m*; 33% *p*).[51] By analyzing the initial reaction products obtained under mild conditions and short periods of time and by observing the products from dichlorobiphenyls, it was concluded that intramolecular phenyl migration predominated over chlorine migration in this aluminum chloride catalyzed isomerization reaction.

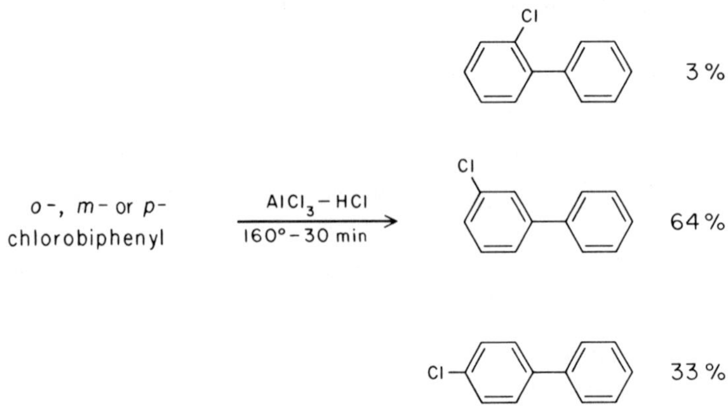


This procedure was used to alter the isomer composition of a technical dichlorobiphenyl mixture (rich in the 2,2′- isomer) to a mixture rich in the 2,3′-isomer showing improved dielectric properties.[52]

When 1,1′-$^{14}$C biphenyl was heated with 10 mol % aluminum chloride and 1 mol % water to 100° for 30 min, the radioactivity was found to be randomly distributed thus indicating intramolecular phenyl migration.[54]

Aluminum chloride induced isomerization reactions were also observed with hydroxychlorobiphenyls (Chapter 4) and isomerizations of chlorobiphenyls were also observed under conditions of photochemical irradiation (Chapter 6).

**Color and Related Reactions**

Spot tests and color reactions have been developed for biphenyl and related aromatic compounds, some of which may be applicable to chlorobiphenyls particularly to components of low chlorine content.

A mixture of sulfuric acid and formaldehyde alone[45] or in the presence of ferric ions[2] has been used for the detection and colorimetric analysis of biphenyl. The latter reagent has recently shown to be useful for the microdetermination of 2-hydroxybiphenyl.[33]

The benzal- or piperonal chloride test for polynuclear hydrocarbons and phenols is applicable to biphenyl.[40] With benzal chloride, in the presence of trifluoroacetic acid a violet color with $\lambda_{max}$ 540 nm is produced.

Electron acceptors such as polynitrofluorenes give colored charge transfer complexes with biphenyl on thin layer plates.[17] Chlorobiphenyls, particularly those with more than two chlorine atoms per molecule, give no or only very weak colors. On the other hand, because of the electron donating nature of the hydroxyl group, chlorohydroxybiphenyls (metabolites) can be located on thin layer chromatograms by electron acceptor spray reagents at a level of ca. 10 µg.[15]

A procedure for the separation of biphenyl from *o*-, *m*-, and *p*-terphenyl by thin layer chromatography has been developed which involves the sulfonation of the mixture and separation of the corresponding sulfonic acids.[14]

**Cyclizations of 2- and 2,2′-Substituted Biphenyls**

*Formation of Dibenzofuran and Related Compounds*

The possible formation of chlorodibenzofurans from chlorobiphenyls is a reaction of considerable toxicological significance. The conditions under which such a cyclization may occur with a number of substituted biphenyls will therefore be briefly discussed here.

The close proximity of the 2 and 2′ substituents in biphenyl facilitates intramolecular cyclization reactions under a variety of conditions.[6] A number of such reactions involving amino, chloro, and hydroxyl substituents lead to substituted dibenzofurans.[32,37] Alkali fusion of 2-chloro-2′-hydroxybiphenyls,[41] hydrolysis of 2,2′-dichloro-3,3′,5,5′-tetranitrobiphenyl[10] (in the original paper sodium nitrite is the reagent used), and dehydration of 2,2′-dihydroxybiphenyls[5,16,25] are examples.

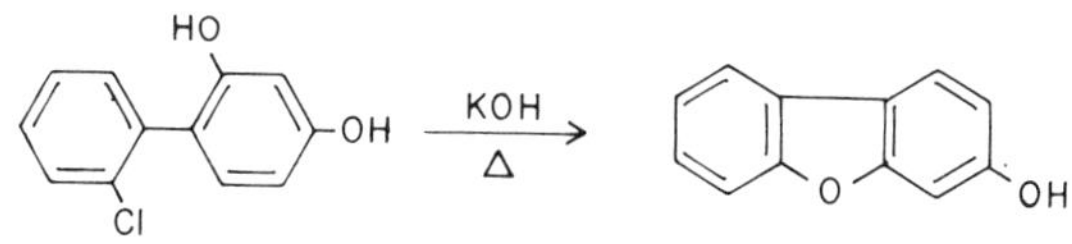


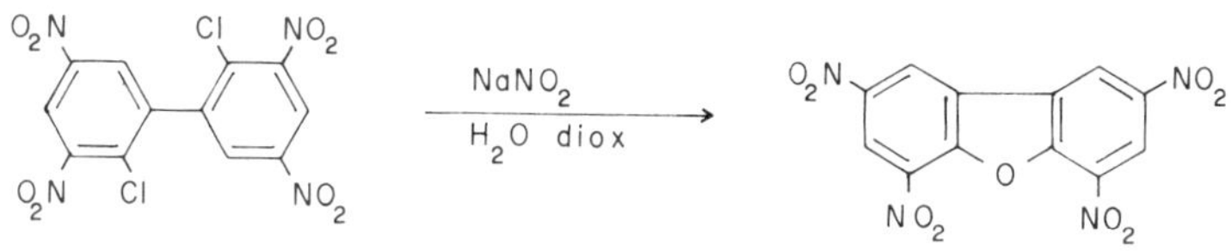


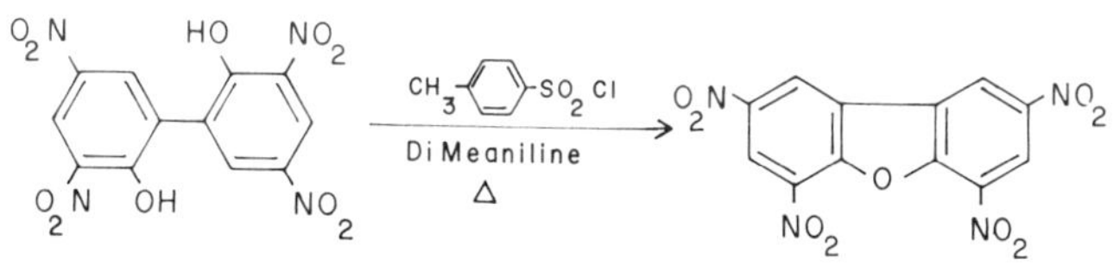


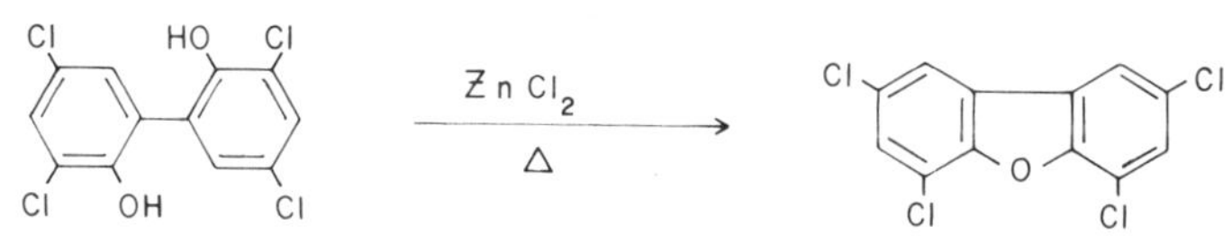


For a discussion of the possible formation of chlorodibenzofurans from chlorobiphenyls by photochemical reaction see Chapter 6.

## REFERENCES

1. *Beilsteins Handbuch der Organischen Chemie,* Springer, Berlin, Vol. 5, Hauptwerk, 579; Ergänzungsband I, 272; Ergänzungsband II, 483; Ergänzungsband III, 1736.
2. **Bellen, Z. and Utnik, A.,** *Chemia Analityczna,* 14, 45, 1969.
3. **Berg, O. W., Diosady, P. L., and Rees, G. A. V.,** *Bull. Environ. Contam. Toxicol.,* 7, 338, 1972.
4. **Binns, F. and Suschitzky, H.,** *J. Chem. Soc.,* (C), 1913, 1971.
5. **Borsche, W. and Scholten, B. G. B.,** *Ber.,* 50, 596, 1917.
6. **Buntrock, R. E. and Taylor, E. C.,** *Chem. Rev.,* 68, 209, 1968.

7. **Buu-Hoi, N. P., Dat Xuong, N., and van Bac, N.,** *Bull. Soc. Chim.,* 2442, 1963.
8. **Case, F. H.,** *J. Am. Chem. Soc.,* 64, 1848, 1942.
9. **Case, F. H.,** *J. Am. Chem. Soc.,* 64, 2225, 1942.
10. **Case, F. H. and Schock, R. U., Jr.,** *J. Am. Chem. Soc.,* 65, 2086, 1943.
11. **Colbert, J. C., Fox, D. W., and Matuszak, C.,** *J. Am. Chem. Soc.,* 77, 2447, 1955.
12. **Copin, A., Severin, M., and Evrard, J.,** *J. Chromatogr.,* 68, 89, 1972.
13. **Caubere, P. and Gorny, B.,** *J. Organometal. Chem.,* 37, 401, 1972.
14. **Denti, E., Luboz, M. P., and Massaglia, A.,** *J. Chromatogr.,* 11, 339, 1963.
15. **Heacock, R. A. and Hutzinger, O.,** submitted to *J. Chromatogr.*
16. **Hutzinger, O.,** unpublished results.
17. **Hutzinger, O. and Heacock, R. A.,** submitted to *Mikrochim Acta.*
18. **Hutzinger, O., Jamieson, W. D., Safe, S., and Zitko, V.,** *J. Assoc. Offic. Anal. Chem.,* 56, 982, 1973.
19. **Hutzinger, O., Safe, S., and Zitko, V.,** *Bull. Environ. Contam. Toxicol.,* 6, 209, 1971.
20. **Hutzinger, O., Safe, S., and Zitko, V.,** *Int. J. Environ. Anal. Chem.,* 2, 95, 1972.
21. **Hutzinger, O., Safe, S., and Zitko, V.,** in preparation.
22. **Jensen, S. and Widmark, G.,** Swedish Report at the Organization for Economic Cooperation and Development, Pesticide Conference, Scotland, 1967.
23. **Kaufmann, T., Henkler, H., and Zengel, H.,** *Angew. Chem., Int. Ed.,* 1, 214, 1962.
24. **Kovar, V. and Marhold, J.,** Czechoslovakian Patent 117,301, 1966; *Chem. Abstr.,* 65, 16895, 1966.
25. **Kruber, O.,** *Ber.,* 65, 1382, 1932.
26. **Langille, K. R. and Peach, M. E.,** *J. Fluorine Chem.,* 1, 407, 1971/72.
27. **LeFevre, R. J. W. and Turner, E. E.,** *J. Chem. Soc.,* 245, 1928.
28. **Levy, L. A.,** *Synthesis,* 170, 1973.
29. **Mack, W.,** *Angew. Chem. Int. Ed.,* 6, 1083, 1967.
30. Monsanto Chem. Co., British Patent 669,756, 1952; *Chem. Abstr.,* 47, 1742, 1953.
31. **Osawa, E. and Yoshida, Z.,** *Bull. Chem. Soc. Jap.,* 40, 1996, 1967.
32. **Parham, W. E.,** Dibenzofuran, in *Heterocyclic Compounds,* Vol. 2, Elderfield, R. L., Ed., Wiley, New York, 1951, 123.
33. **Rajzman, A.,** *Analyst,* 95, 490, 1970.
34. **Reeve, W., Mutchler, J. P., and Liotta, C. L.,** *Can. J. Chem.,* 44, 575, 1966.
35. **Reynolds, L. M.,** *Bull. Environ. Contam. Toxicol.,* 4, 128, 1969.
36. **Rocklin, A. L.,** *J. Org. Chem.,* 21, 1478, 1956.
37. **Rodd, E. H.,** *The Chemistry of Carbon Compounds,* Vol. IV,A., Elsevier, Amsterdam, 1957, 192.
38. **Safe, S. and Hutzinger, O.,** *J. Chem. Soc.,* Perkin Trans I, 686, 1972.
39. **Sandroni, S., Beaudet, C., and Juppe, G.,** *Communaute Eur. Energ. At. Euratom (Rapp),* 1969, Eur-4215.e; *Chem. Abstr.,* 71, 70202, 1969.
40. **Sawicki, E., Miller, R., Stanley, T., and Hauser, T.,** *Anal. Chem.,* 30, 1130, 1958.
41. **Schimmelschmidt, K.,** *Ann. Chem.,* 566, 184, 1950.
42. **Sharnin, G. P., Moisak, I. E., Gryazin, E. E., and Soponova, T. N.,** *Izv. Vyssh. Uchebn. Zaved., Khim. Khim. Tekhnol.,* 12, 907, 1969; *Chem. Abstr.,* 72, 21186, 1970.
43. **Sharnin, G. P. and Falyakhov, I. F.,** *Izv. Vyssh. Uchebn. Zaved., Khim. Khim., Tekhnol.,*12, 1057, 1969; *Chem. Abstr.,* 72, 31346, 1970.
44. **Shaw, F. R. and Turner, E. E.,** *J. Chem. Soc.,* 285, 1932.
45. **Silverman, L. and Bradshaw, W.,** *Anal. Chem.,* 27, 96, 1955.
46. **Slaugh, L. H. and Raley, J. H.,** *J. Org. Chem.,* 32, 369, 1967.
47. **Smith, E.,** U.S. Patent 3,480,673, 1969; *Chem. Abstr.,* 72, 31426, 1970.
48. **Vernon, C. C., Rebernak, A., and Ruwe, H. H.,** *J. Am. Chem. Soc.,* 54, 4456, 1932.
49. **Vuillemenot, J. and Grimaud, E.,** Belgian Patent 643,938, 1963; *Chem. Abstr.,* 63, 11430, 1965.
50. **Weingarten, H.,** *J. Org. Chem.,* 26, 730, 1961.
51. **Weingarten, H.,** *J. Org. Chem.,* 27, 2024, 1962.
52. **Weingarten, H.,** U.S. Patent 3,068,297, 1962; *Chem. Abstr.,* 58, 11274, 1963.
53. **White, J. and Adams, R.,** *J. Am. Chem. Soc.,* 54, 2104, 1932.
54. **Wynberg, H. and Wolf, A. P.,** *J. Am. Chem. Soc.,* 85, 3308, 1963.
55. **Zalkind, Y. S. and Belikova, M. V.,** *Zh. Obshch. Khim.,* 8, 1918, 1938; *Chem. Abstr.,* 33, 5833, 1939, National Research Council of Canada Technical Translation, Ottawa, 1970.

Chapter 6

# PHOTODEGRADATION OF CHLOROBIPHENYLS

## Introduction

In view of the remarkable chemical stability of polychlorinated biphenyls, environmental breakdown initiated by the photochemically active part of the solar spectrum is of particular interest. Until recently, little attention has been paid to the photochemistry of chloroaromatic compounds since it was generally assumed[56,66,67] that, because of the increasing bond strength C-I through C-F, little photochemical cleavage of the C-Cl bond will occur. There is experimental evidence for and against C-Cl bond cleavage[49,56] depending on the conditions used (e.g., wavelength of irradiation, solvent, substituents, etc.).

Photochemical degradation was shown to be a possible major route of environmental breakdown for a number of pesticides including chloroaromatic compounds[9–12,15,41,50–52,57–59] but simple compounds in which the breaking of the C-Cl bond is the main reaction have not been studied until relatively recently.[14] For instance, in the photochemical breakdown of DDT and related compounds, cleavage of the aromatic C-Cl bond is usually not involved.[32,39,40,43,53]

Because of lack of other structural features, photochemical conditions leading to C-Cl fission are of main interest in PCB photochemistry. Although it has been speculated that photochemical degradation of PCB may be one route of environmental breakdown of these compounds, experimental evidence was not obtained until 1971 when it was shown[62] that 2,2′,4,4′,6,6′-hexachlorobiphenyl photolyzes rather readily in organic solvents when irradiated at 310 nm to give products which are formed by stepwise loss of chlorine, rearrangement, and condensation (Table 1; Figure 1). In addition, "polar" products were also observed. Sub-

TABLE 1

**GLC and Mass Spectral Data for Hexachlorobiphenyl Photolysis Products**

| | Retention time[a] (min) | m/e of $M^+$ | Chlorine[b] content |
|---|---|---|---|
| TLC band I | | | |
| Peak 1 | 5.6 | –[c] | 0 |
| Peak 2 | 8.2 | –[c] | 0 |
| Peak 3 | 9.9 | 222 | 2 |
| Peak 4 | 13.5 | 256 | 3 |
| Peak 5 | 14.1 | 290 (256) | 4, (3[tr]) |
| Peak 6 | 19.0 | 324 (290) | 5, (4[tr]) |
| Peak 7 | 29.7 | 358 | 6 |
| Peak 8 | 56.9 | 304 | 2 |
| TLC band II | | | |
| Peak 1 | 19.6 | 290 | 4 |
| Peak 2 | 26.8 | 324 | 5 |
| Peak 3 | 35.2[d] | 358 | 6 |

[a]Conditions: instrument, Hewlett-Packard 5750; column, 6% QF-1 + 4% SE30 on "Chromosorb" (AW) 60–80 mesh 5′, ¼″ outside diameter, glass; injection port temperature, 220°C; column temperature, 175°C; flame ionization detector temperature and collection port temperature, 240°C; helium flow rate, 40 ml/min
[b]From isotope peak distribution; tr, trace.
[c]Hydrocarbon peaks, $M^+$ not identified.
[d]Corresponding to starting material.

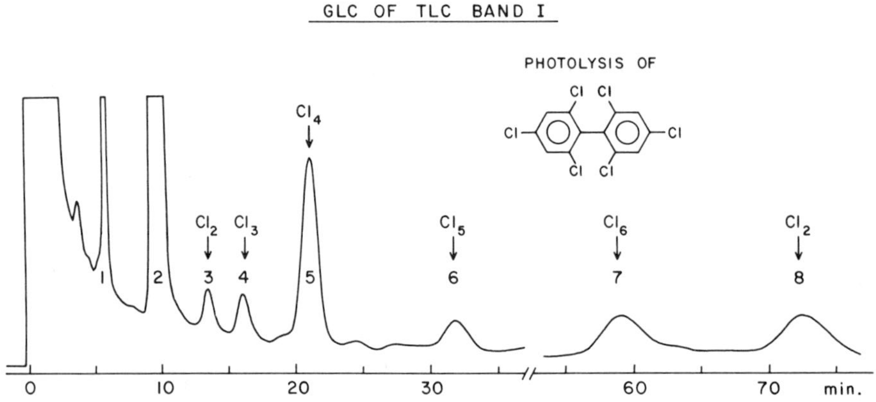


FIGURE 1. Gas chromatogram of TLC band I from hexachlorobiphenyl irradiation experiment; $Cl_2$, $Cl_3$ etc. indicate chlorine content of peak. Further details and GC conditions are given in Table 1.

sequently, this photochemical reactivity has been proven to be quite general with a number of pure chlorobiphenyls (e.g., Figure 2) as well as with commercial PCB preparations and employing laboratory UV sources and natural sunlight.[3,24-26,29,46,47,61]

The lability of decachlorobiphenyl had previously been shown by the fact that its luminescence yield could not be measured due to the photochemical instability.[20] An analytical method for the identification of PCB's is based on the change in GC patterns of irradiated PCB fractions and hence their photochemical reactivity.[22]

Compounds related to the chlorobiphenyls such as chlorodibenzodioxins[17] and chlorodibenzofurans[20] have recently shown to readily undergo photodecomposition in organic solvents. General information on experimental approaches and theory of photochemical reactions can be found in the reviews on the photochemistry of pesticides mentioned above, in reviews on the photochemistry of natural products,[64] drugs and related substances,[55] in books on the chemical action of light,[4,18,19] ultraviolet radiation in general,[33] photochemistry in particular,[6,35,65] and in review series.[5,8,48]

## Conditions of Irradiation

### *Choice of Irradiation Sources*

It has been repeatedly pointed out[58,59] that irradiation with low-pressure mercury lamps which are commonly used UV sources in photochemical studies[6] is of academic interest only, as far as environmental photochemical breakdown studies are concerned. The bulk of the radiation emitted by these lamps is at 2547 Å which is of considerably shorter wavelength (higher energy) than radiation received from the sun at the earth's surface (for practical purposes > 295 nm). A number of pesticides which photolyze at 254 nm are stable to sunlight (for examples see References 57 and 58). The observation that the rate of photochemical oxidation of organic pollutants in waste increases severalfold when the wavelength of irradiation is reduced to from > 250 to ca. 250 nm[7] is of interest in this regard.

In the recent studies on the photochemical reactions of chlorobiphenyls, high energy radiation UV sources have been used[24,25] but increasingly, lamps with emission of radiation > 290 are being employed.[3,24,26,29,46,61,62]

Long-wavelength UV fluorescent lamps ("blacklight," e.g., General Electric F 40 BL) closely simulate natural sunlight in the critical photochemically active wavelength region above 290 nm.[9,15] Irradiations with these lamps were found to produce similar results to sunlight in herbicide degradation studies[15] and in the formation of photochemical smog.[34] An identical spectrum of products was recently obtained when thin films of a number of chlorobiphenyls were irradiated with "blacklight" lamps and sunlight.[26]

### *Physical State of the Irradiated Compound – Laboratory Models for Natural Conditions*

Pesticide photodegradation studies have been carried out in the gas-phase, in solution (water or organic solvents), and in the solid state (thin films on glass plates or adsorbed on soil or silica).

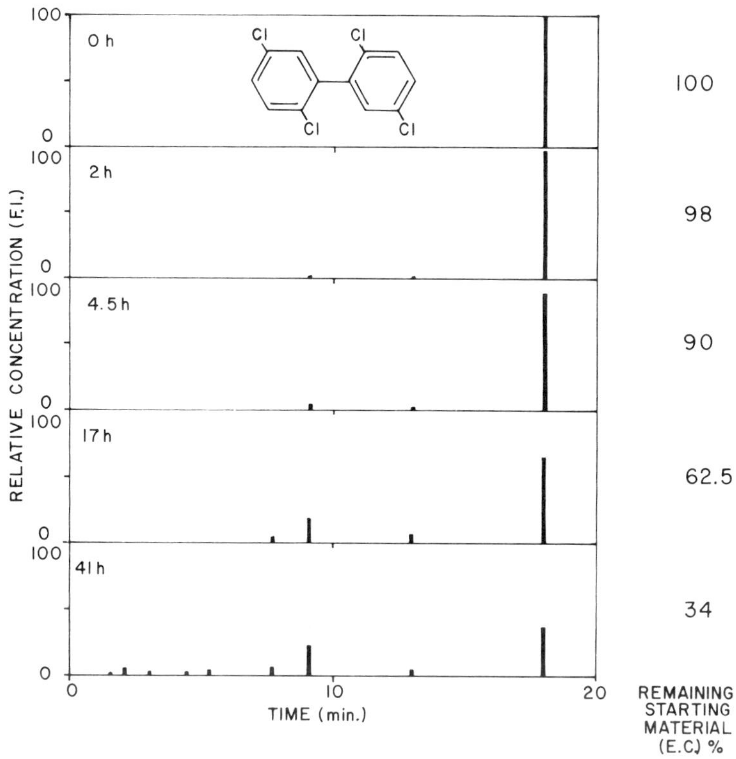


FIGURE 2. Gas chromatograms and % starting material remaining from tetrachlorobiphenyl irradiation experiment. Irradiation conditions: 2,2′,5,5′-tetrachlorobiphenyl (50 mg) in hexane (300 ml) was irradiated under $N_2$ at $\lambda_{max}$ 310 nm and aliquots taken at times indicated. For the gas chromatograms a flame ionization detector was used and for the quantitative determinations on electron capture detector. (From *Environmental Health Perspectives.* With permission.)

Attention has been paid in these studies to simulating natural conditions and an effort has been made to create environmental conditions including controlled irradiation in a weathering chamber.[68]

*Gas-phase irradiation of chlorobiphenyls* – Irradiation of chlorobiphenyls in the gas phase would be highly relevant, particularly for the more volatile components with low chlorine content, but appropriate data have not yet become available. In preliminary gas-phase irradiations of 2,2′,4,4′,6,6′-hexachlorobiphenyl with a mercury lamp and pyrex filter at 50° only the unchanged starting material could be recovered.[25]

A number of decomposition products including polar materials were observed when the vapor phase of a refluxing suspension of 2,2′,5,5′-tetrachlorobiphenyl and water was exposed to UV irradiation (General Electric "Sun-lamps").[29] A vapor-phase photoreactor designed to simulate environmental conditions and to eliminate "wall effects" has recently been described.[16,42]

*Irradiations of chlorobiphenyls in solution* – Because of the low solubility of chlorobiphenyls in water,[70] irradiations in true solutions have to be carried out in organic solvents. Such experiments cannot be directly related to environmental conditions[59] although it has been pointed out[17,59] that organic hydrogen donors such as oil films and cuticular waxes are abundantly present in the environment. It has been shown with other chloroaromatic compounds such as pesticides that reductive dechlorination is the main reaction in nonpolar solvents such as hexane whereas in water or alcohols replacement of the chlorine by a hydroxy group also occurs.[11,50,57] Reductive dechlorination is the main initial photoreaction of chlorobiphenyls in organic solvents and the rate of dechlorination is faster in hydroxylic solvents such as methanol[25] or iso-propanol[46] than in nonpolar solvent, an observation which holds true also for other chloroaromatic compounds.[28] For instance, even chlorobenzene loses chlorine rapidly on irradiation in iso-propanol[49] and chloronitrobenzene is dechlorinated in ethanol but not in benzene.[56] Complete dechlorination of a technical PCB preparation was observed in 15 min in alkaline iso-propanol solution and using a mercury lamp as the UV source,[47] biphenyl and sodium chloride were identified in the reaction mixture. Since the rate of photodegradation can be changed significantly by photosensitizers, solvents must be free of impurities. Much higher rates of photochemical degradation of PCB in commercial versus purified methanol for instance have been observed.[13]

In perfluorohydrocarbon solution, isomerization and dechlorination with concomitant chlorination to yield more highly chlorinated biphenyls are observed.[25]

Experiments on the photodecomposition of chlorobiphenyls in water are difficult not only because of the low solubility of these compounds but also because of the rapid, irreversible adsorption of PCB onto glass walls.[23] Attempts have been made to increase the solubility of chlorobiphenyls by the addition of dioxane[29] and surfactant (Tween 80)[26] to the aqueous mixture but even there a more or less stable suspension on not a true solution was obtained.

In an attempt to simulate irradiation in an aqueous environment of a similarly isoluble class of compounds (polycyclic hydrocarbons), benzo(a)pyrene has been irradiated adsorbed onto calcium carbonate in aqueous suspension.[2]

*Irradiations of chlorobiphenyls in the solid state* – Photochemical decomposition of solid, adsorbed species is probably of importance in pesticide degradation on agricultural land and appropriately the decomposition of many pesticides has been studied as thin films or adsorbed onto soil particles or silica.[31] However, irradiation of chlorobiphenyls adsorbed on soil particles or similar adsorbents have only limited validity since PCB's are mainly found in the aquatic environment and are not usually associated with soils as are many pesticides. Aerial fallout on the other hand may contribute to PCB contamination of soils.[45a] Thin films of chlorobiphenyls have been exposed to UV light or sunlight[25,26,29] to investigate the photochemical behavior of these compounds in the solid state. Initially, glass dishes loosely covered with transparent plastic material were used to expose a number of chlorobiphenyls to sunlight. However, since large quantities of material disappeared presumably by evaporation, closed quartz tubes which keep volatile compounds in and impurities out were used in subsequent experiments. There are disadvantages to using thin films of chlorobiphenyls in irradiation experiments. UV radiation will not penetrate deeply into the solid film which makes photolysis in the solid state inefficient and dependent on an extremely thin uniform layer. Also natural levels may never give concentrations of PCB high enough to allow close proximity of a significant number of molecules to allow certain reactions (e.g., dimerization) to occur. Attempts were made to approach natural conditions more closely by adding water and iso-octane (as organic hydrogen donor) to the bottom of the quartz tube. With the more volatile chlorobiphenyls in particular, gas phase irradiation occurs also in the quartz tubes when exposed to the light heat of the summer sun. 4-Chlorobiphenyl, which was found to be photostable when irradiated at 310 nm in iso-propanol under nitrogen,[46] decomposed completely when exposed to sunlight for 2 months as a thin film (50 mg) coated on the inside of a quartz tube (40 x 8 cm) in the presence of water and iso-octane.[26] Obviously, evaporation and mixing of the content must have taken place to a certain extent.

### Theoretical Aspects

#### *Correlation of Absorption Spectra to Photochemical Activity*

UV radiation is effective in initiating photochemical change only if absorbed directly or indirectly by the target molecule. The UV spectra and their change with the structure of chlorobiphenyls are therefore of particular interest[37] (see Chapter 10). Most chlorobiphenyls show some absorption at wavelength > 280 nm. Compounds with less than two chlorine atoms ortho to the phenyl-phenyl bond absorb in this region as tailing from a main peak at ca. 240 to 250 nm and compounds with two or more chlorine atoms in ortho positions exhibit weak but distinct bands at > 270. From the evidence available it is not clear whether this absorption is sufficient to explain the photochemical lability of chlorobiphenyls. The photochemical activity of pesticides without significant absorption at $\lambda > 290$ nm has been explained either by the presence of a small quantity of high energy UV radiation in sunlight (ozone "window"), by the change of the spectrum of absorbed species or by photosensitization by singlet oxygen, atmospheric pollutants, or natural products.

No study has been reported as to the relationship of chlorine content and the effect of positional isomerism on photochemical activity. Preliminary experiments with individual isomers[29] and PCB mixtures[22,29] indicate that higher chlorinated biphenyls disappear faster than those with lower chlorine content on irradiation (Table 2; Figure 3).

TABLE 2

**Relative Photochemical Labilities of Some* Chlorobiphenyl Isomers**

| Compound | Starting material remaining after 24 hr at 310 nm in hexane |
|---|---|
| Tetrachlorobiphenyls | % |
| 3,3′,4,4′- | 32 |
| 2,2′,6,6′- | 29 |
| 2,2′,5,5′- | 33 |
| Hexachlorobiphenyl | |
| 2,2′,4,4′,5,5′- | 3.8 |
| Octachlorobiphenyl | |
| 2,2′,3,3′,4,4′,5,5′- | <1 |

*A 0.1% solution of each chlorobiphenyl was irradiated under nitrogen for 24 hr at λmax 310 nm.

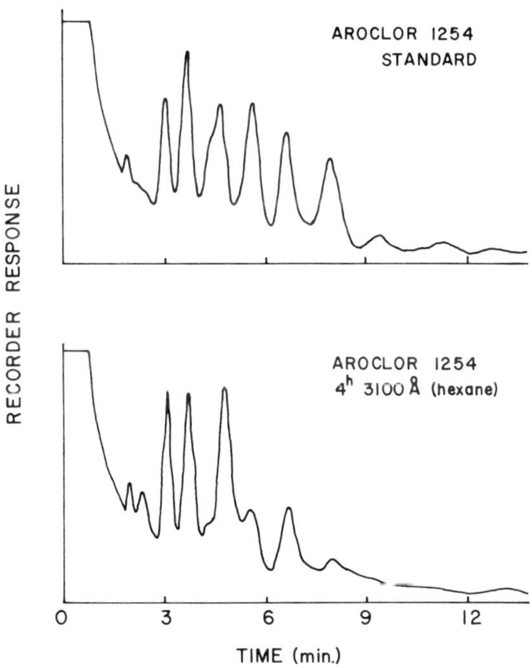


FIGURE 3. Gas chromatograms of Aroclor 1254 before and after irradiation. 6′ x 0.5″ Column packed with 3% SE-30 on Chromosorb W, oven temperature 200°. (From *Environmental Health Perspectives*. With permission.)

*Mechanism of Photochemical Degradation*

Little information on the mechanistic aspect of chlorobiphenyl photochemistry is available and so far detailed studies have only been reported for 4,4′-dichlorobiphenyl.[46] In iso-propanol under nitrogen, 4,4′-dichlorobiphenyl decomposes when irradiated at 310 nm to form only HCl and 4-chlorobiphenyl which is photostable under these conditions. In the presence of oxygen, a number of unidentified products were also formed. The primary photochemical process appears to be the formation of an aryl radical (Figure 4) and may involve a triplet electronic state since this reaction was sensitized by benzophenone ($E_t$ 69.0) and triphenylene ($E_t$ 67.0) (the triplet energy ($E_t$) of 4,4′-dichlorobiphenyl is 62.5 kcal/mol) and this reaction was slowed down in the presence of oxygen, an efficient triplet quencher. Chlorine abstraction from 4,4′-dichlorobiphenyl by a radical species (such as postulated in DDT-photolysis; see Reference 40) is unlikely since in a chain mechanism radicals would not be able to differentiate between 4,4′-dichlorobiphenyl and 4-chlorobiphenyl. The quantum yield for the disappearance of 4,4′-dichlorobiphenyl was calculated as 0.002 by comparison with the photoreaction of benzophenone.

*Photosensitization and Quenching*

Photosensitizers may accelerate environmental photodecomposition, may lead to different products or make photodegradation possible. The interactions of photosensitizers with pesticides have been studied[31,40,52,60] but to date only one report deals with the effect of photosensitizers and quenchers on chlorobiphenyls.[46] In this study, it was shown that tryptophan, diethylaniline, benzophenone, and triphenylene sensitize the photoreaction of 4,4′-dichlorobiphenyl under nitrogen. In air, tryptophane and aromatic amines are effective but not benzophenone. The reaction is quenched by *n*-hexylmercaptan, di-*n*-butyl sulfide, and *t*-butyl-disulfide.

**Photoproducts Formed from Chlorobiphenyls**

So far, the photodegradation of the chlorobiphenyls depicted in Figure 5 have been studied as have some commercial PCB mixtures.

*Reductive Dechlorination*

Products formed by reductive dechlorination are predominant particularly in organic sol-

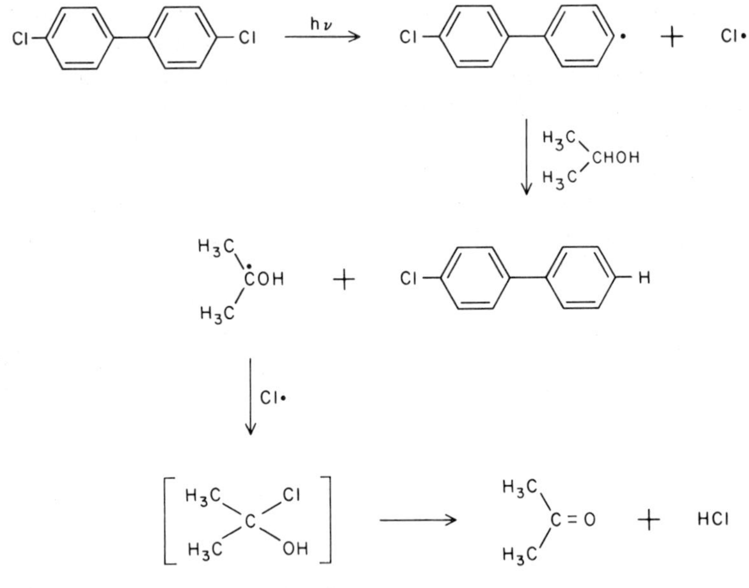

FIGURE 4. Proposed mechanism for the photochemical dechlorination of 4,4′-dichlorobiphenyl in isopropanol. (Courtesy of Dr. Miller.)

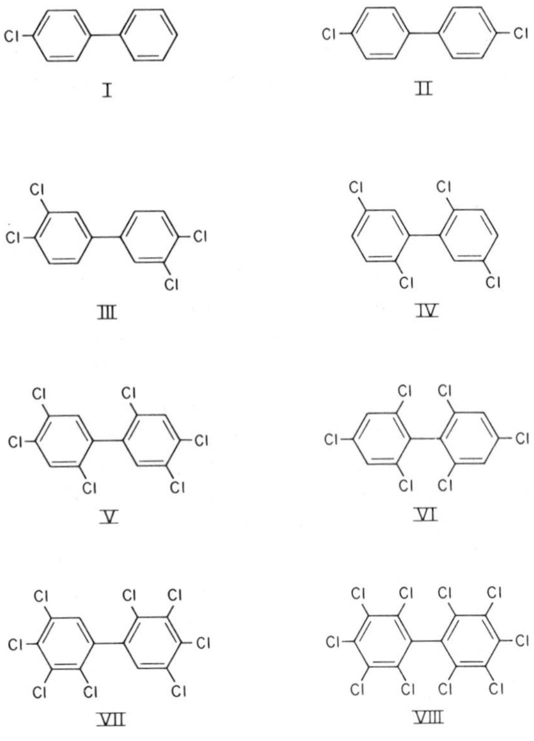

FIGURE 5. Structures of chlorobiphenyls which have been used in photodegradation studies.

vents.[3,25,29,46,47,61,62] This reaction has previously been observed to occur with chloroaromatic pesticides[14,50] and related chloroaromatic compounds.[17,28] In thin films[25,26,29] and fluorocarbon solvents[25] dechlorination occurs but concomitant chlorination as well as rearrangement is also observed.

*4-Chlorobiphenyl (I)* – was found to be photostable when irradiated in iso-propanol under $N_2$ at 310 nm.[46] An ion corresponding to the dechlorination product biphenyl, as well as the starting material, however, was observed in the mass spectrum of TLC fraction from the exposure of 4-chlorobiphenyl to sunlight for 1 month as thin film in a quartz tube in the presence of water and iso-octane.[26] After 2 months, complete decomposition was indicated from the mass spectrometric results.

*4,4′-Dichlorobiphenyl (II)* – irradiated in solution under similar conditions gave 4-monochlorobiphenyl (I) as the main product.[46,61] I, on the other hand, was not found among the photoproducts when II was exposed to sunlight as thin film.

*3,3′,4,4′-Tetrachlorobiphenyl (III)* – irradiation in hexane at 300 nm shows successive dechlorination to give a trichloro and a dichloro derivative, probably 3,4,4′-trichlorobiphenyl and 4,4′-dichlorobiphenyl.[61] A trichlorobiphenyl derivative was among the products formed on exposure of III as thin film to sunlight.[26]

*2,2′,5,5′-Tetrachlorobiphenyl (IV)* – showed also reductive dechlorination when irradiated in hexane[29] and methanol.[25] A dichlorobiphenyl was the main chlorinated intermediate in the photochemical decomposition of IV in hexane (Figure 6). A trichlorobiphenyl was among the products formed on exposure of IV as thin film to sunlight.[26]

*2,2′,4,4′,5,5′-Hexachlorobiphenyl (V)* – irradiation of this compound in solution has not yet been described. On exposure of a thin film of this compound to sunlight, tetra- and pentachlorobiphenyls are formed among other products.[26]

*2,2′,4,4′6,6′-Hexachlorobiphenyl (VI)* – irradiated in organic solvents shows stepwise dechlorination[25,62] (see Table 1 and Figure 1). 4,4′-Dichlorobiphenyl and a pentachlorobiphenyl (probably the 2,2′,4,4′6-isomer) were among the main products. A tri-, tetra-, and monochlorobiphenyl as well as biphenyl was also detected in the reaction mixture.[25]

*2,2′,3,3′,4,4′,5,5′-Octachlorobiphenyl (VII)* – exposed to blacklight UV lamps as thin film gave dechlorinated biphenyls containing 4 to 7 chlorine atoms (Figure 7).

*Decachlorobiphenyl (VIII)* – was recovered unchanged after exposure to blacklight UV lamps.[26] Since the photolability of VIII in solution was previously noted,[20] lack of observable breakdown products is most likely due to the fact that thin layers (a large surface area) of this compound cannot be obtained due to its low solubility; few and relatively large needles are formed on evaporating the solvent from the tubes.

*Commercial PCB mixtures* – show reductive dechlorination on irradiation in organic solvents[22,24,29] which is indicated by the change in the GC-pattern (Figure 3). These results and data from irradiation of trapped fractions[22] indicated that chlorobiphenyls with higher chlorine content decompose more readily. In iso-propanol, in the presence of alkali, dechlorination of PCB mixtures is rapid with biphenyl and sodium chloride being the identifiable end-products.[47]

*Isomerization, Condensation, and Chlorination*

Aluminum chloride catalyzed isomerizations of biphenyl[72] and chlorobiphenyls[71] are known, and photoisomerizations of methyl[1] and dimethylbiphenyl[38] have been reported. It is also known that photochemical hydrogen isotope exchange occurs in aromatic compounds.[36]

The mechanism of photoinduced isomerization of chlorobiphenyls may be different to the cases cited above and could proceed via phenyl-phenyl carbon bond or carbon chlorine bond breakage. Although no definite compounds were isolated and no structures assigned, evidence points to isomerization reactions occurring in fluorocarbon solvents.[25]

Condensation products are observed mainly when irradiation is carried out in thin films,[26]

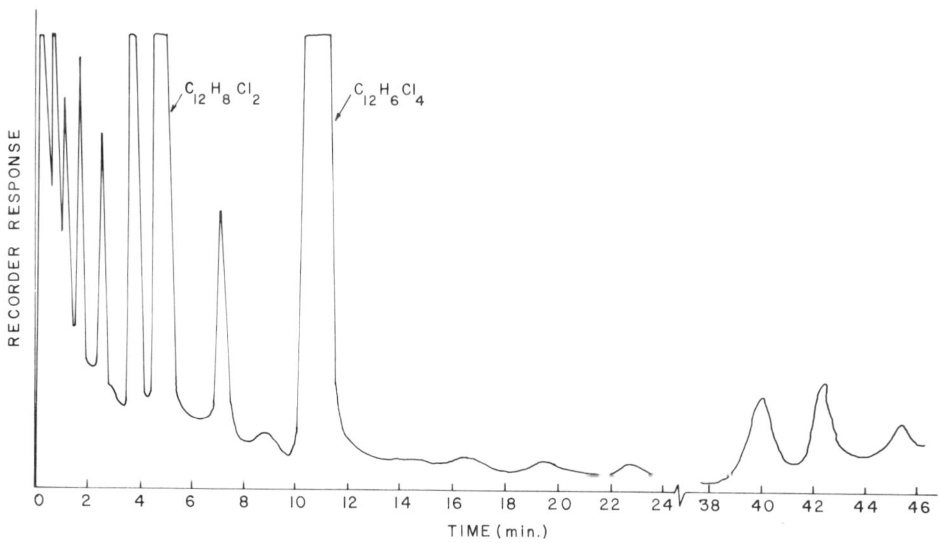


FIGURE 6. GC-MS data for the 41h sample of the 2,2′,5,5′-tetrachlorobiphenyl irradiation experiment (see Figure 2). 8′ x 0.25″ Column packed with SE-30 on Chromosorb W, oven temperature isothermal at 150° at 36 min beginning of programming at 4°/min. A number of further peaks were observed after 46 min. (From *Environmental Health Perspectives.* With permission.)

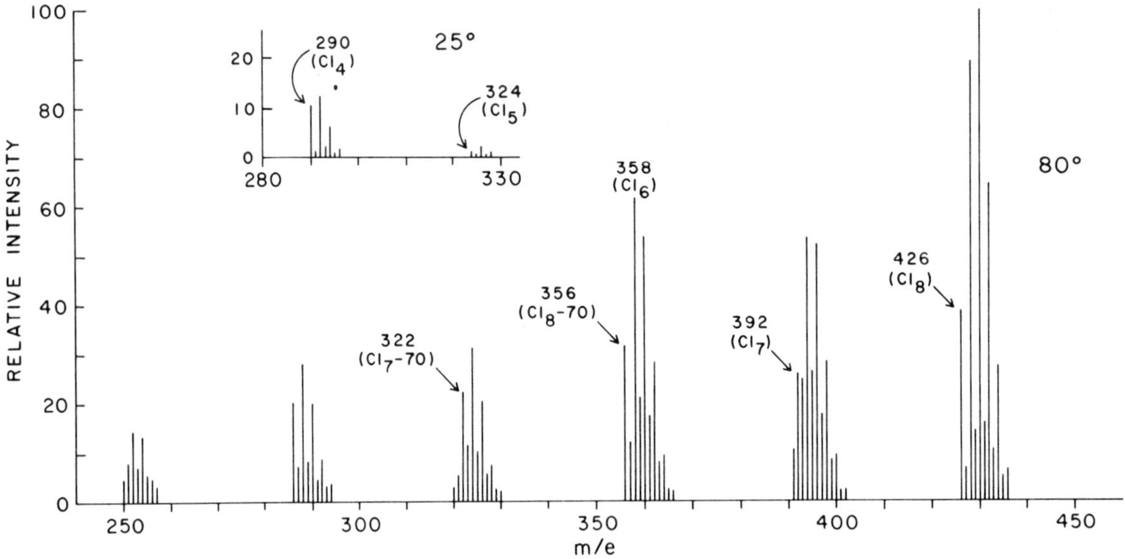


FIGURE 7. 70 eV Mass spectra of 2,2′,3,3′,4,4′,5,5′- octachlorobiphenyl photolysis products. $Cl_8$, $Cl_7$, etc. indicate octachlorobiphenyl, heptachlorobiphenyl, etc. The spectra were taken on the same sample at different probe temperatures as indicated.

presumably because of the proximity of chlorobiphenyl molecules to the photochemically generated aryl radical. For instance, a number of compounds whose structures are best rationalized as chlorinated quaterphenyls are formed on irradiating 2,2′,5,5′-tetrachlorobiphenyl (Figure 8) and 2,2′,4,4′,5,5′-hexachlorobiphenyl (Figure 9) as well as Aroclor 1254 (*m/e* 680; 646; 612 and 578) with blacklight fluorescent UV lamps or sunlight. The formation of chlorinated terphenyls from 3,3′,4,4′-tetrachlorobiphenyl (Figure 10) is indicated by the mass spectra of some of the photoproducts. In one instance (2,2′,5,5′-tetrachlorobiphenyl exposed to sun as thin film[26]), a compound is formed which appears to be a cyclic structure of the type depicted in Figure 11. This product may have formed from quaterphenyls.[63] The presence of chlorinated terphenyls was also indicated in the mixture obtained by irradiating Aroclor 1254 as thin films by perchlorination-gas chromatographic techniques.[27,30,73] However, in this instance, contamination of the starting material, Aroclor 1254, with chlorinated terphenyls was not rigorously excluded. Although dimerization involving C-C bond formation is a well-known photochemical process,[45,69] and condensations of chlorophenols via C-O bonds[44] and DDT via side chain addition have been observed,[21] condensations of aromatic nuclei have not been commonly observed with environmental chemicals.

In thin films[25,26] and solutions in fluorocarbon solvents[25] dechlorination is accompanied by chlorination to give chlorobiphenyls of higher chlorine content than the starting material (for example see Figure 12).

*Formation of Products Containing Oxygen*

In hydroxylic organic solvents, irradiation of chlorobiphenyls yields, in addition to the dechlorinated species, polar oxygenated products.[3,25,62] Little information on the exact structure of these compounds is available. Mass spectra were reported for products containing oxygen obtained from the irradiation of Aroclor 1254 in dioxane-water in the presence of sodium bicarbonate.[29] Figure 13 indicates species formed by the addition of the elements of water to components of the Aroclor 1254 mixture. On the other hand, irradiation of Aroclor 1254 as thin film in the presence of water gave products whose mass spectra (Figure 14) indicate formation of hydroxylated species.

Because of the toxicity of chlorinated dibenzofurans, the possibility of their photochemical formation (Figure 15) is an item of special importance. Reports given in recent conferences[3,13] would, in fact, indicate that in model

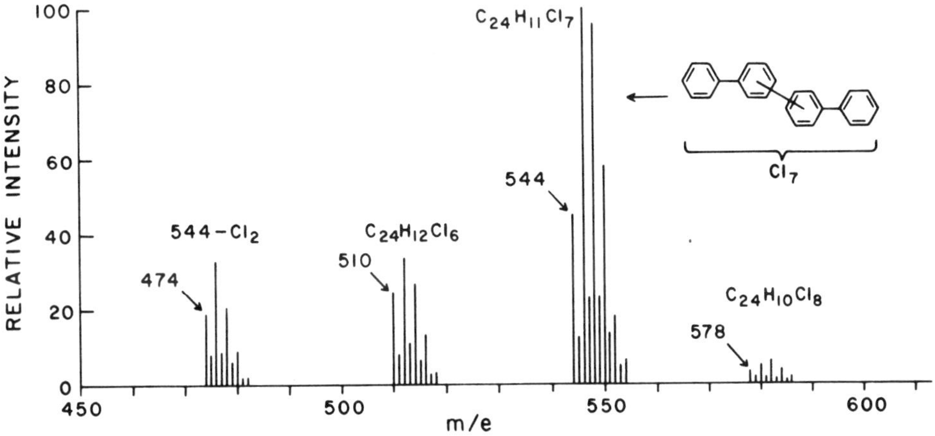


FIGURE 8. 70 eV Mass spectrum of TLC fraction from irradiated 2,2′,5,5′-tetrachlorobiphenyl. Quaterphenyl-type structures containing six, seven, and eight chlorine atoms are indicated.

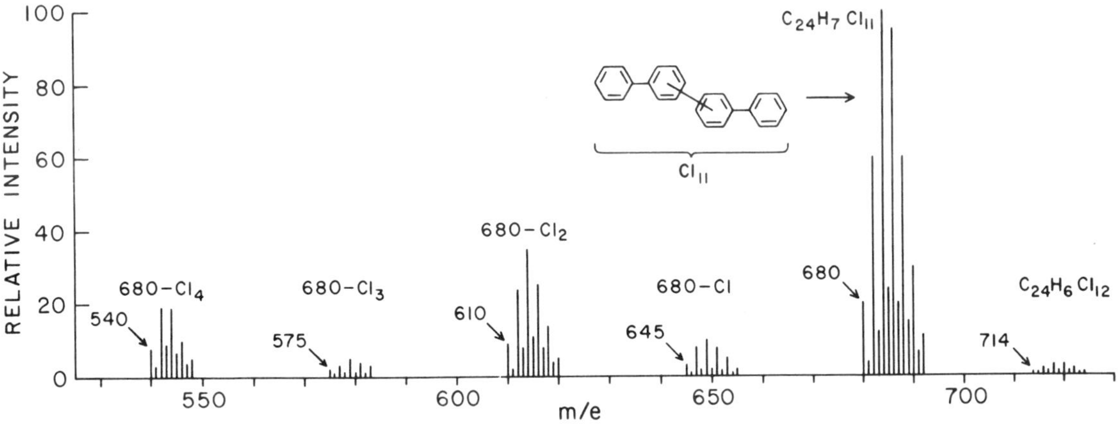


FIGURE 9. 70 eV Mass spectrum of TLC fraction from irradiated 2,2′,4,4′,5,5′-hexachlorobiphenyl. Quaterphenyl-type structures containing 11 and 12 chlorine atoms are indicated.

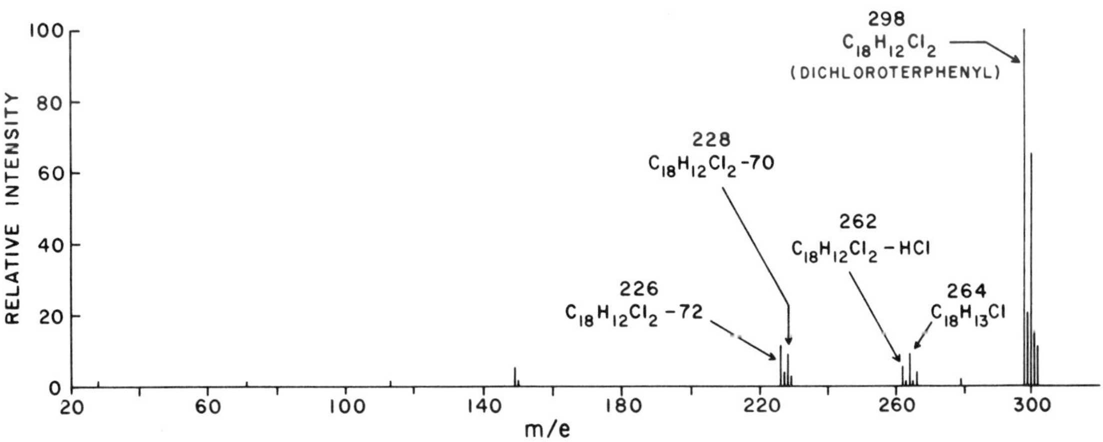


FIGURE 10. 70 eV Mass spectrum of TLC fraction from irradiated 3,3′,4,4′-tetrachlorobiphenyl. Mono-and dichloroterphenyl-type structures are indicated.

irradiations compounds of this type are formed. In other experiments, chlorodibenzofurans could not be detected. No benzofuran was formed from 2,2′-dichlorobiphenyl[54] and irradiation of Aroclor 1254 as thin film[26] gave no chlorodibenzofuran as indicated by a sensitive perchlorination-GC-procedure.[27,30] Of four chlorobiphenyls of different chlorine content irradiated in aqueous suspension,[26] only 4,4′-dichlorobiphenyl gave traces of a product whose mass spectrum ($M^+$ 236) is compatible with a chlorodibenzofuran ($C_{12}H_6Cl_2O$). Loss of chlorine was apparently not involved in this reaction. Chlorodibenzofurans are themselves photolabile[28] and an assessment of their possible environmental accumulation due to photochemical formation from chlorobiphenyls cannot be made at this point.

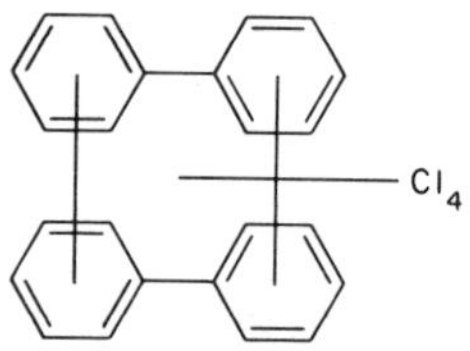


FIGURE 11. Possible structure for a photoproduct from 2,2′,5,5′-tetrachlorobiphenyl (m/e 440).

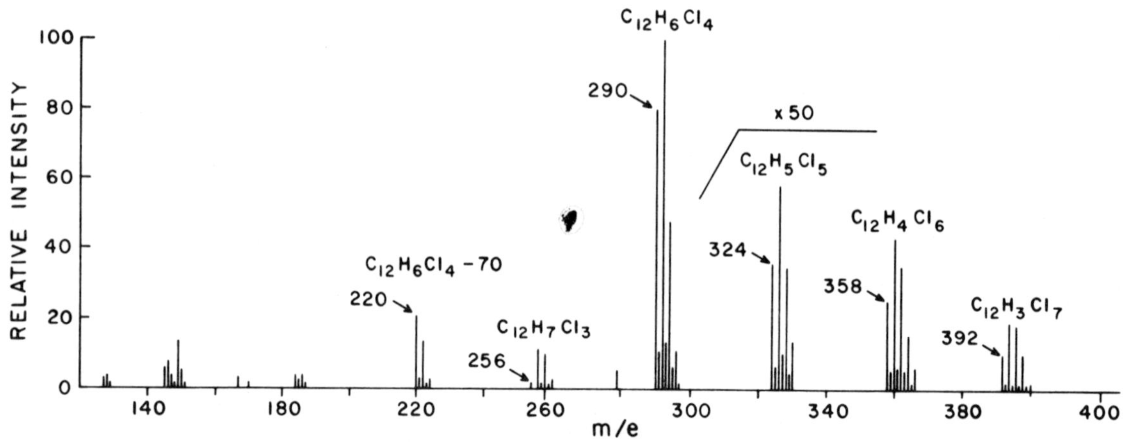


FIGURE 12. 70 eV spectrum of TLC band A from irradiated 3,3′,4,4′-tetrachlorobiphenyl. The presence of penta-, hexa-, hepta- as well as trichlorobiphenyls is shown.

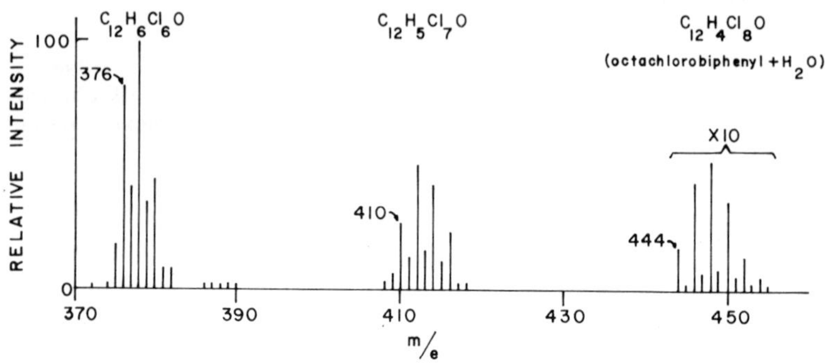


FIGURE 13. Partial mass spectrum (70 eV; probe temperature 35°) of a TLC fraction from Aroclor 1254 irradiation experiment. The irradiation was carried out at $\lambda_{max}$ 310 nm in a suspension (dioxane-water) in the presence of sodium bicarbonate. The chlorine isotope pattern for the $Cl_6$ and $Cl_7$ compounds are distorted due to impurities or fragment ions. Compounds with molecular compositions corresponding to octachloro-, heptachloro- and chlorobiphenyls plus the elements of water are indicated. (From *Environmental Health Perspectives.* With permission.)

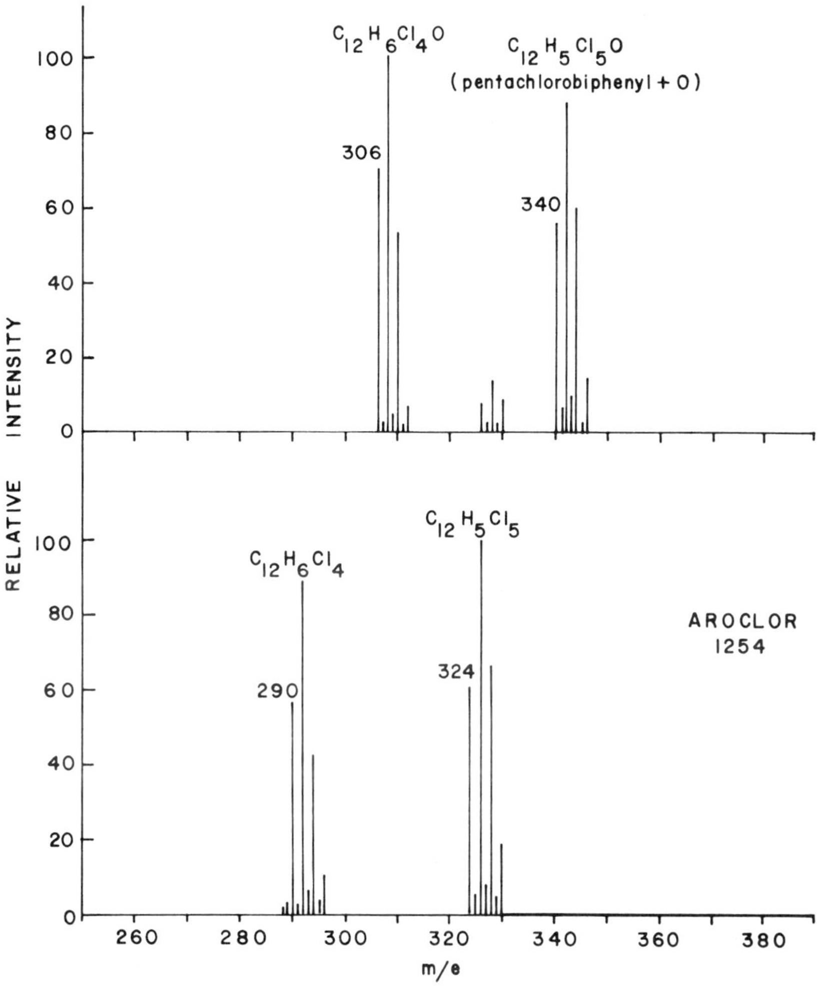


FIGURE 14. Partial mass spectra (70 eV; probe temperature 35°) of: TOP; TLC fraction from Aroclor 1254 irradiation experiment. The irradiation was carried out with "blacklight" fluorescence lamps and a thin film of Aroclor in the presence of water. Compounds with molecular compositions corresponding to pentachloro- and tetrachloro-hydroxybiphenyls are shown. BOTTOM; Standard Aroclor 1254 (ions below *m/e* 340 not shown). (From *Environmental Health Perspectives.* With permission.)

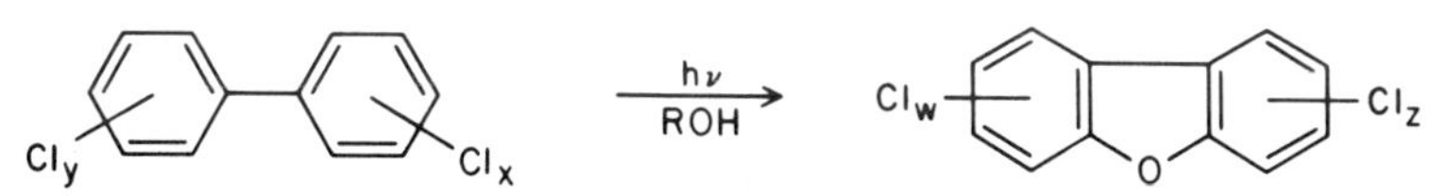


FIGURE 15. Formation of chlorobiphenyls from polychlorinated biphenyls as a possible photochemical reaction.

*Formation of Polymeric Products*

Prolonged irradiation of chlorobiphenyls results in the formation of yellow gums of polymeric character.[29] These products give complicated mass spectra ("picket fence" spectra) presumably due to thermal degradation of the polymer in the ion source. The infrared spectra (e.g., Figure 16) are indicative of the types of groups present but no further information as to their structures is available. Polymers showing similar properties are also obtained from solvents alone which have not been rigorously purified.

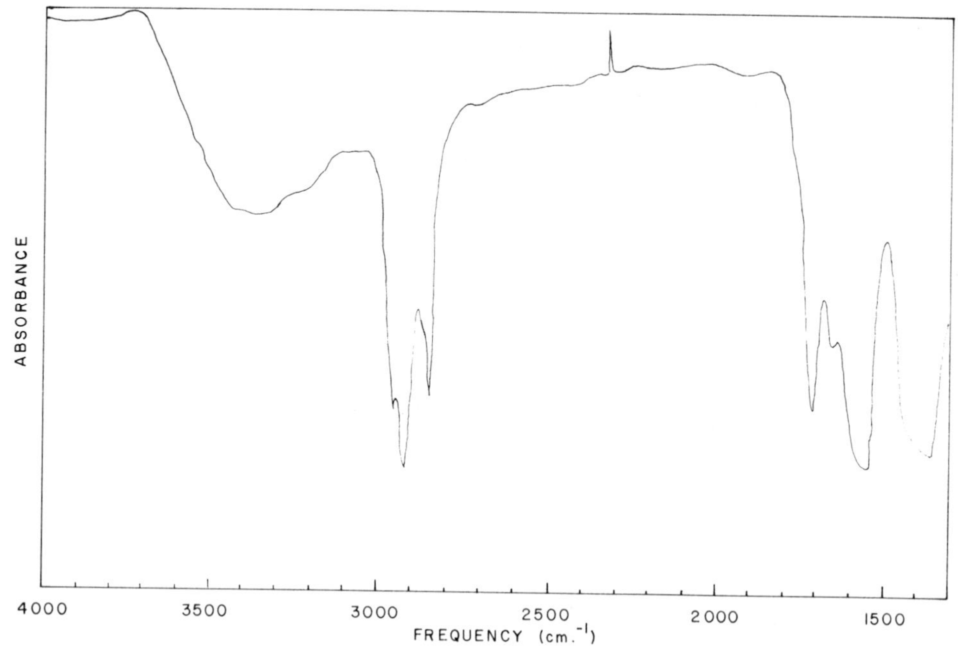


FIGURE 16. Partial infrared spectrum (thin film) of polar TLC fraction from Aroclor 1254 irradiation experiment (irradiation conditions of Figure 13). (From *Environmental Health Perspectives.* With permission.)

## REFERENCES


1. **Abramovich, R. A. and Takaya, T.,** *Chem. Commun.,* 1369, 1969.
2. **Andelman, J. B. and Suess, M. J.,** The photodecomposition of 3,4-benzpyrene sorbed on calcium carbonate, in *Organic Compounds in Aquatic Environments,* Faust, S. D. and Hunter, J. V., Eds., Marcel Dekker, New York, 1971, 439.
3. **Andersson, K., Nilsson, C. A., Norstrom, A., and Rappe, C.,** PCB Conference II, Stockholm, 1972.
4. **Bowen, E. J.,** *Chemical Aspects of Light,* Oxford University Press, Oxford, 1946.
5. **Bryce-Smith, D.,** *Photochemistry,* Chemical Society Specialist Periodical Report, Vol. 1–3, The Chemical Society, London, 1970, 1971, and 1972.
6. **Calvert, G. J. and Pitts, J. N., Jr.,** *Photochemistry,* Wiley, New York, 1966.
7. **Cha, C. Y. and Smith, J. M.,** *Ind. Eng. Chem. Process Des. Develop.,* 11, 451, 1972.
8. **Chapman, O. L., Ed.,** *Organic Photochemistry,* Vol. 1–3, Marcel Dekker, New York, 1967–1973.
9. **Crosby, D. G.,** *Residue Rev.,* 25, 1, 1969.
10. **Crosby, D. G.,** in *Pesticides in the Soil,* Symposium at Michigan State University, East Lansing, 1970, 86.
11. **Crosby, D. G.,** The photodecomposition of pesticides in water, *Advan. Chem. Series,* 111, 173, 1972.
12. **Crosby, D. G.,** Environmental photooxidation of pesticides, in *Degradation of Synthetic Organic Molecules in the Biosphere,* National Academy of Sciences, Washington, D.C., 1972, 260.
13. **Crosby, D. G., Moilanen, K., and Wong, A.,** Conference on Chlorinated Dibenzodioxins and Dibenzofurans, Research Triangle Park, N.C., April 1973 and personal communication.
14. **Crosby, D. G. and Hamadmad, N.,** *J. Agric. Food Chem.,* 19, 1171, 1971.
15. **Crosby, D. G. and Li, M. Y.,** in *Degradation of Herbicides,* Kearney, P. C. and Kaufman, D. D., Eds., Marcel Dekker, New York, 1969, 321.
16. **Crosby, D. G. and Moilanen, K. W.,** Presentation at 163rd National Meeting, ACS, Boston, Mass., April 11, 1972.
17. **Crosby, D. G., Wong, A. S., Plimmer, J. R., and Woolson, E. A.,** *Science,* 173, 748, 1971.
18. **Dhar, N. R.,** *The Chemical Action of Light,* Blakie, London, 1931.
19. **Ellis, C., Wells, A. A., and Heyroth, F. F.,** *The Chemical Action of Ultraviolet Rays,* Reinhold, New York, 1941.
20. **Ermolaev, V. L. and Sveshnikova, E. B.,** *Optika i Spektroskopiya,* 21, 134, 1966; *Chem. Abstr.,* 65, 14670, 1966.
21. **Fleck, E. E.,** *J. Am. Chem. Soc.,* 71, 1034, 1949.
22. **Hannan, E. J., Bills, D. D., and Herring, J. L.,** *J. Agric. Food Chem.,* 21, 87, 1973.

23. **Hattula, M. L. and Karlog, O.,** *Dansk. Tidsskr. Farm.,* 45, 259, 1971.
24. **Herring, J. L., Hannan, E. J., and Bills, D. D.,** *Bull. Environ. Contam. Toxicol.,* 8, 153, 1972.
25. **Hustert, K. and Korte, F.,** *Chemosphere,* 1, 7, 1972.
26. **Hutzinger, O., Jamieson, W. D., Safe, S., MacNeil, J. D., and Zitko, V.,** *PCB in the Environment,* Marcel Dekker, New York, in press.
27. **Hutzinger, O., Jamieson, W. D., Safe, S., and Zitko, V.,** *J. Assoc. Offic. Anal. Chem.,* 56, 982, 1973.
28. **Hutzinger, O., Safe, S., Wentzell, B. R., and Zitko, V.,** *Environ. Health Perspect.,* 2, 267, 1973.
29. **Hutzinger, O., Safe, S., and Zitko, V.,** *Environ. Health Perspect.,* 1, 15, 1972.
30. **Hutzinger, O., Safe, S., and Zitko, V.,** *Intern. J. Environ. Anal. Chem.,* 2, 95, 1972.
31. **Ivie, G. W. and Casida, J. E.,** *J. Agric. Food Chem.,* 19, 405, 1971.
32. **Kerner, I., Klein, W., and Korte, F.,** *Tetrahedron,* 28, 1575, 1972.
33. **Koller, L. R.,** *Ultraviolet Radiation,* Wiley & Sons, New York, 1965.
34. **Laity, J. L.,** *Environ. Sci. Technol.,* 5, 1218, 1971.
35. **Lamola, A. A. and Turro, J. N.,** *Energy Transfer and Organic Photochemistry* (Technique of Organic Chemistry, Vol. XIV), Interscience, New York, 1969.
36. **Lodder, G. and Havinga, E.,** *Tetrahedron,* 28, 5583, 1972.
37. **MacNeil, J. D., Safe, S., and Hutzinger, O.,** in preparation.
38. **Mende, U., Laseter, J. L., and Griffin, G. W.,** *Tetrahedron Lett.,* 3747, 1970.
39. **Miller, L. L. and Narang, R. S.,** *Science,* 169, 368, 1970.
40. **Miller, L. L., Narang, R. S., and Nordblom, G. D.,** *J. Org. Chem.,* 38, 340, 1973.
41. **Mitchell, L. C.,** *J. Assoc. Offic. Anal. Chem.,* 44, 643, 1961.
42. **Moilanen, K. W. and Crosby, D. G.,** Presentation at 165th National Meeting ACS, Dallas, Texas, April, 1973.
43. **Mosier, A. R., Guenzi, W. D., and Miller, L. L.,** *Science,* 164, 1083, 1969.
44. **Munakata, K. and Kuwahara, M.,** *Residue Rev.,* 25, 13, 1969.
45. **Mustafa, A.,** *Chem. Rev.,* 51, 1, 1952.
45a. **Nisbet, C. T. and Sarofim, A. F.,** *Environ. Health Perspect.,* 1, 21, 1972.
46. **Nordblom, J. D. and Miller, L. L.,** *Environ. Sci. Technol.,* in press.
47. **Nishiwaki, T., Ninomiya, A., Yamanaka, S., and Anda, K.,** *Nippon Kagaku Kaishi,* 2225, 1972; *Chem. Abstr.,* 78, 29339, 1973.
48. **Noyes, W. A., Jr., Hammond, G. S., and Pitts, J. N., Jr.,** *Advances in Photochemistry,* Vol. 1–8, Interscience, New York, 1963–1971.
49. **Pinkey, J. T. and Rigby, R. D. G.,** *Tetrahedron Lett.,* 1267, 1969.
50. **Plimmer, J. R.,** *Residue Rev.,* 33, 47, 1970.
51. **Plimmer, J. R.,** Photochemistry of pesticides: A discussion of the influence of some environmental factors, in *Pesticide Chemistry,* Vol. VI, Proc. 2nd IUPAC Congress (1971), Tahori, A. S., Ed., Gordon and Breach, New York, 1972, 47.
52. **Plimmer, J. R.,** Principles of photodecomposition of pesticides, in *Degradation of Synthetic Organic Molecules in the Biosphere,* National Academy of Sciences, Washington, D.C., 1972, 279.
53. **Plimmer, J. R., Klingbiel, V. I., and Hummer, B. E.,** *Science,* 167, 67, 1970.
54. **Pomerantz, I.,** personal communication.
55. **Reid, S. T.,** *Fortschritte Arzneimittelforsch,* 11, 48, 1968.
56. **Robinson, J. E. and Vernon, J. M.,** *J. Chem. Soc.,* (C), 3363, 1971.
57. **Rosen, J. D.,** Photodecomposition of organic pesticides, in *Organic Compounds in Aquatic Environments,* Faust, S. D., and Hunter, J. V., Eds., Marcel Dekker, New York, 1971, 425.
58. **Rosen, J. D.,** The photochemistry of several pesticides, in *Environmental Toxicology of Pesticides,* Matsumura, F., Boush, C. M., and Misato, T., Eds., Academic Press, New York, 1972, 435.
59. **Rosen, J. D.,** *Environ. Qual. Saf.,* 1, 85, 1972.
60. **Rosen, J. D., Siewierski, M., and Winnett, G.,** *J. Agric. Food Chem.,* 18, 494, 1970.
61. **Ruzo, L. O., Zabik, M. J., and Schuetz, R. D.,** *Bull. Environ. Contam. Toxicol.,* 8, 217, 1972.
62. **Safe, S. and Hutzinger, O.,** *Nature,* 232, 641, 1971.
63. **Sato, T. and Morita, T.,** *Bull. Chem. Soc. Jap.,* 45, 1548, 1972.
64. **Schaffner, K.,** *Fortschr. Chem. Org. Naturst.,* 22, 1, 1964.
65. **Schönberg, A.,** *Preparative Organic Photochemistry,* Springer-Verlag, New York, 1968.
66. **Sharma, S. K. and Kharasch, N.,** *Angew. Chem. Intern. Ed.,* 1, 36, 1968.
67. **Stenberg, V. I.,** *Organic Photochemistry,* Chapman, O. L., Ed., Marcel Dekker, New York, 1967, 127.
68. **Thomas, R. O. and Fishbein, L.,** *Chem. Anal.,* 55, 90, 1966.
69. **Trecker, D. J.,** Photodimerisations, in *Organic Photochemistry,* Vol. 2, Chapman, O. L., Ed., Marcel Dekker, New York, 1969, 63.
70. **Wallnöfer, P. R., Königer, M., and Hutzinger, O.,** *Research Notes,* Analabs Inc., in press.
71. **Weingarten, H.,** *J. Org. Chem.,* 27, 2024, 1962.
72. **Wynberg, H. and Wolf, A. P.,** *J. Am. Chem. Soc.,* 85, 3308, 1963.
73. **Zitko, V., Hutzinger, O., Jamieson, W. D., and Choi, P. M. K.,** *Bull. Environ. Contam. Toxicol.,* 7, 200, 1972.

Chapter 7

# METABOLISM OF CHLOROBIPHENYLS

## Introduction

The metabolic alteration of chlorobiphenyls is being actively investigated in many laboratories because of possible relationships of metabolism to toxicity and storage in the body on the one hand and environmental breakdown in soils mud and water on the other. A reasonably complete picture has not yet emerged but it seems fairly certain, as it is with related compounds such as chlorobenzenes, that the ease of metabolism decreases with increasing chlorine content. In addition, certain structural features such as positional isomerison influence the metabolic behavior, but again, available data are only preliminary.

General information on the metabolism of compounds related to chlorobiphenyls by microorganisms.[13,21,24,47–49] and animals[27,44,54] is available.

## Metabolic Change of Chlorobiphenyls by Animals

### *Studies with Technical PCB Mixtures*

Metabolism of certain components of technical PCB mixtures is indicated by the fact that GC-peaks with lower retention time are diminished or absent in tissue extracts from PCB-treated animals. Studies with mice,[57] rats[5,26,56] (Figure 1; Table 1), birds[3,4,15,37] (Figure 2; Table 2), and cows[19] (Figure 3) indicate that peaks with retention times corresponding to di-,tri-, and tetrachlorobiphenyls are mainly affected. Since it was shown that all components of Aroclor 1254 were absorbed at the same rate in quail[3] and a number of mono- to hexachlorobiphenyls were equally retained (> 90%) by rats,[2] the correlation of disappearance of peaks with metabolism seems justified.

Feeding experiments with fish indicate much lower metabolizing activity. Only after prolonged dietary exposure with Aroclor 1254 (>200 days) and additional control feed period (>200 days), is a reduction or disappearance of the first two peaks observed (Figure 4).[58]

Attempts have been made to correlate the structure of the PCB components with metabolic activity by comparing retention times of peaks (e.g., those disappearing fast or not at all) with known standards or by GC-mass spectrometry. Figure 1 and Table 1 show that the accumulation of the more highly chlorinated biphenyls in rats[5] is not uniform; a pentachlorobiphenyl corresponding to peak 33 being selectively retained whereas

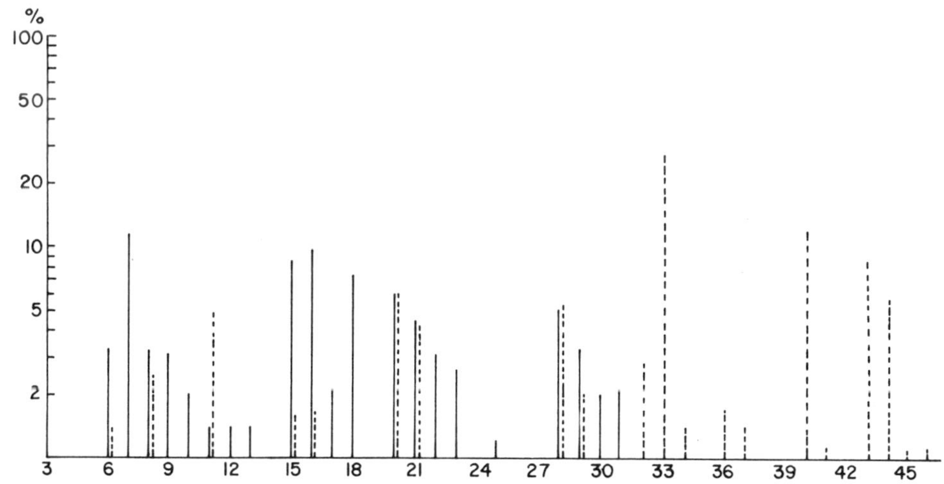


FIGURE 1. Relative intensity (in %) of peaks 3–46 in gas chromatogram of Aroclor 1248 ( – ) and material extracted from rats 4 weeks after administration (- - -). Peak numbers correspond to those of Table 1. (From *Arch. Toxicol.*, 30, 207, 1973. With permission.)

TABLE 1

**Composition of Aroclor 1248 Used for Feeding and Material Isolated from Fatty Tissue of Rats**

| Peak number* | No. of Cl | % In Aroclor 1248 | % In fatty tissue |
|---|---|---|---|
| 1 | | | |
| 2 | | | |
| 3 | | | |
| 4 | 2 | 0,4 | |
| 5 | 3 | 1,0 | 0,9 |
| 6 | 2 | 3,2 | 1,4 |
| 7 | 3 | 11,2 | 0,9 |
| 8 | 3 | 3,2 | 2,4 |
| 9 | 3 | 3,1 | 0,5 |
| 10 | 3 | 2,0 | 0,2 |
| 11/12 | 3,4 | 1,4 | 4,8 |
| 13 | 3,4 | 1,4 | 0,8 |
| 14 | 3,4 | 0,8 | 0 |
| 15 | 3 | 8,5 | 1,6 |
| 16 | 3 | 9,7 | 1,6 |
| 17 | 3 | 2,1 | 0,8 |
| 18 | 3,4 | 7,4 | 0,5 |
| 19 | 3,4 | 7,4 | – |
| 20 | 3,4 | 6,0 | 7,0 |
| 21 | 3,4 | 4,5 | 4,1 |
| 22 | 4 | 3,1 | 0,3 |
| 23 | 4 | 2,6 | 0,4 |
| 24 | 3,4,5 | 0,8 | – |
| 25 | 4,5 | 1,2 | 1,0 |
| 26 | 4,5 | 1,0 | – |
| 27 | 4 | 0,2 | – |
| 28 | 4 | 5,0 | 5,3 |
| 29 | 4 | 3,3 | 2,0 |
| 30 | 4 | 2,0 | 0,4 |
| 31 | 4 | 2,1 | 0,5 |
| 32 | 4,5 | 1,0 | 2,9 |
| 33 | 4,5 | 0,5 | 29,0 |
| 34 | 4,5 | 0,5 | 1,4 |
| 35 | 4,5 | 0,8 | – |
| 36 | 5 | 1,0 | 1,7 |
| 37 | 5 | 0,4 | 1,4 |
| 38 | 6 | <0,1 | 0,1 |
| 39 | 4 | <0,1 | 0,1 |
| 40 | 6 | 0,2 | 11,0 |
| 41/42 | 5 | 0,3 | 1,1 |
| 43 | 5,6 | 0,2 | 8,0 |
| 44 | 6 | <0,1 | 5,3 |
| 45 | 6 | 0,2 | 0,9 |
| 46 | 6,7 | (0,7) | (1,1) |

*Conditions of Figure 1.

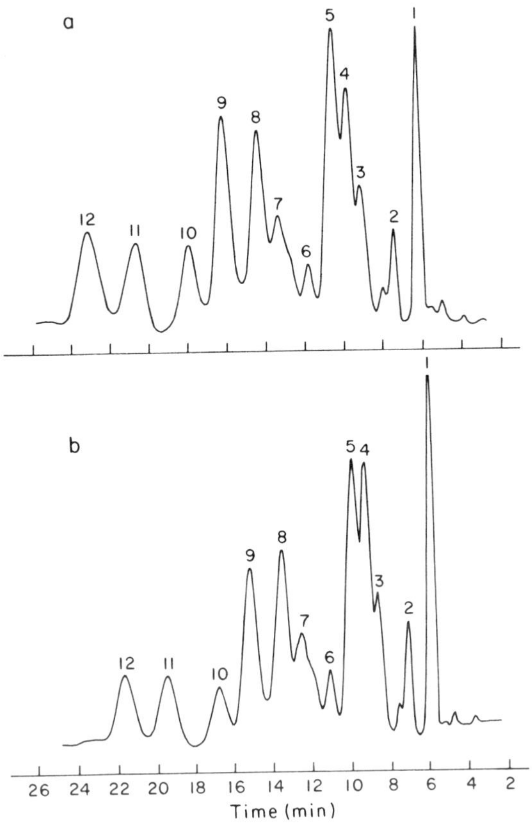


FIGURE 2. Total ion current chromatograms of Aroclor 1254 and quail extracts. GC-conditions: 9′ x 0.25″ column, 5% OV-17, oven temperature 220, helium flow-rate 40 ml/min. (a) Aroclor 1254 standard (S of Table 2); (b) Aroclor 1254 standard, partitioned between acetonitrile and hexane (= treatment of quail carcass extracts)(P of Table 2); (c) Carcass extract of quail administered a single dose of Aroclor 1254 at 500 mg/kg (extract C of Table 2); (d) Carcass extract of quail sacrificed after 14 days' dietary dosage at 300 ppm of Aroclor 1254 (extract D of Table 2); (e) Carcass extract of quail sacrificed after 14 days' dietary dosage at 300 ppm of Aroclor 1254 followed by 14 days of untreated food (extract E of Table 2); (f) Carcass extract of quail sacrificed after 14 days' dietary dosage of 300 ppm of Aroclor 1254 followed by 42 days of untreated food (extract F of Table 2).

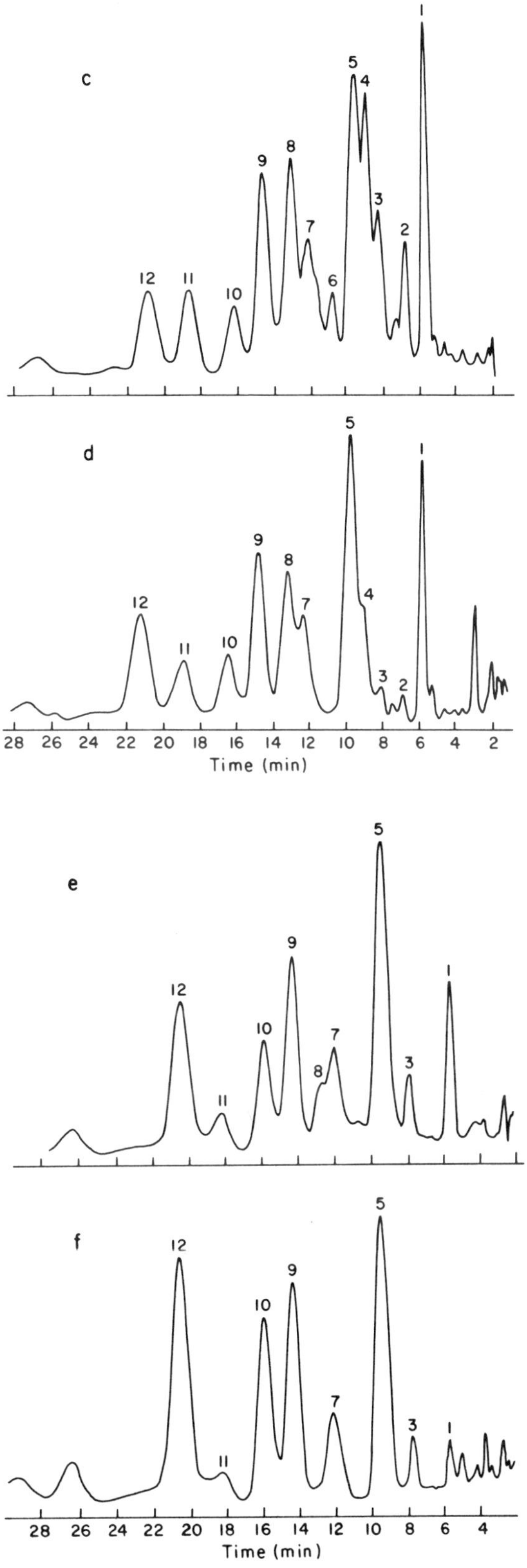


(From *J. Chromatog.*, 75, 219, 1973. With permission.)

pentachlorobiphenyls (peaks 25, 26, 35, and 42) diminish. Among the hexachlorobiphenyls, all peaks except 38 are significantly increased, particularly peaks 40 and 44. The total amounts of tri- and tetrachlorobiphenyls are significantly decreased but here also differences in ease of metabolism are obvious. Tetrachlorobiphenyl 11 is enriched and peaks 20, 21, and 28 remain the same. In the trichlorobiphenyl series, only peak 8 proves to be relatively persistent.

TABLE 2

**Relative Peak Heights[a] of Aroclor 1254 and Material Isolated from Quail**

| Sample | | Peak No. 1 | 2 | 3 | 4 | 5 | 6 | 7 | 8 | 9 | 10 | 11 | 12 |
|---|---|---|---|---|---|---|---|---|---|---|---|---|---|
| | $R_x$[b] | 0.49 | 0.58 | 0.69 | 0.78 | 0.83 | 0.94 | 1.06 | 1.13 | 1.27 | 1.40 | 1.61 | 1.81 |
| | No. of Cl[c] | 4 | 4 | 4,5 | 5 | 5 | 5 | 5 | 6 | 6 | 6 | 6 | 6 |
| Aroclor[d] 1254 (S) | | 101 | 34 | 49 | 80 | 100 | 23 | 39 | 68 | 73 | 30 | 30 | 34 |
| Aroclor[e] 1254 (P) | | 125 | 48 | 57 | 98 | 100 | 32 | 44 | 71 | 65 | 27 | 31 | 31 |
| Extract C[f] | | 118 | 48 | 58 | 96 | 100 | 32 | 49 | 75 | 70 | 28 | 33 | 33 |
| Extract D[g] | | 93 | 15 | 15 | 45 | 100 | 0 | 39 | 51 | 63 | 28 | 22 | 39 |
| Extract E[h] | | 54 | 0 | 24 | 0 | 100 | 0 | 33 | 23 | 63 | 36 | 12 | 48 |
| Extract F[i] | | 25 | 0 | 26 | 0 | 100 | 0 | 34 | 0 | 77 | 66 | 14 | 86 |

[a] Peak No. 5 = 100%, total ion current chromatogram.
[b] *p,p'*-DDE = 1.0.
[c] No. of Cl in PCB.
[d] S = standard.
[e] P = standard partitioned by acetonitrile-hexane and eluted on Florisil.
[f] Carcass of quail administered 1 capsule of Aroclor 1254, 500 mg/kg body weight.
[g] Carcass of quail sacrificed at day 14.
[h] Carcass of quail sacrificed at day 28.
[i] Carcass of quail sacrificed at day 56.

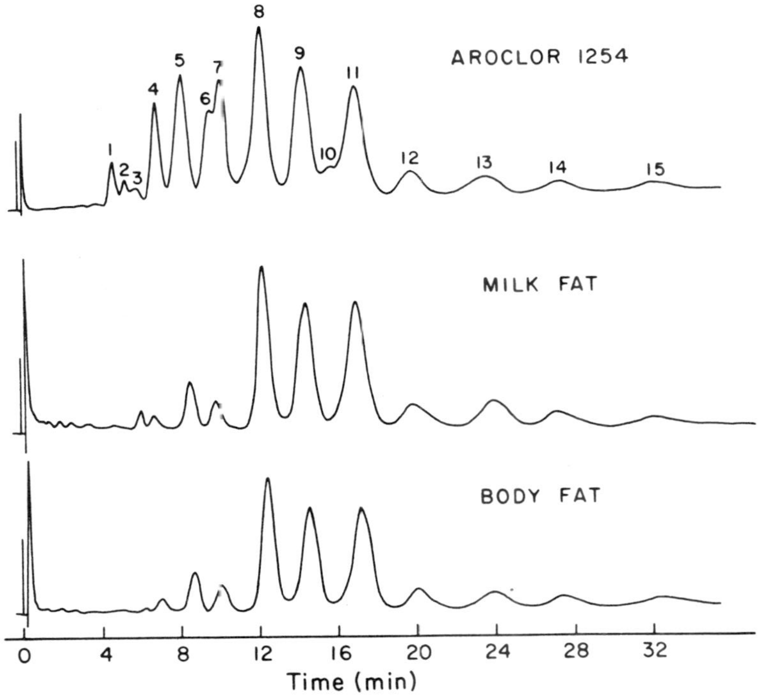


FIGURE 3. Gas chromatogram of Aroclor 1254 standard and milk and body fat residue samples from a cow fed Aroclor 1254. The retention times of the following peaks were identical to the standard compound in parentheses 1,(2,2′,5,5′-tetrachlorobiphenyl); 2,(2,2′,3,5′-tetrachlorobiphenyl); 4,(2,3′,4′,5-tetrachlorobiphenyl); 6,(2,2′,3,4,5′-pentachlorobiphenyl); 9,(2,2′,4,4′,5,5′-hexachlorobiphenyl). (From *J. Agr. Food Chem.*, 27, 117, 1973. With permission.)

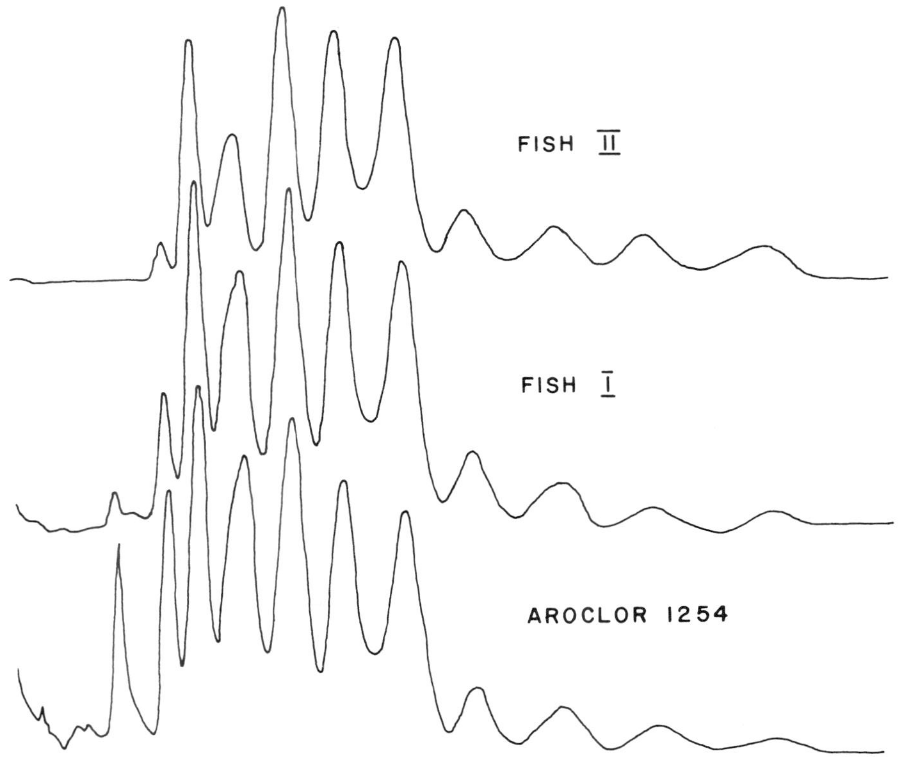


FIGURE 4. PCB patterns in juvenile Atlantic salmon fed 100 $\mu$g/g of Aroclor 1254 in the diet for 244 days and untreated food for additional 243 days. The two upper chromatograms correspond to extracts from two individual fish. A standard chromatogram of Aroclor 1254 is shown at the bottom of the Figure.

In another study with rats,[56] major losses of early eluters were observed but more significantly one later peak was completely lost. This peak was tentatively assigned as being 2,3,3′,4′,6-pentachlorobiphenyl by comparison (GC and infrared) with authentic material.

The change of peak intensity of Aroclor 1254 during a feeding study with quail[3] is shown in Figure 2 and Table 2. All early peaks decrease markedly during the experiment. Of the later eluters, peaks 4, 6, and 8 are completely lost. Most significantly, the authors of this study report dechlorination and isomerization to take place. However, these reactions can be much easier observed with individual chlorobiphenyls and confirmation has to await such experiments.

The diminishing of early eluting peaks in milk and body fat of cows fed Aroclor 1254 is shown in Figure 3.[19] Compounds with retention times identical to some of the peaks are indicated in the legend.

*Studies with Individual Chlorobiphenyls*

The intestinal absorption of biphenyl and 19 mono- to hexachlorobiphenyls in rats has been studied by monitoring fecal excretion after the feeding of individual compounds.[2] The percentage retained was >90 in all cases (Table 3). The lowest retention was shown by the 3,3′,4,4′-tetra- and 3,3′,4-trichlorobiphenyl which have obvious structural similarities. Interestingly, 3,3′,4,4′-tetrachlorobiphenyl was the least oil soluble of 11 mono- to octachlorobiphenyls tested.[16] The solubility of this compound in water, on the other hand, is relatively high compared to other tetrachlorobiphenyls examined.[52]

The distribution of $^{14}$C-labeled 2,2′,4,5,5′-pentachlorobiphenyl was studied in the mouse after

TABLE 3

**Uptake of Chlorobiphenyls by Rats**

| Compound | Percentage retained* | | | | | |
|---|---|---|---|---|---|---|
| | 5 mg/kg | S.D. | 50 mg/kg | S.D. | 100 mg/kg | S.D. |
| Biphenyl | 99.5 | 0.11 | 98.8 | 0.25 | 98.2 | 0.30 |
| Monochlorobiphenyls | | | | | | |
| 2- | 98.9 | 0.45 | 98.0 | 0.20 | 98.4 | 0.15 |
| 3- | 98.0 | 1.01 | 96.6 | 0.10 | 97.3 | 0.40 |
| 4- | 97.3 | 1.73 | 96.3 | 0.40 | 96.4 | 0.42 |
| Dichlorobiphenyls | | | | | | |
| 2,6- | | | 95.0 | 0.96 | 94.9 | 0.75 |
| 2,2′- | 95.0 | 2.42 | 97.4 | 0.20 | 98.2 | 2.10 |
| 2,4- | | | 95.7 | 0.40 | 97.1 | 0.35 |
| 2,3′- | | | 96.8 | 0.51 | 97.9 | 0.28 |
| 2,4′- | 97.2 | 3.38 | 95.5 | 1.20 | 97.6 | 0.78 |
| 3,4′- | 95.7 | 3.45 | 94.3 | 0.86 | 97.1 | 0.63 |
| 4,4′- | 95.4 | 3.05 | 95.6 | 1.12 | 97.4 | 1.10 |
| Trichlorobiphenyls | | | | | | |
| 2,2′,5- | | | 95.1 | 3.00 | 94.3 | 3.18 |
| 2′,3,5- | | | 95.6 | 1.05 | 95.0 | 1.12 |
| 3,3′,4- | | | 93.5 | 2.85 | 90.3 | 3.58 |
| 2,4,4′- | | | 92.6 | 0.86 | 94.0 | 1.50 |
| Tetrachlorobiphenyls | | | | | | |
| 2,2′,5,5′- | 96.0 | 1.75 | | | | |
| 2,3,4,5- | 97.3 | 0.86 | | | | |
| 3,3′,4,4′- | 91.8 | 0.57 | | | | |
| Pentachlorobiphenyls | | | | | | |
| 2,2′,4,5,5′- | 95.1 | 0.57 | | | | |
| Hexachlorobiphenyls | | | | | | |
| 2,2′,4,4′,5,5′- | 95.3 | 1.11 | | | | |

$$\text{*Percentage retained} = \frac{\text{(amount fed)} - \text{(amount excreted)}}{\text{amount fed}} \times 100\%$$

intravenous injection.[6] Within 20 min, most of the radioactivity from the blood had entered the tissues with fat and liver containing most of the activity. Details of the excretion patterns and tissue distribution are shown in Tables 4 and 5. No attempt has been reported yet on the nature of possible metabolites.

Preliminary data from experiments in which $^{14}C$-labeled 2,2′,4,4′-tetrachlorobiphenyl was administered to mice and a mixture of 2,2′,3,4′- and 2,3′,4,4′-tetrachlorobiphenyl ($^{14}C$) to quail have been reported.[41] The excretion of the 2,2′,4,4′-tetrachlorobiphenyl in mice was found to be similar to the pentachloro derivative mentioned above. The most striking observation in the distribution pattern in mice was the accumulation in the adrenal cortex and the corpora lutea of the ovaries. Fat and liver were highest in activity in both species but quantitative data have not been reported. In this study, polar metabolites resembling hydroxybiphenyls have been detected.

In the first study on the structure of a chlorobiphenyl metabolite reported, Block and Cornish in 1959 administered 4-chlorobiphenyl (I) to rabbits via stomach tube.[7] The compound was metabolized to the extent of 64% with the glucuronide of 4-chloro-4′-hydroxybiphenyl accounting for 50%, the corresponding etherial sulfate for 11% and the free phenol for 3% (Figure 5). This distribution was in contrast to the parent compound biphenyl for which the corresponding values (4-hydroxybiphenyl) were 26%, 13%, and 24%. No mercapturic acid derivatives were found. The metabolic behavior of a number of unlabeled chlorobiphenyls (compounds I – XI; Table 6) in rats after intraperitoneal injection was recently reported.[30,31] In this study, urine (after acid hydrolysis) and feces were extracted and these extracts analyzed by thin layer chromatography and mass spectrometry. In those cases in which enough material was recovered, the structures were

TABLE 4

**Excretion of Radioactivity in Urine and Faeces of Mice Dosed with $^{14}C$-2,2′,4,5,5′-Pentachlorobiphenyl**

| Days after dosing | Feces | Urine | Total | Cumulative total |
|---|---|---|---|---|
| 1 | 20.3 | 0.23 | 20.53 | 20.5 |
| 2 | 9.9 | 0.11 | 10.01 | 30.5 |
| 3 | 4.1 | 0.12 | 4.22 | 34.8 |
| 4 | 7.0 | 0.13 | 7.13 | 41.9 |
| 5 | 3.6 | 0.17 | 3.77 | 45.7 |
| 6–7 | 5.4 | 0.16 | 5.56 | 51.2 |
| 8 | 3.6 | 0.13 | 3.73 | 55.0 |
| 9–10 | 6.4 | 0.11 | 6.51 | 61.5 |
| 11 | 2.3 | 0.07 | 2.37 | 63.8 |
| 12 | 1.8 | 0.10 | 1.90 | 65.7 |
| 13 | 2.5 | 0.08 | 2.58 | 68.3 |
| 14 | 2.2 | 0.09 | 2.29 | 70.6 |
| 15 | 2.0 | 0.09 | 2.09 | 72.7 |
| 16 | 2.3 | 0.06 | 2.36 | 75.1 |
| 17–18 | 2.4 | 0.12 | 2.52 | 77.6 |
| 19 | 1.6 | 0.06 | 1.66 | 79.2 |
| Total | 77.4 | 1.8 | 79.2 | |

TABLE 5

**Distribution of Radioactivity in Tissues After a Single Dose of $^{14}C$ 2,2′,4,5,5′-Pentachlorobiphenyl[a]**

| Dose, μCi/mouse | 10[b] | 10[b] | 10[b] | 5 | 5 | | 5 | 5 | 10[b] |
|---|---|---|---|---|---|---|---|---|---|
| Route | iv | iv | iv | oral | oral | | oral | oral | oral |
| Time after dosing | 20 min | 1 hr | 4 hr | 1 day | 4 days | | 8 days | 16 days | 35 days |
| Sex | M | F | M | M | M | F | M | M | F |
| Blood | 7.90 | 3.50 | 3.60 | | | | 0.43 | 0.18 | |
| red cells | | | | 0.40 | 0.16 | | | | |
| plasma | | | | 1.30 | 2.76 | 0.45 | | | |
| Heart | 12.50 | 9.90 | 13.00 | 2.18 | 0.50 | 0.37 | 0.41 | 0.57 | |
| Lungs | 27.00 | 14.80 | 5.70 | | 3.10 | 3.15 | 1.90 | 0.22 | 2.28 |
| Liver | 50.90 | 33.40 | 12.20 | 6.50 | 2.80 | 1.80 | 3.60 | 0.78 | 0.64 |
| Kidneys | 19.30 | 9.80 | 8.40 | 8.60 | 1.70 | 1.50 | 1.00 | 0.50 | 0.76 |
| Fat | 4.30 | 13.30 | 11.90 | 27.90 | 10.70 | 14.00 | 13.40 | 8.20 | 3.30 |
| Muscle | 7.80 | 5.60 | 1.73 | | 1.95 | 0.94 | 1.51 | 2.65 | 0.29 |
| Brain | 5.03 | 6.52 | 3.80 | 0.97 | 0.18 | 0.23 | 0.21 | 0.11 | 0.08 |

[a]Figures in the Table represent μCi/100 mg tissue.
[b]Results have been halved to bring in line with 5 μCi experiments.

TABLE 6

**Polar Metabolites of Chlorobiphenyls Found in Different Animal Species**

| Chlorobiphenyl administered | | | Metabolite found | | | |
|---|---|---|---|---|---|---|
| | | | | Monohydroxy* compound | Dihydroxy* compound | |
| No. | Name | Formula | Animal species used | $M^{+\cdot}$ = m/e chlorobiphenyl +16 | $M^{+\cdot}$ = m/e chlorobiphenyl +32 | Other derivative |
| I | 4-Chlorobiphenyl | 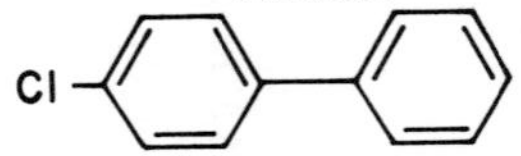 | rat | + | + | – |
| | | | pigeon | + | – | – |
| | | | trout | – | – | – |
| | | | rabbit | + | – | – |
| II | 4,4′-Dichlorobiphenyl | 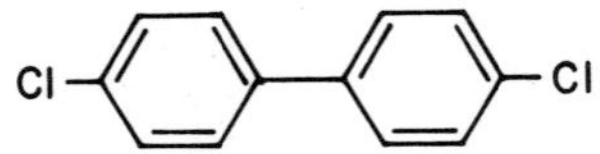 | rat | + | +† | – |
| | | | pigeon | + | – | – |
| | | | trout | – | – | – |
| III | 2,2′-Dichlorobiphenyl | 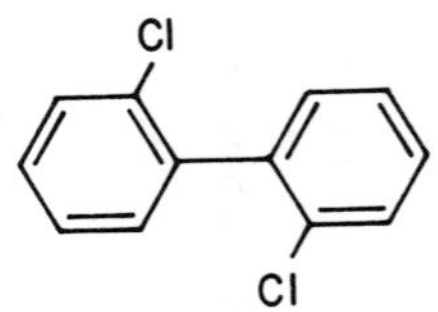 | rat | + | + | – |
| IV | 2,4′-Dichlorobiphenyl | 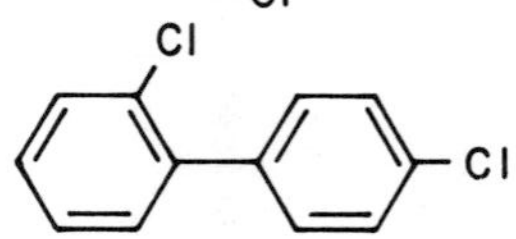 | rat | + | – | – |
| V | 2,2′,4,4′-Tetrachloro-biphenyl | 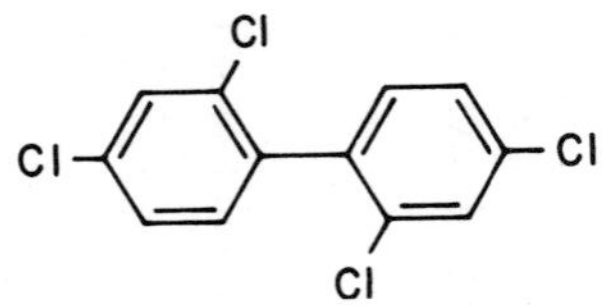 | rat | + | – | – |
| VI | 2,2′,5,5′-Tetrachloro-biphenyl | 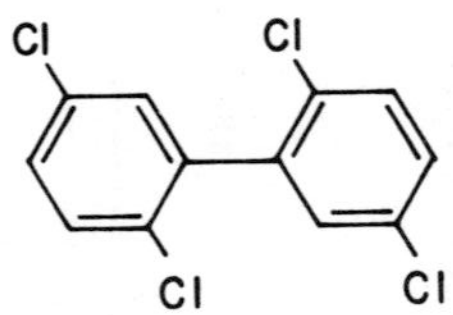 | rat | + | – | – |
| | | | pigeon | + | – | – |
| | | | trout | – | – | – |
| | | | rabbit | + | – | Dihydrodihydroxy-tetrachlorobiphenyl |

TABLE 6 (continued)

**Polar Metabolites of Chlorobiphenyls Found in Different Animal Species**

| Chlorobiphenyl administered: No. | Chlorobiphenyl administered: Name | Formula | Animal species used | Metabolite found: Monohydroxy* compound $M^{+\cdot}$ = m/e chlorobiphenyl +16 | Metabolite found: Dihydroxy* compound $M^{+\cdot}$ = m/e chlorobiphenyl +32 | Other derivative |
|---|---|---|---|---|---|---|
| VII | 3,3′,4,4′-Tetrachloro-biphenyl | 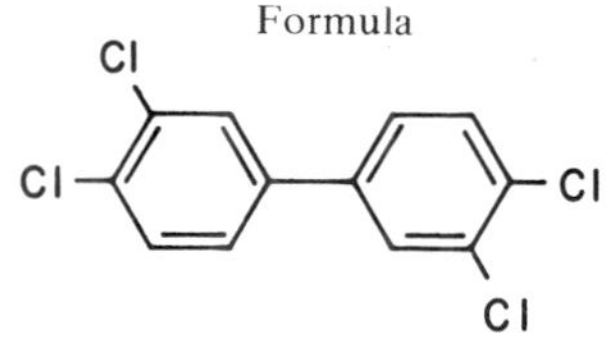 | rat | – | – | – |
| VIII | 2,2′,4,4′,5,5′-Hexa-chlorobiphenyl | 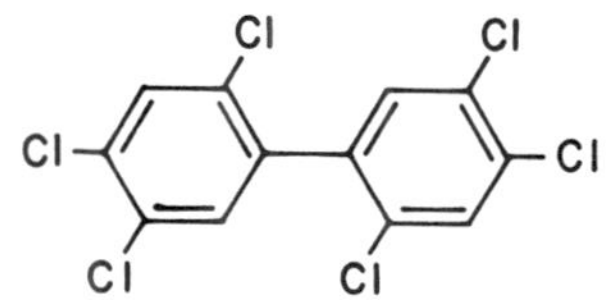 | rat | – | – | – |
| | | | pigeon | – | – | – |
| | | | trout | – | – | – |
| | | | rabbit | + | – | Hydroxypentachlorobiphenyl hydroxymethoxypenta-chlorobiphenyl |
| IX | 2,2′,4,4′,6,6′-Hexa-chlorobiphenyl | 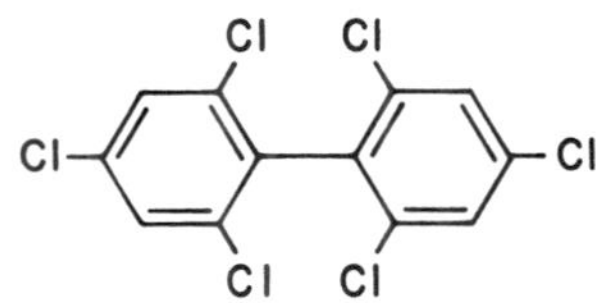 | rat | – | – | – |
| X | 2,2′,3,3′,5,5′-Hexa-chlorobiphenyl | 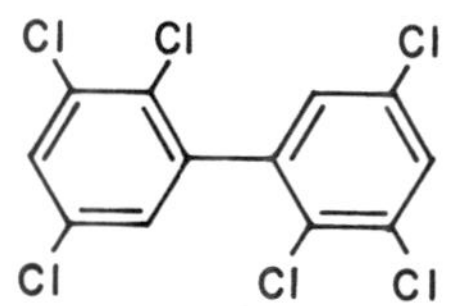 | rat | – | – | – |
| XI | 2,2′,3,3′,4,4′,5,5′-Octa-chlorobiphenyl | 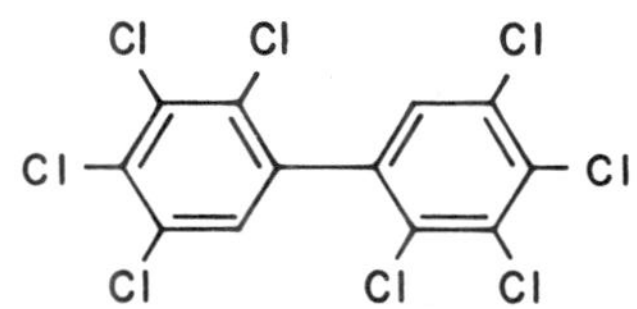 | rat | – | – | – |

*Including conjugates
†Detected by dansylation procedure

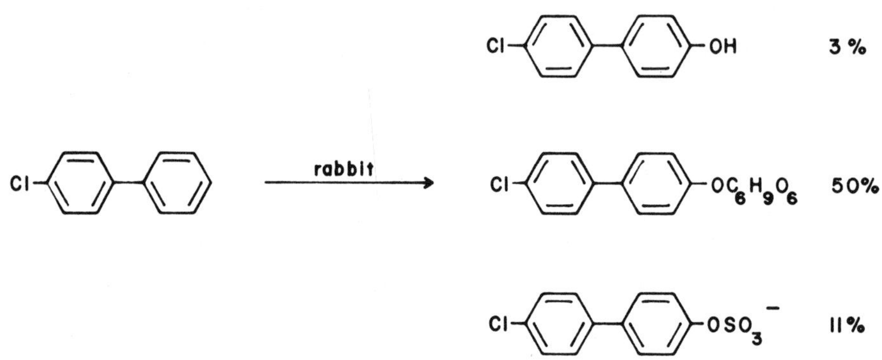


FIGURE 5. Metabolism of 4-chlorobiphenyl in the rabbit.

further investigated by 220 MHz NMR spectroscopy and confirmed by synthesis.[48a]

The results of this study are summarized together with some other data in Table 6. With the method used in this study, metabolites were not found for hexa- and octachlorobiphenyls (decachlorobiphenyl was not soluble enough in oil for injection into rats).

Of the three tetrachlorobiphenyls (V – VII) investigated, the 2,2′4,4′- and the 2,2′,5,5′- derivative both gave a monohydroxy derivative whereas no such metabolite could be detected with the method used for the 3,3′,4,4′-tetrachlorobiphenyl.

All three dichlorobiphenyls (II – IV) gave a monohydroxy derivative and the 2,2′-isomer also gave a dihydroxydichlorobiphenyl. A dihydroxy-derivative was also observed for the 4,4′-isomer employing a sensitive dansylation procedure, however, the 2,4′-isomer was not analyzed in such fashion. The monohydroxyderivative of 4,4′-dichlorobiphenyl (II) was shown to be 4,4′-dichloro-3-hydroxybiphenyl (XIII; Figure 6) by proton magnetic resonance and comparison with authentic material synthesized by an unambiguous route.

The only monochlorobiphenyl studied, 4-chlorobiphenyl (I), gave 4-chloro-4-hydroxybiphenyl (XII; Figure 6) and a dihydroxychlorobiphenyl of unknown structure.

In the same study, the chlorobiphenyls I, II, VI, and VIII were also fed to Carneau pigeons and brook trout. 4-Chlorobiphenyl (I), 4,4′-di-chlorobiphenyl, and 2,2′,5,5′-tetrachlorobiphenyl gave a monohydroxylated derivative in the birds, whereas 2,2′,4,4′,5,5′-hexachlorobiphenyl did not give a hydroxylated metabolite in the pigeons. Hydroxy derivatives could not be detected for any of the four compounds in brook trout.

Tissues of rats were not analyzed in this

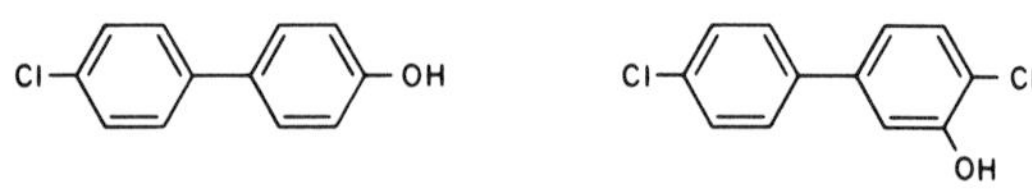


XII XIII

FIGURE 6. Structures of two metabolites isolated from rats.

investigation and the information on nonpolar metabolites is therefore limited to those excreted in the feces. Generally, dechlorination does not seem to have taken place; in only two instances (3,3′,4,4′-tetrachlorobiphenyl VII and 2,2′,3,3′,-4,4′,5,5′-octachlorobiphenyl XI) were traces of substances found whose mass spectra indicated loss of a chlorine atom (a tri- and a heptachlorobiphenyl, respectively).

When 2,2′,5,5′-tetrachlorobiphenyl (VI) was fed to a rabbit, a dihydroxy-dihydrotetrachlorobiphenyl was found to be the main water soluble metabolite in the urine.[20] Two monohydroxy derivatives of VI were also found in free and conjugated forms in the urine. In contrast, urine of rabbits fed with 2,2′4,4′5,5′-hexachlorobiphenyl contained a hexachlorohydroxybiphenyl, a pentachlorohydroxybiphenyl, and a pentachlorohydroxy-methoxybiphenyl as shown by high resolution mass spectrometric techniques.[29]

*Studies with Related Compounds*

Since very few studies on the metabolism of chlorobiphenyls have been reported, the behavior of some related compounds will be described briefly.

Chlorobiphenyls will provide interesting substrates for the study of labile intermediates (e.g., arene oxides) or special reactions (e.g., NIH-shift) known to occur with aromatic hydrocarbons[14,22,33]

*Biphenyl*

It was shown early that biphenyl is metabolized to 4-hydroxybiphenyl in the dog[36] and hydroxybiphenyls were subsequently shown to be the main mammalian metabolites for this compound.[46,54] The excretion of a significant portion of these hydroxylated derivatives is often in the form of etherial sulfates[7] or glucuronides.[7,42,53] In the rat, diphenylmercapturic acid was formed.[53] Compounds isolated as metabolites of biphenyl in rat or rabbit are depicted in Figure 7.

Species differences in the hydroxylation of biphenyl by liver microsomal enzymes have been observed[11,12,55] and are reported in Table 7.

Houseflies and bowflies were shown to convert biphenyl to 4-hydroxybiphenyl.[35]

Solubilizers such as Tween 80 are being used frequently in the study of the metabolism of water-insoluble compounds. It may be interesting to note that Tween 80 (0.25 to 2.5%) was found to inhibit the hydroxylation of biphenyl by liver microsomal preparations.[8]

2-Hydroxybiphenyl was shown to be metabolized to 2,5-dihydroxybiphenyl (Figure 8) in the rat.[17]

*Chlorobenzenes and Chloronaphthalenes*

Products observed in the mammalian metabolism of chlorobenzene[25,38,44,54] are depicted in Figure 9. In line with the current thinking on the metabolism of aromatic compounds in general, 3,4-chlorobenzene oxide has been suggested as the key intermediate in the metabolism of chlorobenzene.[38]

Similar products are obtained from the more highly chlorinated benzenes but, in general, the ease of metabolism decreases with increasing chlorine content. Some comparative data for various chlorobenzenes[54] are shown in Table 8. Recently,[40] evidence was presented for the formation of phenolic and dechlorinated products from hexachlorobenzene in rats. Similarly to the chlorobenzenes, studies on the metabolism of chloronaphthalenes in rabbits[10] show that mono-, di, and to a lesser extent, tetrachloronaphthalene were converted to phenolic compounds, their glucuronides or sulfates or to mercapturic acid derivatives. Penta-, hepta-, and octachloronaphthalenes apparently were not metabolized by the rabbit. The suggestion was made that a high degree of chlorination may interfere with the formation of a 1,2-dihydro-1,2-diol type intermediate.

**Metabolic Change of Chlorobiphenyls by Microorganisms**

*Studies Involving Mixed Microbial Populations*

Very few studies on the metabolic alteration of chlorobiphenyls by microorganisms have been

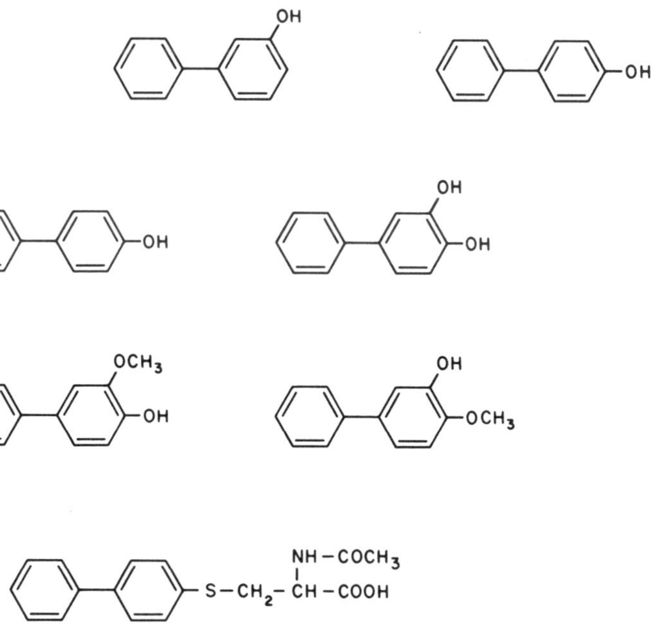


FIGURE 7. Compounds isolated as metabolites of biphenyl in rat or rabbit.

TABLE 7

**Occurrence of Biphenyl Hydroxylase in the Livers of Various Adult Species**

| Species | Product observed: 2-hydroxybiphenyl | Product observed: 4-hydroxybiphenyl |
|---|---|---|
| Man | – | + |
| Pig | – | + |
| Fox | – | + |
| Rabbit* | – | + |
| Guinea pig | – | + |
| Rat* | – | + |
| Cat | + | + |
| Hamster | + | + |
| Mouse | + | + |
| Coypu | + | + |
| Hen | + | + |
| Frog | + | + |

*Young animals also produce the 2-isomer.

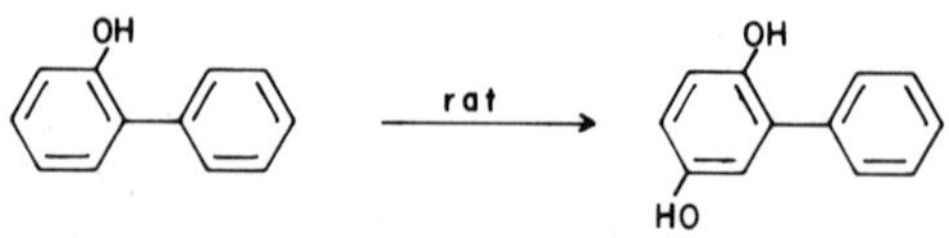


FIGURE 8. Metabolism of 2-hydroxybiphenyl in the rat.

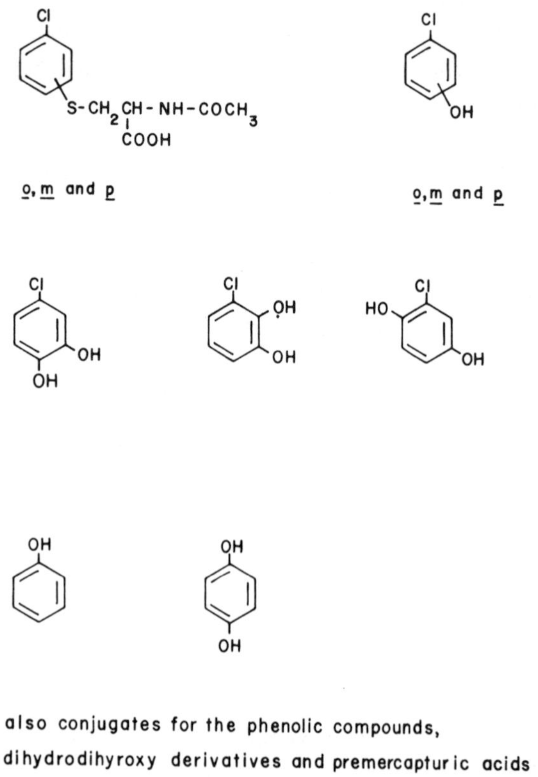


FIGURE 9. Compounds observed in the mammalian metabolism of chlorobenzene.

reported in the literature. Metabolites obtained by microbial action on the parent compound biphenyl will therefore be included in this discussion.

The microbial degradation of a number of commercial PCB preparations was studied using activated sludge under aerobic conditions in a semicontinuous operation.[43] Similarly to the results in animal metabolism studies, degradation of compounds with low chlorine contents was easiest (Table 9). A typical gas chromatogram showing loss or reduction of the early eluting peaks from Aroclor 1242 is presented in Figure 10. Analysis of silage containing Arclor 1254 indicated that no preferential anaerobic degradation of any of its components had occurred after several months storage during which normal fermentation had taken place.[18]

In a study on the persistence of Aroclor 1254 in six California soils under laboratory conditions,[32] it was shown that, depending on the soil type, no change, some change, or considerable change of the Aroclor 1254 peak pattern had occurred after 1 year of incubation at 30° and 40% soil moisture content (Figure 11). However, under the experimental conditions used, loss of some of the Aroclor components may have been due to evaporation. In parallel experiments, DDT was added to the same soils and, with the possible exception of loamy sand, PCB and DDT+DDE residues were found to be equally persistent.

In a preliminary communication,[34] the degradation of single chlorobiphenyls in soil and soil amended with cattle manure was studied after 1 month incubation under flooded conditions in the laboratory. The compounds were 4,4′-dichloro, 2,2′,5,5′-, 2,2′,6,6′-, 3,3′,4,4′-, and 3,3′,5,5′-tetrachloro, 2,2′,4,5,5′-pentachloro and 2,2′,4,4′, 5,5′- and 2,2′,4,4′,6,6′-hexachlorobiphenyl. No indication of any metabolism was found for any of the compounds. It is interesting to note that *p,p*′-DDT degrades almost completely under these conditions.

*Studies Involving Pure Cultures*

The products and intermediates obtained on incubating biphenyl with a number of microorganisms are shown in Figure 12. Pure cultures of

TABLE 8

**The Metabolism of Chlorinated Benzenes in Rabbits**

| Compound | Time of excretion days | % of dose excreted as: Mercapturic acid | Monophenols | Catechols | Total O-conjugates | Eliminated unchanged |
|---|---|---|---|---|---|---|
| Chlorobenzene | 1–2 | 25 | 2–3 | 27 | 47 | in 27, expired air |
| Dichlorobenzene | | | | | | |
| 1,2- | 5 | 5 | 40 | 4 | 69 | – |
| 1,3- | 5 | 11 | 25 | 3 | 37 | – |
| 1,4- | 5 | 0 | 35 | 0(6)[a] | 63 | – |
| Trichlorobenzene | | | | | | |
| 1,2,3- | 5 | 0.3 | 78 | tr.[b] | 62 | 0) |
| 1,2,4- | 5 | 0.4 | 42 | tr.[b] | 38 | 0) |
| 1,3,5- | 8 | 0 | 9 | 0 | 23 | 10) in |
| Tetrachlorobenzene | | | | | | |
| 1,2,3,4- | 6 | 0 | 43 | tr.[b] | 34 | 5) feces |
| 1,2,3,5- | 6 | 0 | 5 | 0 | 8 | 14) |
| 1,2,4,5- | 6 | 0 | 2 | 0 | 5 | 16) |
| Pentachlorobenzene | 6 | 0 | <1 | 0 | 9 | 5 |
| Hexachlorobenzene | 6 | 0 | 0 | 0 | 0 | 6 |

[a]This figure is for a hydroquinone
[b]tr = trace

TABLE 9

**Degradation of PCB by Activated Sludge***

| Substrate | % Chlorine | % Degradation in 47-hr cycle |
|---|---|---|
| Biphenyl | 0 | 100 |
| Aroclor 1221 | 21 | 80.6 ± 5.7 |
| Aroclor 1016 | 41 | 32.9 ± 13.8 |
| Aroclor 1242 | 42 | 26.3 ± 15.5 |
| Aroclor 1254 | 54 | 15.2 ± 37.7 |

*Semicontinuous operation; addition rate: 1 mg/48 hr.

gram-negative bacteria isolated from soil utilized biphenyl as the sole carbon source in a salt medium.[39] 2,3-Dihydroxybiphenyl which was isolated from the culture was cleaved by a particulate fraction from cells grown in biphenyl to give α-hydroxy-β-phenylmuconic semialdehyde. A soluble cell-free extract in turn converted this initial fission product to phenylpyruvate via unidentified intermediates.

Cultures of *Pseudomonas putida* were reported to oxidize biphenyl via 2,3-dihydro-2,3-dihydroxybiphenyl to benzoic acid.[9] The structure of the intermediate dihydro-dihydroxy compound was deduced by acid catalyzed dehydration which gave a mixture of 2- and 3-hydroxybiphenyl.

*Beijerinckia* species were also shown to oxidize biphenyl to a dihydro-dihydroxybiphenyl which, in this case, was identified to be *cis*-2,3-dihydro-2,3-dihydroxy-biphenyl (*cis*-2,3-dihydroxy-1-phenylcyclohexa-4,6-diene).[23] It is interesting to note that in this and other cases[23] of oxidation of aromatic compounds by microorganisms, a *cis*-diol is formed whereas mammals oxidize aromatic hydrocarbons to arene oxides which yield *trans* diols by enzymatic hydration.

*Cis*-2,3-dihydro-2,3-dihydroxybiphenyl is further converted to 2,3-dihydroxybiphenyl by a cell-free extract of the *Beijerinckia* species and cell extracts from wild type strain of *Beijerinckia* oxidized this compound to a yellow product

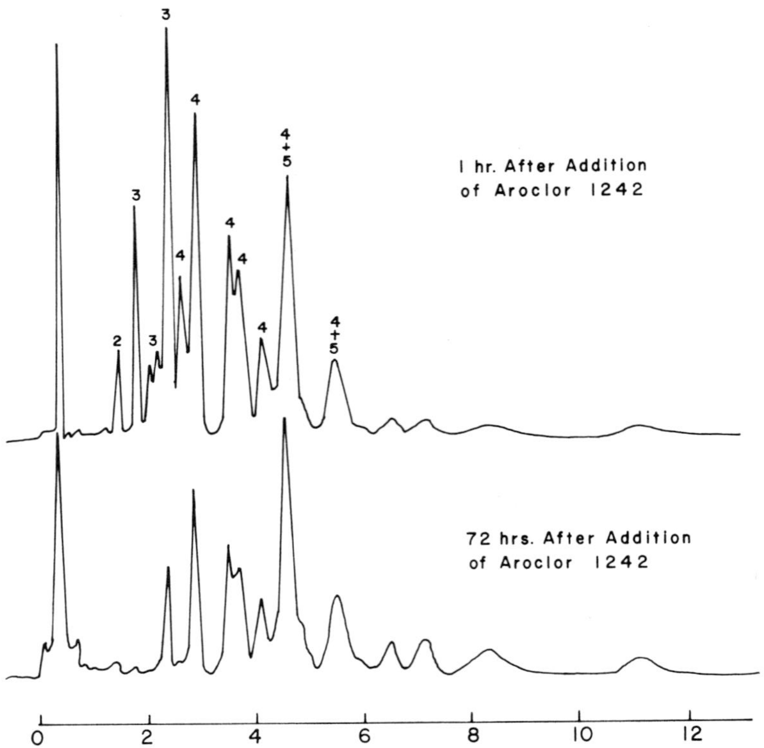


FIGURE 10. Gas chromatogram showing loss or reduction of the early eluting peaks from Aroclor 1242 after treatment in the activated sludge process. (Courtesy of Monsanto Co.)

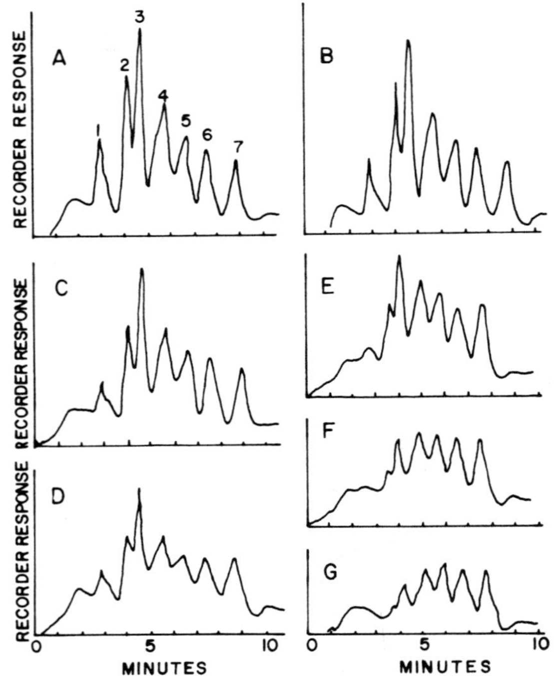


FIGURE 11. Microcoulometric gas chromatograms. (A) Represents 240 ng of Aroclor 1254. Chromatograms B through G each represents the injection of the extractives from 24 mg of soil fortified at 10 ppm and stored for 4 to 12 months at 30°C and a soil moisture of 40 to 45% of saturation. (B) Windy loam after 12 months. Note essential superimposability with A. (C) Linne clay after 12 months. Note relative decreased peak heights in earlier eluting peaks. (D) Mocho silt loam after 12 months. Overall recovery from this soil is about 20% less than from the others. Laveen loamy sand after (E) 4, (F) 8, and (G) 12 months. (From *Bull. Environ. Contam. Toxicol.*, 9, 204, 1973. With permission.)

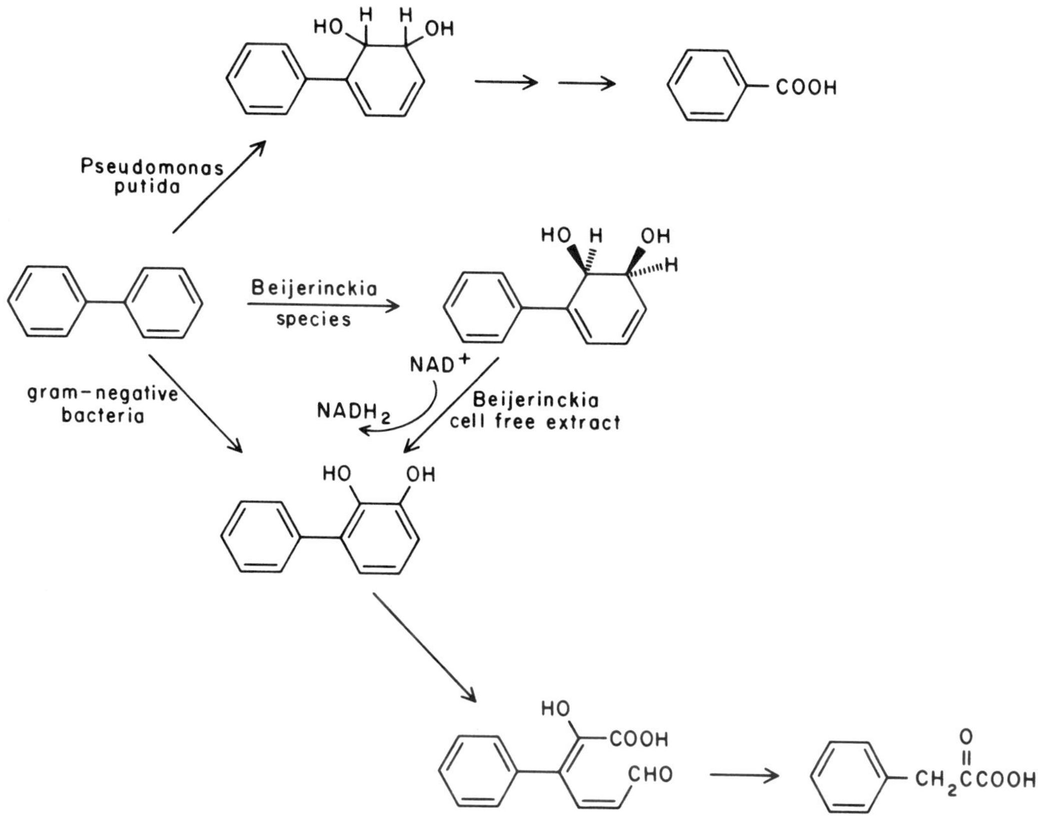


FIGURE 12. Metabolic conversion of biphenyl by several microorganisms.

whose spectral characteristics were identical to those reported for α-hydroxy-β-phenylmuconic semialdehyde.

A hydroxylated biphenyl, the fungicide 2-hydroxybiphenyl, was converted to 2,5-dihydroxy-biphenyl and 2,5-6-trihydroxybiphenyl by a pigmented strain of Mucor (Figure 13).[28]

To date, only two reports on the metabolism of individual chlorobiphenyls by pure cultures of microorganisms appeared in the literature.

The product formed by the action of *Rhizopus japonicus* on titrated 4-chlorobiphenyl in shaken culture was shown to be 4-chloro-4-hydroxy-biphenyl (Figure 14) by mass spectrometry (Chapter 8) proton magnetic resonance spectroscopy (Chapter 9) and comparison with synthetic material (Chapter 4).[50] 4,4′-Dichlorobiphenyl ($^3$H) gave a product whose chromatographic properties indicated the presence of a hydroxy group. Several isomeric dichlorobiphenyls similarly gave products with properties of hydroxybiphenyls when incubated with pure cultures of

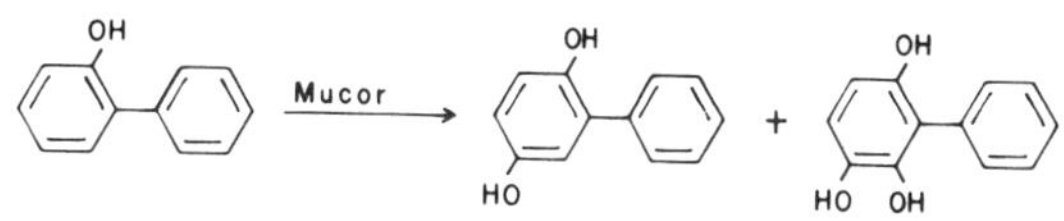


FIGURE 13. Metabolic conversion of 2-hydroxybiphenyl by a pigmented strain of Mucor.

organisms isolated from soil[51] but the structures of these products have not been elucidated yet.

Two species of *Achromobacter* were isolated from sewage effluent using biphenyl and 4-chlorobiphenyl as the sole carbon sources.[1] Suspensions of washed cells of both isolates were able to oxidize a number of aromatic compounds including 2-, 3-, and 4-chlorobiphenyl and 2,2′- and 4,4′-dichlorobiphenyl. Compounds isolated and identified were benzoic acid from biphenyl and 4-chlorobenzoic acid from 4-chlorobiphenyl (Figure 14). Spectroscopic evidence was produced in favor of intermediates produced by cleavage of one benzene ring. No chloride ion was produced by either isolate during the degradation of all chlorobiphenyls tested.

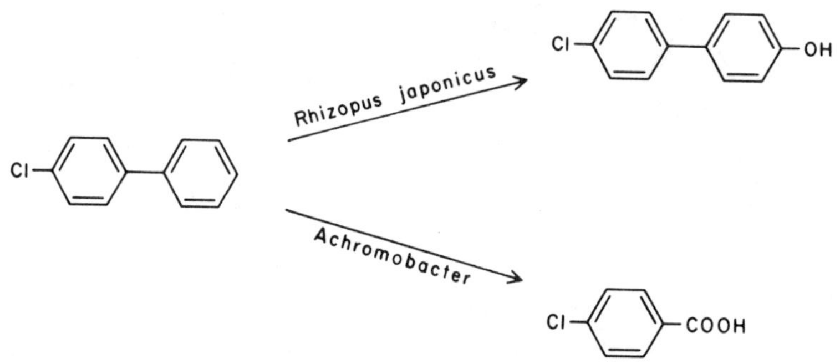


FIGURE 14. Metabolic conversion of 4-chlorobiphenyl by *Rhizopus japonicus* and *Achromobacter.*

## REFERENCES


1. **Ahmed, M. and Focht, D. D.,** *Can. J. Microbiol.,* 19, 47, 1973.
2. **Albro, P. W. and Fishbein, L.,** *Bull. Environ. Contam. Toxicol.,* 8, 26, 1972.
3. **Bagley, G. E. and Cromartie, E.,** *J. Chromatogr.,* 75, 219, 1973.
4. **Bailey, S. and Bunyan, P. J.,** *Nature,* 236, 34, 1972.
5. **Benthe, H. F. and Schmoldt, A.,** *Arch. Toxicol.,* 30, 207, 1973.
6. **Berlin, M., Gage, J. C., and Holm, S.,** PCB Conference II, Stockholm, 1972.
7. **Block, W. D. and Cornish, H. H.,** *J. Biol. Chem.,* 234, 3301, 1959.
8. **Burke, M. D. and Bridges, J. W.,** *Biochem. J.,* 130, 17 p, 1972.
9. **Catelani, D., Sorlini, C., and Treccani, V.,** *Experientia,* 27, 1174, 1971.
10. **Cornish, H. H. and Block, W. D.,** *J. Biol. Chem.,* 231, 583, 1958.
11. **Creaven, P. J., Parke, D. V., and Williams, R. T.,** *Biochem. J.,* 96, 879, 1965.
12. **Creaven, P. J. and Williams, R. T.,** *C. R. Symp. Int. Limites Tolerables Subst. Toxiques Ind.,* 2nd, Paris 1963, 273, 1965.
13. **Dagley, S.,** *Adv. Microb. Physiol.,* 6, 1, 1971.
14. **Daly, J. W., Jerina, D. M., and Witkop, B.,** *Experientia,* 28, 1129, 1972.
15. **DeFreitas, A. S. W., Norstrom, R. T., and Hutzinger, O.,** *PCB in the Environment,* Marcel Dekker, in press.
16. **Ecobichon, D. J., Johnstone, G. J., and Hutzinger, O.,** *Tox. Appl. Pharm.,* in press.
17. **Ernst, W.,** *Arzneimitt. Forsch.,* 15, 632, 1965.
18. **Fries, G. F.,** *Adv. Chem. Ser.,* 111, 256, 1972.
19. **Fries, J. F. Marrow, G. S., and Gordon, C. H.,** *J. Agric. Food Chem.,* 21, 117, 1973.
20. **Gardner, A. M.,** private communication.
21. **Gibson, D. T.,** *Science,* 161, 1093, 1968.
22. **Gibson, D. T.,** Initial reactions in the degradation of aromatic hydrocarbons, in *Degradation of Synthetic Molecules in the Biosphere,* National Academy of Sciences, Washington, D.C., 1972.
23. **Gibson, D. T., Roberts, R. L., Wells, M. C., and Kobal, V. M.,** *Biochem. Biophys. Res. Commun.,* 50, 211, 1973.
24. **Gibson, D. T., Wood, T. M., Chapman, P. J., and Dagley, S.,** *Biotechnol. Bioeng.,* 9, 33, 1967.
25. **Gillham, B. and Young, L.,** *Biochem. J.,* 109, 143, 1968.
26. **Grant, D. L., Phillips, W. E. T., and Villeneuve, D. C.,** *Bull. Environ. Contam. Toxicol.,* 6, 102, 1971.
27. **Hathway, D. E.,** Foreign compound metabolism, *Specialist Periodical Reports,* Vol. 1–2, The Chemical Society, London, 1970, 1972.

28. **Herber, R., Villoutreix, J., and Pierfitte, M.,** *Compt. Rend. Soc. Biol.,* 163, 1657, 1969.
29. **Hutzinger, O., Jamieson, W. D., Safe, S., Paulmann, L., and Ammon, R.,** in preparation.
30. **Hutzinger, O., Nash, D. M., Safe, S., DeFreitas, A. S. W., Norstrom, R. J., Wildish, D. J., and Zitko, V.,** *Science,* 178, 312, 1972.
31. **Hutzinger, O., Safe, S., Ecobichon, D. J., DeFreitas, A. S. W., Norstrom, R. J., MacNeil, J. D., Wildish, D. J., and Zitko, V.,** *PCB in the Environment,* Marcel Dekker, in press.
32. **Iwata, Y., Westlake, W. E., and Gunther, F. A.,** *Bull. Environ. Contam. Toxicol.,* 9, 204, 1973.
33. **Jerina, D. M.,** *Chem. Technol.,* 127, February, 1973.
34. **Johnsen, R. E.,** *PCB Newsletter,* January, 1973.
35. **Jordan, T. W. and Smith, J. N.,** *Int. J. Biochem.,* 1, 139, 1970.
36. **Klingenberg, K.,** *Jahresber. Fortschr. Tierchem.,* 21, 57, 1891.
37. **Koeman, J. H., ten Noever de Brauw, M. C., and DeVos, R. H.,** *Nature,* 221, 1126, 1969.
38. **Lindsay Smith, J. R., Shaw, B. A. J., and Foulkes, D. M.,** *Xenobiotica,* 2, 215, 1972.
39. **Lunt, D. and Evans, W. C.,** *Biochem. J.,* 118, 54P, 1970.
40. **Mehendale, H. M. and Matthews, H. B.,** 165th meeting of the American Chemical Society, Dallas, April 1973, Pesticide Chemistry Paper No. 10.
41. **Melvås, B. and Brandt, I.,** PCB Conference II, Stockholm, 1972.
42. **Millburn, P., Smith, R. L., and Williams, R. T.,** *Biochem. J.,* 105, 1275, 1967.
43. Monsanto Co., Presentation to the Interdepartmental Task Force on PCB, Washington, May 15, 1972.
44. **Parke, D. V.,** *The Biochemistry of Foreign Compounds,* Pergamon Press, Oxford, 1968.
45. **Parke, D. V. and Williams, R. T.,** *Biochem. J.,* 74, 5, 1960.
46. **Raig, P. and Ammon, R.,** *Arzneim. Forsch.,* 22, 1399, 1972.
47. **Ribbons, D. W.,** *Ann. Reports,* 62, 445, 1965.
48. **Rogoff, M. H.,** *Adv. Appl. Microbiol.,* 3, 193, 1961.
48a. **Safe, S. and Hutzinger, O.,** *Experientia,* in press.
49. **Treccani, V.,** *Progr. Ind. Microbiol.,* 4, 1, 1963.
50. **Wallnöfer, P. R., Engelhardt, G., Safe, S., and Hutzinger, O.,** *Chemosphere,* 69, 1973.
51. **Wallnöfer, P. R., Engelhardt, G., Safe, S., and Hutzinger, O.,** unpublished results.
52. **Wallnöfer, P. R., Königer, M., and Hutzinger, O.,** *Research Notes,* Analabs, in press.
53. **West, H. D., Lawson, J. R., Miller, I. H., and Mathura, G. R.,** *Arch. Biochem. Biophys.,* 60, 14, 1956.
54. **Williams, R. T.,** *Detoxication Mechanisms,* Chapman & Hall, London, 1959.
55. **Williams, R. T.,** *Ann. Biol. Clin.,* 23, 7, 1965.
56. **Yip, G. and Link, J. D.,** *PCB Newsletter,* March 30, 1972.
57. **Yoshimura, H. and Oshima, M.,** *Fukuoka Igaku, Zasshi,* 62, 5, 1971; National Research Council of Canada Technical Translation, Ottawa, 1972.
58. **Zitko, V. and Hutzinger, O.,** *PCB in the Environment,* Marcel Dekker, in press.

Chapter 8

# MASS SPECTROMETRY OF CHLOROBIPHENYLS

Mass spectrometry has been extensively used in the detection and confirmation of PCB's in the environment,[27] and in addition the electron impact-induced fragmentation reactions of pure isomeric chlorobiphenyls have also been examined. The mass spectra of chlorinated biphenyls exhibit two important features:

1. Most of the compounds give relatively intense molecular and M-70$^{7\cdot+}$ ionic species and in addition their doubly-charged ions are also observed. Since these species contain a large percentage of the ion current this greatly facilitates identification of PCB residues at low concentration levels.

2. The highly characteristic isotopic distribution of chlorine atoms (i.e., 75.8% $^{35}Cl$ and 24.2% $^{37}Cl$) permits easy recognition of compounds which contain one or more of these atoms. Thus the mass spectrum of 4-chlorobiphenyl exhibits molecular ion species at m/e 188 ($C_{12}H_9{}^{35}Cl$) and m/e 190 ($C_{12}H_9{}^{37}Cl$) with the ratio [m/e 188]/[m/e 190] $\cong$ 3:1 as predicted from the natural isotopic distribution of the two chlorine isotopes. Similarly all the fragment ions which contain a Cl moiety give an analagous isotope distribution pattern. Figure 1 shows the calculated abundance ratios for molecules containing up to 15 chlorine atoms/molecule.[13] The above abundance ratios are highly characteristic and facilitate the identification of PCB's by mass spectrometry.

**Mass Spectra of Individual Chlorobiphenyls**

In order to understand the relatively complex mass spectra of commercial Aroclor mixtures, the mass spectra of the isomeric chlorobiphenyls and several di-, tri-, tetra-, hexa-, and octachlorobiphenyl isomers have been studied in detail.[25,26] In addition, spectra of penta-, hepta-, nona-, and decachlorobiphenyl have also been reported[22] and representative spectra of the chlorobiphenyls are shown (Figures 2 – 6). The primary ion mass spectra of *o*-, *m*-, and *p*-chlorobiphenyl were all similar (Table 1) and the results suggested that their decomposing molecular ions were of similar internal energies with resultant substituent (i.e., Cl) scrambling in the substituted ring. The frag-

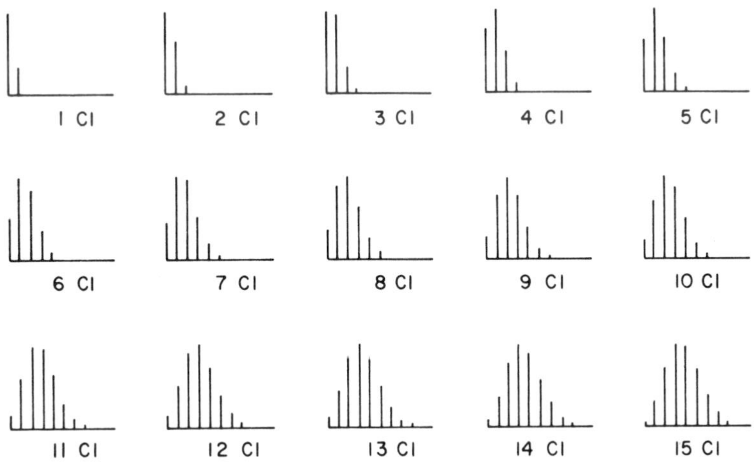


FIGURE 1. Calculated Cl abundance ratios for compound containing 1 to 15 Cl atoms.

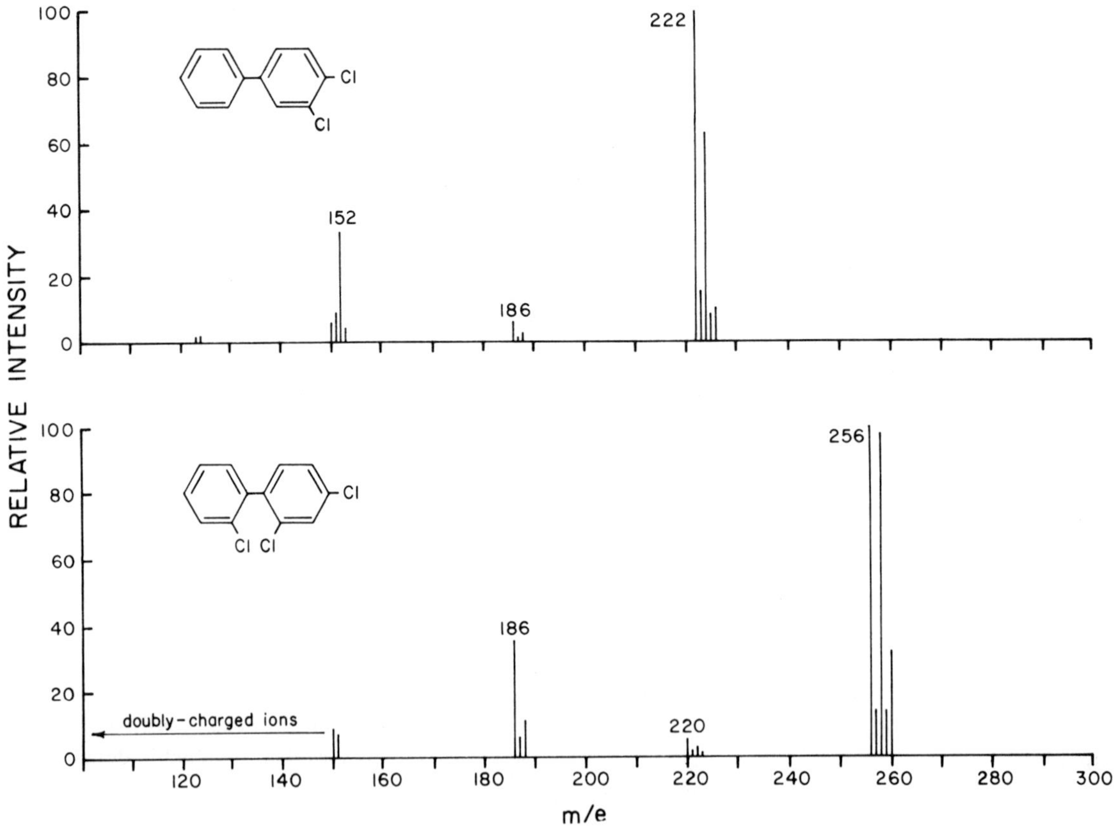


FIGURE 2. Top, mass spectrum 3,4-dichlorobiphenyl; bottom, mass spectrum 2,2′,4-trichlorobiphenyl.

TABLE 1

**Mass Spectral Data for the Mono-, Di-, and Tetrachlorobiphenyl Isomers**

| Compound | Ion Abundances | | | |
|---|---|---|---|---|
| | $M^{+\cdot}$ | M-35 | M-36 | M-70 |
| 2-chloro | 100 | 12 | 27 | – |
| 3-chloro | 100 | 10 | 27 | – |
| 4-chloro | 100 | 12 | 25 | – |
| 4,4′-dichloro | 100 | 1.4 | 6.0 | 38 |
| 3,3′-dichloro | 100 | 1.4 | 6.1 | 38 |
| 3,4-dichloro | 100 | 1.5 | 6.7 | 35 |
| 2,4-dichloro | 100 | 1.5 | 7.2 | 39 |
| 2,2′-dichloro | 100 | 2.8 | 7.5 | 79 |
| 2,6-dichloro | 100 | 2.4 | 6.3 | 35 |
| 3,3′,5,5′-tetrachloro | 100 | 1.0 | 5.0 | 33 |
| 2,3,4,5-tetrachloro | 100 | 1.0 | 4.0 | 38 |
| 2,3,5,6-tetrachloro | 100 | 1.5 | 3.0 | 44 |
| 2,2′,4,4′-tetrachloro | 100 | 4.0 | 2.5 | 56 |
| 3,3′,4,4′-tetrachloro | 100 | 0.5 | 1.0 | 30 |
| 2,2′,5,5′-tetrachloro | 100 | 13 | 5.0 | 71 |
| 2,2′,6,6′-tetrachloro | 100 | 2.0 | 2.0 | 75 |
| 2,3′,4,4′-tetrachloro | 100 | 0.5 | 2.0 | 35 |

mentation pattern of the PCB isomers featured successive expulsion of Cl atoms with the odd-electron species (i.e., $M^{+\cdot}$, $M\text{-}Cl_2{}^{7+\cdot}$, $M\text{-}Cl_4{}^{7+\cdot}$ etc.) being more abundant than the even-electron ions (i.e., $M\text{-}Cl^{7+}$, $M\text{-}Cl_3{}^{7+}$, $M\text{-}Cl_5{}^{7+}$ etc). In addition, the elimination of $Cl_2$ from the odd-electron ions was also observed and this elimination from the molecular ions is accompanied by the appropriate metastable ions. A general fragmentation scheme is shown (Equation 1) as well as fragmentation schemes for the di-, tetra-, and hexachlorobiphenyl isomers (Equations 2 – 4) and it should also be noted that for the less chlorinated isomers the expulsion of HCl from many of the ions was also noted. The mass spectral and metastable ion data for a series of isomeric mono-, di-, tetra-, and hexachlorobiphenyls are given (Table 1) and the data showed some marked differences in the spectra of the various isomers. In addition, the ion abundance data obtained for other chlorinated biphenyls are shown (Table 2). The mass spectral and metastable ion data (Table 3) suggested that the decomposing molecular ions for the 4,4′-,

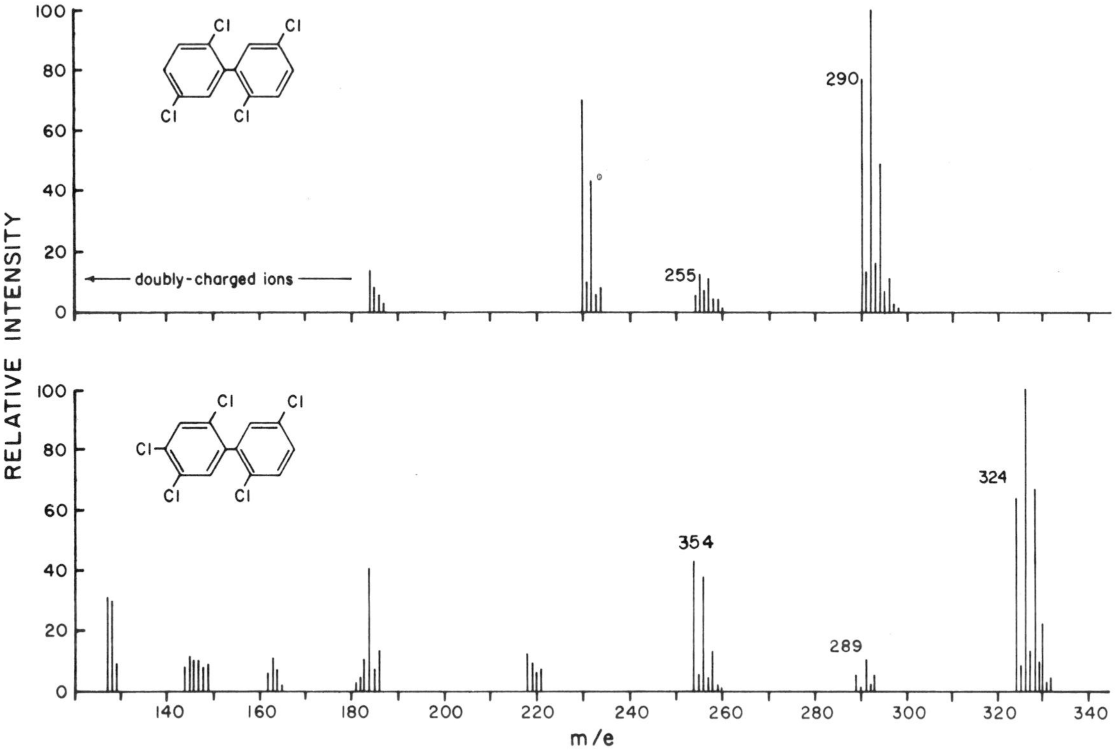


FIGURE 3. Top, mass spectrum of 2,2′,5,5′-tetrachlorobiphenyl; bottom, mass spectrum of 2,2′,4,5,5′-pentachlorobiphenyl.

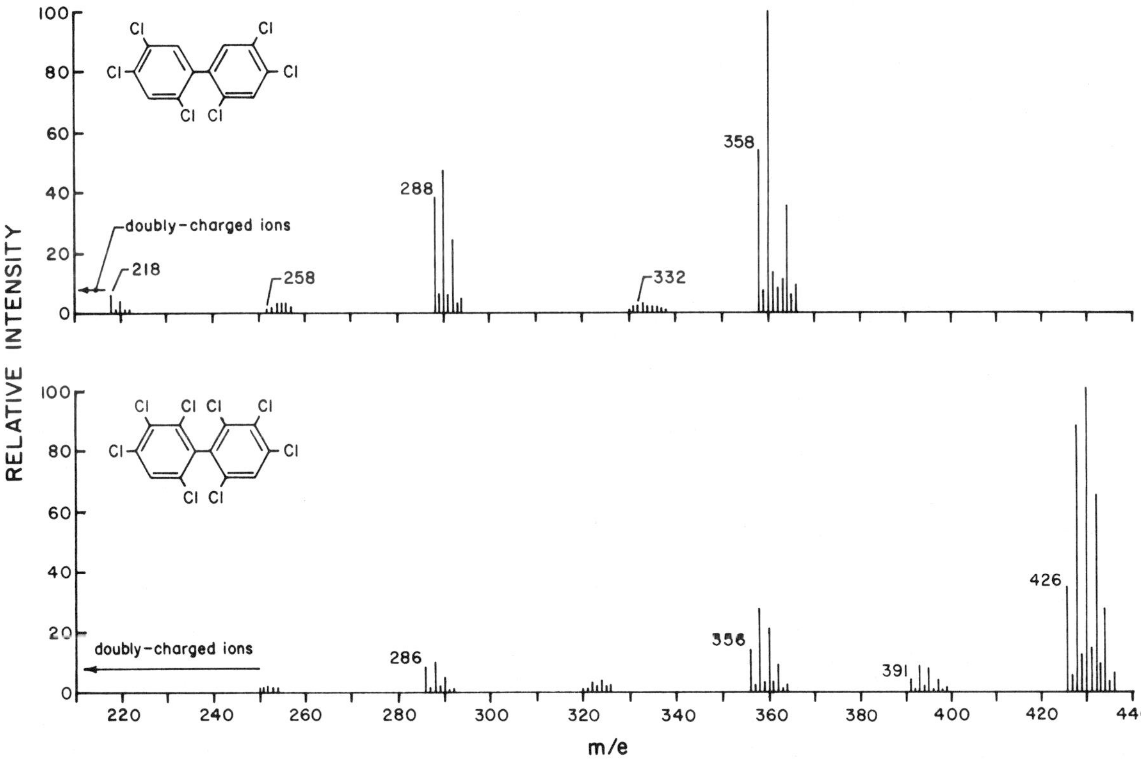


FIGURE 4. Top, mass spectrum of 2,2′,4,4′,5,5′-hexachlorobiphenyl; bottom, mass spectrum of 2,2′,4,4′,5,5′,6,6′-octachlorobiphenyl.

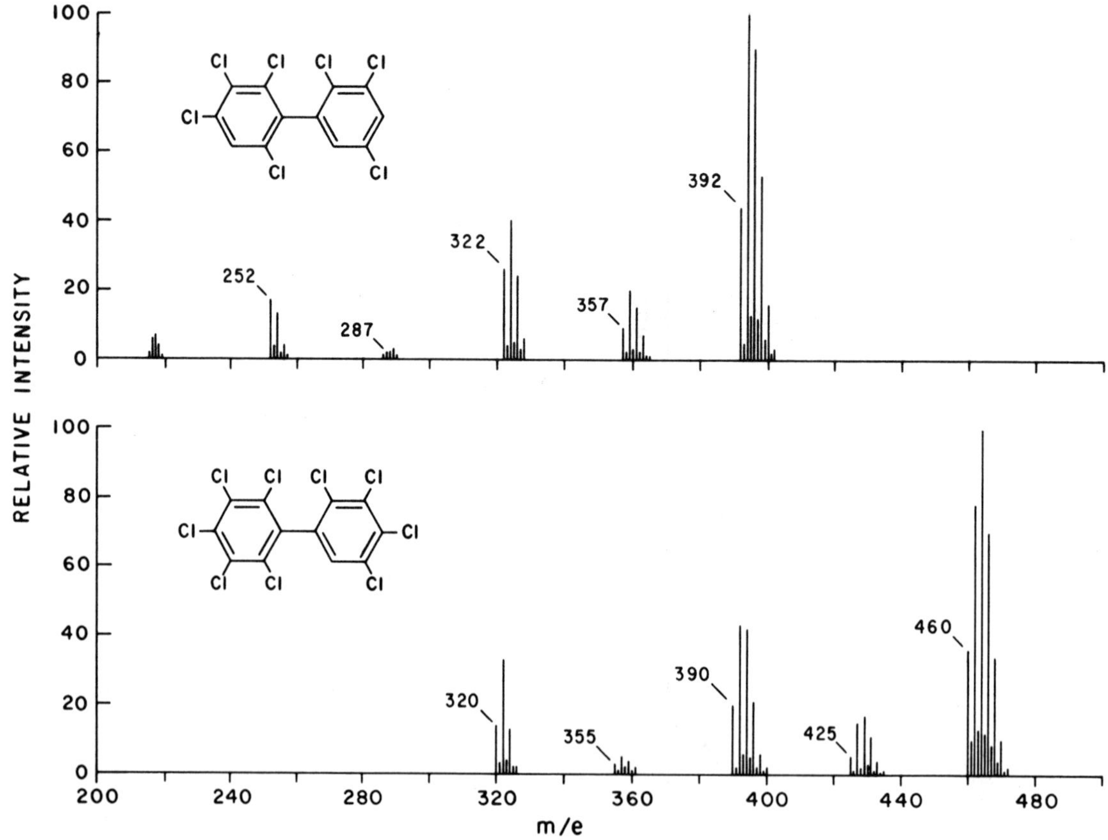


FIGURE 5. Top, mass spectrum of 2,2′,3,3′,4,6,6′-heptachlorobiphenyl; bottom, mass spectrum of 2,2′,3,3′,4,4′,5,5′,6-nonachlorobiphenyl

3,3′-, 3,4, and 2,4-dichlorobiphenyls were energetically similar and markedly different from the 2,2′ and 2,6-dichlorobiphenyl isomers. The 3,3′,5,5′-, 2,3,4,5-, 3,3′,4,4′-, and 2,3′4,4′-tetrachlorobiphenyl isomers also gave similar mass spectra and the metastable ion data suggested that their respective decomposing molecular ions were energetically similar to apparent chlorine scrambling over the entire biphenyl nucleus. The (metastable ion)/(daughter ion) ratios for the 2,2′,4,4′-, 2,3,5,6-, 2,2′,5,5′-, and 2,2′,6,6′-tetrachlorobiphenyls were all different from the above and clearly have retained some substituent identity in their decomposing molecular ions. All the di- and tetrachlorobiphenyl isomers which have retained their substituent identity contain two or more chloro substituents ortho the biphenyl ring junction. Other spectroscopic studies have shown that there is considerable steric hindrance to rotation with *ortho* substituted biphenyl moieties and in addition this steric effect also prevents coplanarity between the two phenyl rings. Thus, it was evident that this ground state effect is still retained by the highly energetic ions produced upon electron impact.

The ion kinetic energy (IKE) spectrum of the di-, tetra-, and hexachlorobiphenyl isomers (II, IV, and V) have also been reported (Table 4).[24] IKE spectroscopy is a technique which records the ionic decompositions which occur in the first field-free region of the mass spectrometer prior to the electric sector.[3,7] The main ion beam which passes through the electric sector is transmitted at specific electric sector voltage (i.e., 100% E). However, ions which have undergone unimolecular decompositions in the first field-free region (i.e., $m_1^+ \rightarrow m_2^+ + m_3$) are less energetic than the unreacted species and are not transmitted through the electric sector at 100% E. However, if the electric sector voltage is scanned from 100% to 0 E, these less energetic daughter ions (i.e., $m_2^+$) are transmitted at a specific fractional energy, $m_2/m_1$ x 100% E. These ionic species thus pass through the electric sector exit slit (i.e., B slit) and can be

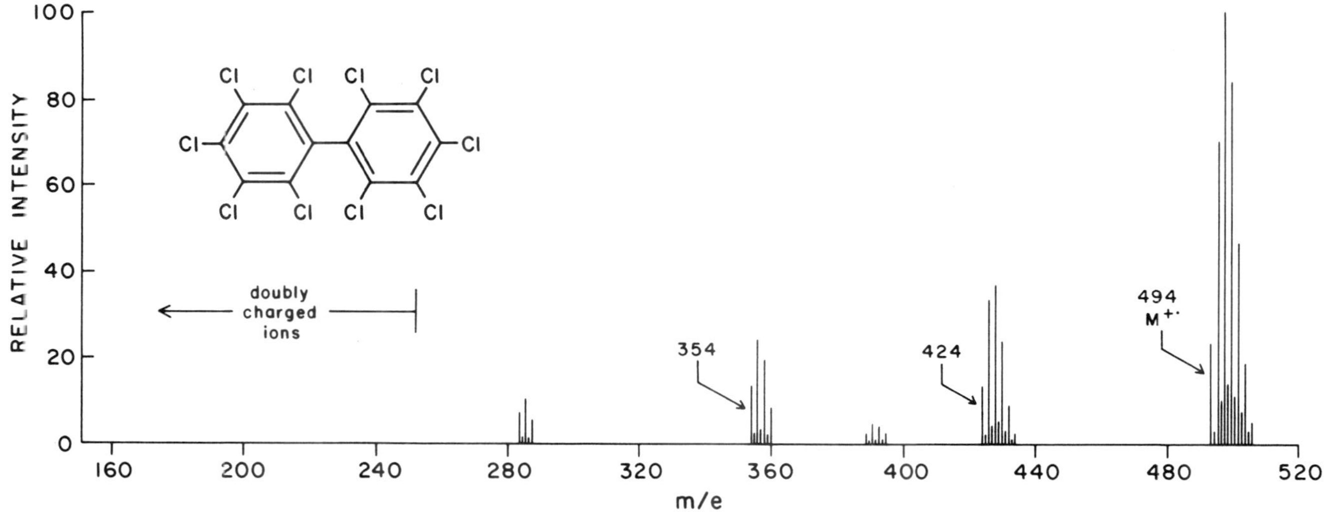


FIGURE 6. Mass spectrum of decachlorobiphenyl.

$$\left[C_{12}H_xCl_y\right]^{+\cdot} \xrightarrow{-Cl} \left[C_{12}H_xCl_{y-1}\right]^{+} \xrightarrow{-Cl} \left[C_{12}H_xCl_{y-2}\right]^{+\cdot} \quad \left(C_{12}H_xCl_y \xrightarrow{-Cl_2} C_{12}H_xCl_{y-2}\right)$$

$$\xrightarrow{-Cl} \left[C_{12}H_xCl_{y-3}\right]^{+} \xrightarrow{-Cl} \left[C_{12}H_xCl_{y4}\right]^{+\cdot}$$

Equation I

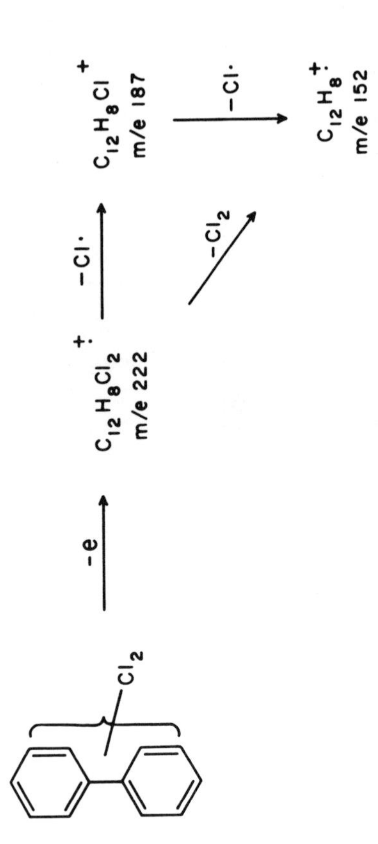


Equation 2

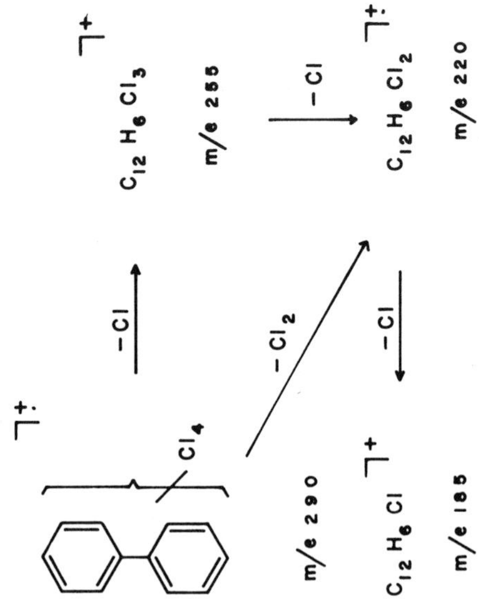


Equation 3

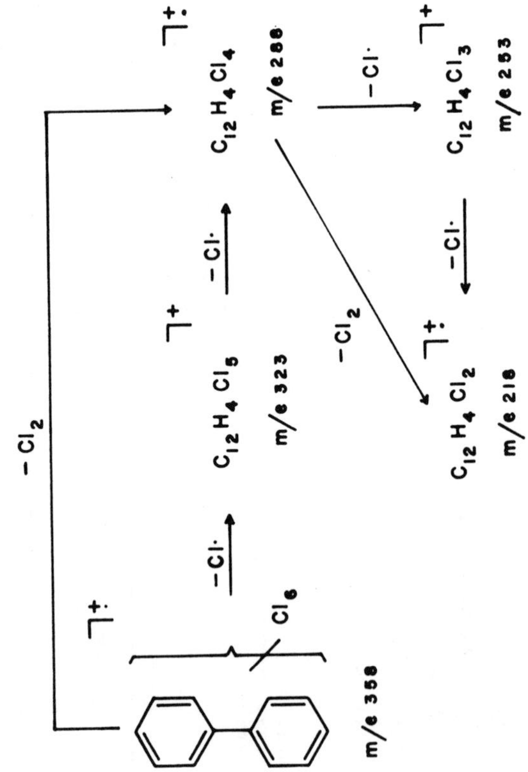


Equation 4

TABLE 2

**Mass Spectral Data of Tri-, Hexa-, Octa-, and Deca-chlorobiphenyls**

| Mass | Relative ion intensities | | | | | | | | | | |
|---|---|---|---|---|---|---|---|---|---|---|---|
| | IIIa | IIIb | Va | Vb | Vc | Vd | Ve | VIa | VIb | VIc | VII |
| $M^{+\cdot}$ | 100 | 100 | 100 | 100 | 100 | 100 | 100 | 100 | 100 | 100 | 100 |
| $M^{+\cdot}$-35 | 1 | 1 | 2 | 1 | 2 | 1 | 5 | 3 | 6 | 8 | 1 |
| $M^{+\cdot}$-36 | 5 | 4 | 3 | 1 | 1 | 15 | 2 | 11 | 11 | 6 | ~0 |
| $M^{+\cdot}$-70 | 36 | 35 | 52 | 56 | 54 | 31 | 36 | 38 | 53 | 45 | 65 |
| $M^{+\cdot}$-71 | 2 | 2 | ~0 | ~0 | ~0 | ~0 | ~0 | 0 | ~0 | ~0 | ~0 |
| $M^{+\cdot}$-105 | 6 | 7 | 5 | 7 | 3 | 3 | 3 | 3 | 4 | 4 | 8 |
| $M^{+\cdot}$-106 | 10 | 9 | 7 | 5 | 7 | 6 | 5 | 2 | 4 | 3 | ~0 |
| $M^{+\cdot}$-140 | | | 22 | 19 | 22 | 15 | 17 | 24 | 33 | 29 | 5° |
| $M^{+\cdot}$-141 | | | 2 | 2 | 3 | 3 | 2 | 1 | 2 | ~1 | ~0 |
| $M^{+\cdot}$-175 | | | * | * | * | * | * | 3 | 4 | 3 | 9 |
| $M^{+\cdot}$-176 | | | | | | | | 3 | 5 | 2 | ~0 |
| $M^{+\cdot}$-210 | | | | | | | | * | * | * | 32 |
| $M^{+\cdot}$-211 | | | | | | | | | | | ~0 |
| $M^{+\cdot}$-245 | | | | | | | | | | | |
| $M^{+\cdot}$-246 | | | | | | | | | | | |

*Doubly charged ions make measurement difficult below this mass.

TABLE 3

**(Metastable Ion)/(Daughter Ion) Ratios for the Di- and Tetrachlorobiphenyl Isomers**

| Compound | Ratios | | | |
|---|---|---|---|---|
| | $\frac{[m_1^*]}{[m/e\ 187]} \times 10^3$ | | $\frac{[m_2^*]}{[m/e\ 152]} \times 10^5$ | |
| | 70 eV | 20 eV | 70 eV | 20 eV |
| (a) 4,4′-dichloro | 15.5 | 13.5 | 2.1 | 1.7 |
| 3,3′-dichloro | 15.5 | 12.0 | 2.2 | 1.5 |
| 3,4-dichloro | 16.0 | 13.0 | 2.3 | 1.5 |
| 2,4-dichloro | 17.0 | 14.0 | 2.3 | 1.7 |
| 2,2′-dichloro | 5.1 | 4.0 | 180 | 110 |
| 2,6-dichloro | 100 | 27.0 | 100 | 100 |
| | $\frac{[m_5^*]}{[m/e\ 185]} \times 10^3$ | | $\frac{[m_4^*]}{[m/e\ 290]} \times 10^3$ | |
| | 70 eV | 20 eV | 70 eV | 20 eV |
| (b) 3,3′,5,5′-tetrachloro | 6.7 | 6.6 | 0.75 | 0.75 |
| 2,3,4,5-tetrachloro | 6.7 | 6.9 | 0.8 | 0.8 |
| 2,3,5,6-tetrachloro | 6.8 | 6.8 | 0.2 | 0.2 |
| 2,2′,4,4′-tetrachloro | 6.6 | 6.6 | <0.01 | <0.01 |
| 3,3′,4,4′-tetrachloro | 6.6 | 6.6 | 0.8 | 0.8 |
| 2,2,5,5-tetrachloro | 6.5 | 6.9 | <0.01 | <0.01 |
| 2,2′,6,6′-tetrachloro | 6.5 | 6.7 | <0.01 | <0.01 |
| 2,3,4,4′-tetrachloro | 6.4 | 6.6 | 0.8 | 0.8 |

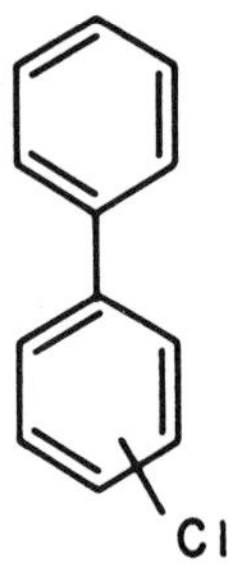

I

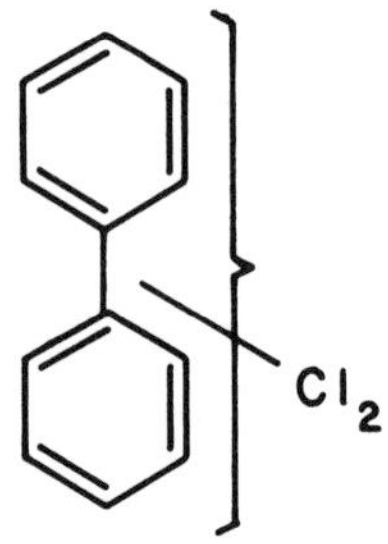

II

a; 4,4′
b; 3,3′
c; 3,4
d; 2,4
e; 2,2′
f; 2,6

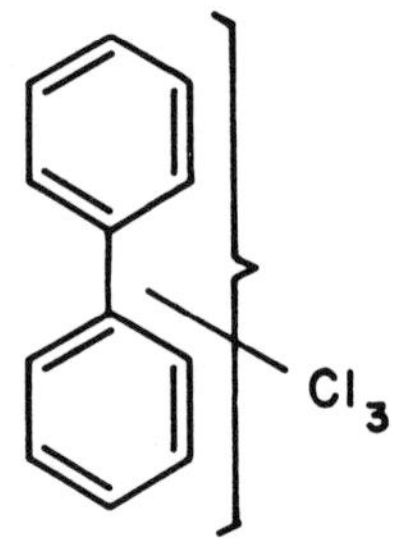

III

a; 2,4,6
b; 2,4,4′

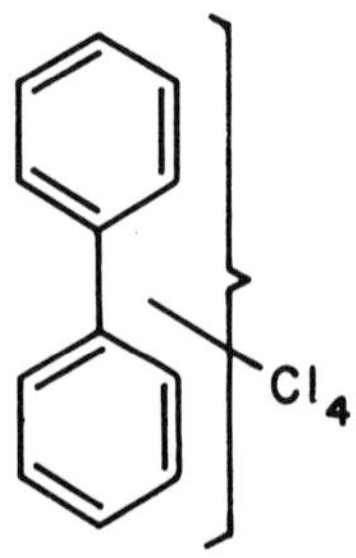

IV

a; 3,3′,5,5′
b; 2,3,4,5
c; 2,3,5,6
d; 2,2′,4,4′
e; 3,3′,4,4′
f; 2,2′,5,5′
g; 2,2′,6,6′
h; 2,3′,4,4′

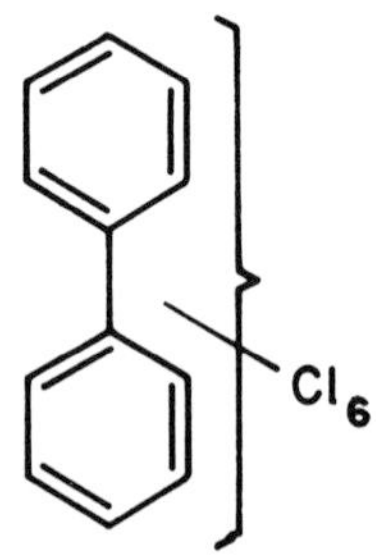

V

a; 2,2′,4,4′,5,5′
b; 2,2′,4,4′,6,6′
c; 2,2′,3,3′,5,5′
d; 3,3′,4,4′,5,5′
e; 2,2′,3,4,4′,5

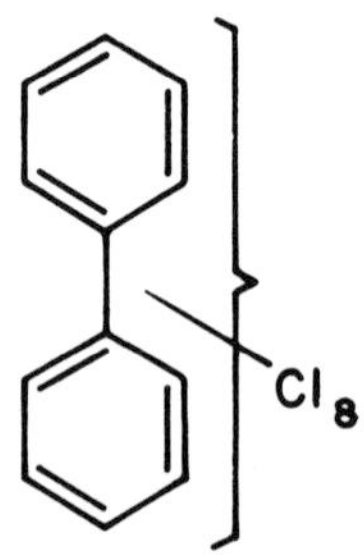

VI

a; 2,2′,3,3′,4,4′,6,6′
b; 2,2′,3,3′,4,4′,5,5′
c; 2,2′,3,3′,5,5′,6,6′

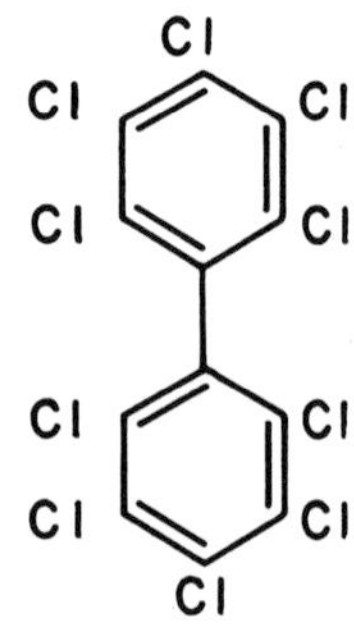

VII

duly detected and recorded. The position of the IKE peaks can then be readily correlated with possible fragmentation pathways. The results which have been reported for the PCB isomers (Table 4, Figures 7, 8, and 9) confirm the fragmentation pathways previously deduced from their corresponding primary ion mass spectra and metastable ion data. The data clearly indicated that the relative peak abundances were different for each isomer thus indicating possible analytical applications for this technique in the structure analysis of PCB isomers. Since the molecular ions of both the di- and tetrachlorobiphenyl isomers fragment with loss of both Cl· and $Cl_2$ the ratios of these two ionic decomposition peaks (i.e., [0.842 E]/[0.685 E], for the dichloro isomers and [0.879 E]/[0.761 E] for the tetrachloro isomers) are a measure of the relative reactivity of their corresponding decomposing molecular ions. The ratios obtained were different for all the isomers thus indicating energetically dissimilar decomposing molecular ions and incomplete substituent randomization. These results thus explain why the spectra of all the isomers are different and show a contrast in the results obtained from the first and second field-free regions of the mass spectrometer.

TABLE 4

**Ion Kinetic Energy Data for the Di-, Tetra-, and Hexachlorobiphenyl Isomers**

| Compound | Relative Peak Areas x $10^3$* | | | | | |
|---|---|---|---|---|---|---|
| | 0.842 E | 0.815 E | 0.685 E | [0.842 E]/[0.685 E] | | |
| (a) 4,4′-dichloro | 10.8 | 10.7 | 12.9 | 0.840 | | |
| 3,3′-dichloro | 11.4 | 13.0 | 11.2 | 0.985 | | |
| 3,4-dichloro | 8.2 | 9.1 | 8.7 | 0.945 | | |
| 2,4-dichloro | 14.4 | 15.0 | 12.0 | 1.16 | | |
| 2,2′-dichloro | 46.0 | 29.0 | 0.9 | 51.0 | | |
| 2,6-dichloro | 22.8 | 22.4 | 8.4 | 2.72 | | |
| | 0.879 E | 0.862 E | 0.840 E | 0.810 E | 0.761 E | [0.879 E]/[0.761 E] |
| (b) 3,3′,5,5′-tetrachloro | 7.9 | 9.6 | 11.9 | 6.4 | 12.3 | 0.64 |
| 2,3,4,5-tetrachloro | 11.1 | 13.9 | 11.7 | 6.4 | 15.0 | 0.74 |
| 2,3,5,6-tetrachloro | 16.4 | 17.4 | 11.9 | 6.4 | 9.3 | 1.77 |
| 2,2′,4,4′-tetrachloro | 31.4 | 23.0 | 12.3 | 7.0 | 4.6 | 6.83 |
| 3,3′,4,4′-tetrachloro | 5.2 | 6.8 | 11.6 | 6.4 | 13.0 | 0.40 |
| 2,2′,5,5′-tetrachloro | 39.0 | 25.2 | 12.0 | 7.0 | 2.2 | 17.7 |
| 2,2′,6,6′-tetrachloro | 23.0 | 20.4 | 11.6 | 6.6 | 3.3 | 6.96 |
| 2,3,4,4′-tetrachloro | 9.4 | 10.7 | 11.8 | 6.7 | 12.2 | 0.77 |
| | 0.895 E | 0.877 E | 0.860 E | 0.838 E | 0.803 E | |
| (c) 2,2′,4,4′,5,5′-hexachloro | 28.4 | ** | 9.8 | 3.7 | 10.6 | |
| 2,2′,4,4′,6,6′-hexachloro | 21.2 | ** | 10.0 | 4.6 | 12.0 | |
| 2,2′,3,3′,5,5′-hexachloro | 41.5 | ** | 9.0 | 3.0 | 4.0 | |
| 3,3′,4,4′,5,5′-hexachloro | 9.6 | 12.0 | 8.8 | 2.6 | 15.8 | |
| 2,2,3,4,4′,5-hexachloro | 41.5 | ** | 9.6 | 3.7 | 5.5 | |

*Total in beam at 1.0 E taken as bask peak
**Shoulder

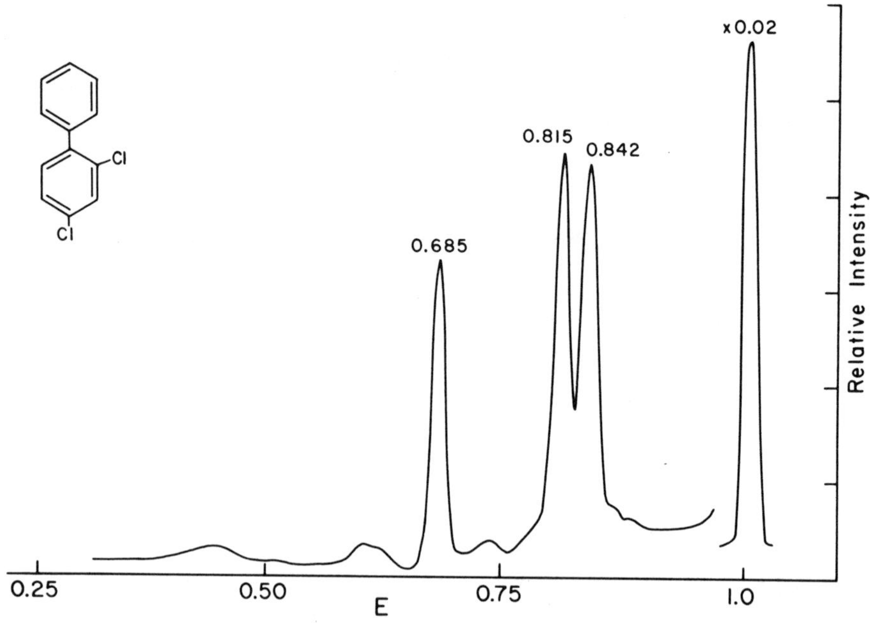


FIGURE 7. Ion kinetic energy spectrum of 2,4-dichlorobiphenyl.

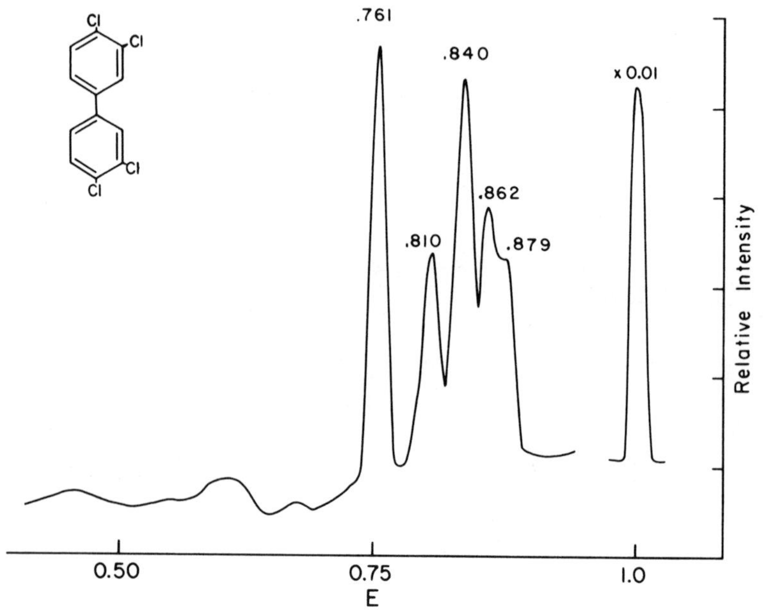


FIGURE 8. Ion kinetic energy spectrum of 3,3′,4,4′-tetrachlorobiphenyl.

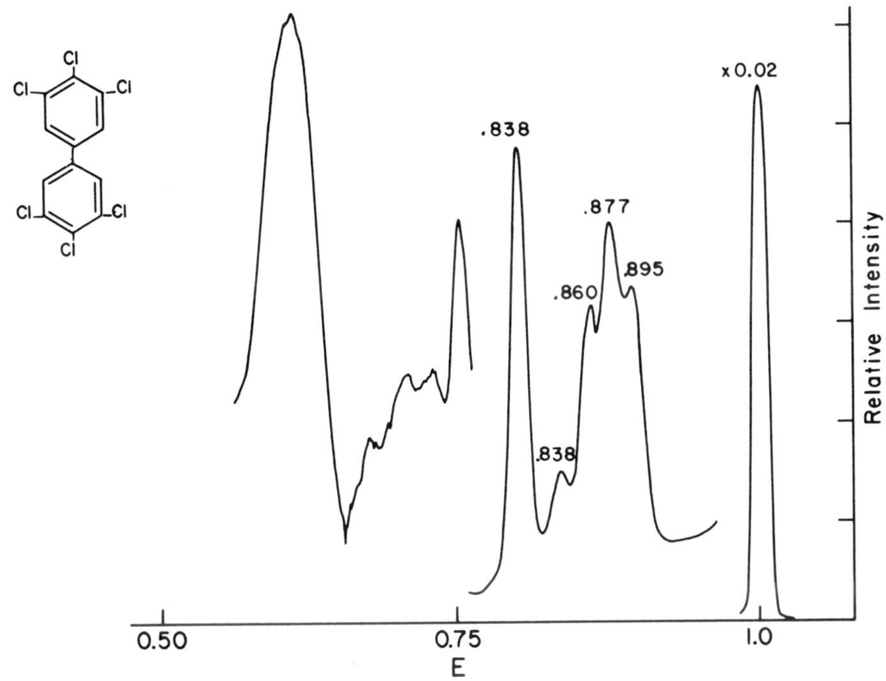

FIGURE 9. Ion kinetic energy spectrum of 3,3',4,4',5,5'-hexachlorobiphenyl.

**Mass Spectra of Commercial Aroclor Samples**

Commercial Aroclor preparations are sold by several companies and are graded according to their % chlorine (see Table 2 Chapter 2). The analysis of several commercial Aroclors has been reported[2,21,29,30,35] and the various GC fractions have been analyzed using mass spectrometry to determine the molecular weight and elemental composition of the various fractions. Since analysis of environmental samples by mass spectrometry is useful in confirming the presence of PCB's it was of interest to examine the mass spectra of the commercial Aroclor samples (see Figure 10). The spectra of the PCB mixtures are distinctive and reflect the chlorine content of these samples with the higher molecular weight species giving much more abundant ions in the chlorine-rich preparation (i.e., Aroclor 1254) than in the lower chlorinated Aroclor mixture. These mass spectra reflected the data obtained for the pure PCB isomers since the spectra clearly exhibited strong molecular ions and their corresponding odd-electron (i.e., $M\text{-}Cl_2{}^{7+\cdot}$) species were also relatively abundant.

The detection of PCB's in the environment[17] has been confirmed by mass spectrometry in a number of studies in which the appropriate GC peaks have been shown to be composed of chlorinated biphenyl isomers(s).[1,4,18,20,33] The identification of PCB's in pesticide mixtures was readily effected using GC/MS linked with appropriate computer facilities.[5,6,31] PCB's have previously been shown to have gas chromatographic properties similar to several chlorinated insecticides. Several PCB peaks overlap the region of the chromatogram in which the DDT degradation product *p,p'*-DDD is eluted; however, with computer assisted GC/MS analysis the spectra of *p,p'*-DDD ($M^{+\cdot}$ 318), hexachlorobiphenyl ($M^{+\cdot}$ 358) and a heptachlorobiphenyl ($M^{+\cdot}$ 392) were obtained with only minimal cross-contamination of the spectra.

The relatively large mass deficiencies of both the chlorine isotopes ($^{35}Cl$ = −32 mmu, $^{37}Cl$ = −34.1 mmu) cause the exact masses of ions which contain chlorine to be usually less than other ions of the same nominal *m/e* value which contain only isotopes of C, H, N, or O, the elements most commonly present in samples of biological origin (e.g., Table 5). Using the high resolution photoplate technique,[16] this permits the detection of trace quantities of organochlorine compounds (e.g., pesticides or pollutants) in the presence of large quantities of coextracted natural products which are primarily composed of the elements C, H, N, and O. Typically the ion beam from a particular

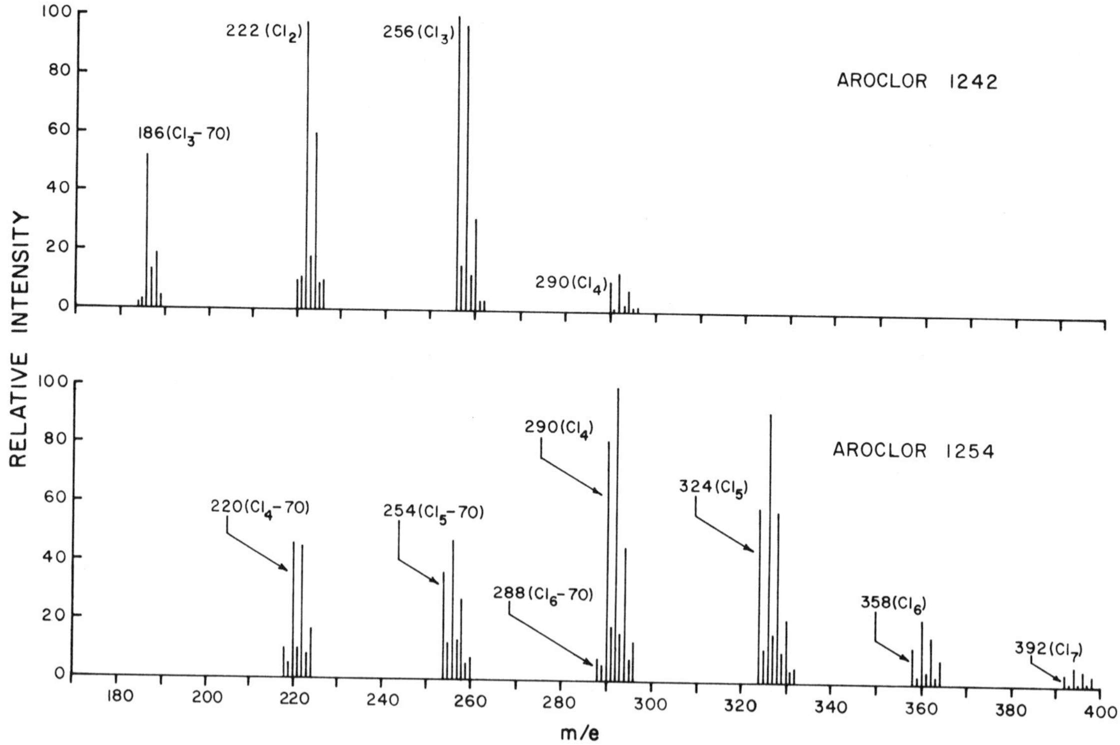


FIGURE 10. Top, mass spectrum of Aroclor 1242; bottom, mass spectrum of Aroclor 1254.

sample is allowed to impinge on a photographic plate for specific exposure periods. In addition, a standard can be appropriately superimposed on a plate in order to facilitate high resolution mass measurements. Recent results[16] are illustrated in Figure 11 in which tetrachlorobiphenyl has been detected in a crude hexane extract of human brain tissue. Using this technique, PCB's have been detected in marine wildlife, human tissue and breast milk samples.[16]

*Mass Spectra of Photolysis and Metabolic Products of Chlorobiphenyls*

The photolysis of pure PCB isomers and Aroclor mixtures yielded a large number of products which were conveniently analyzed by MS and GC/MS techniques. The first such study was reported by Safe and Hutzinger[28] in which they reported that the photolysis of 2,2′,4,4′,6,6′-hexachlorobiphenyl resulted in the formation of a number of dechlorinated products whose composition was determined by their mass spectra (see Table 6). Other groups have since reported similar results.[8,10,14,23]

A number of isomeric chlorobiphenyls as well as Aroclor 1254[14] have been photolyzed under a variety of conditions and the photoproducts were examined by MS analysis. Photolysis in nonhydroxylic solvents (i.e., hexane) gave the usual range of dechlorinated photoproducts, however, under aerobic conditions in the presence of hydroxylic solvents a number of oxygenated species were identified. Examination of the polar products from the Aroclor 1254 irradiation by MS analysis indicated formation of both hydrolysis and hydrated products (see Figure 12)[14].

Hutzinger and co-workers have investigated the metabolism of several PCB isomers in fish, rats, and pigeons.[11,12] The products were purified by chromatography and analyzed by mass spectrometry. The mass spectral analysis showed conversion of mono-, di-, and tetrachlorobiphenyl isomers into their corresponding mono-hydroxy derivatives in the rat and the pigeons and traces of dihydroxy metabolite were also detected by MS (Figure 13). It was interesting to note that no hydroxylation of any of the isomers was observed with the fish nor was 2,2′,4,4′,5,5′-hexachlorobiphenyl oxidized by any of the animals used in the study. Recent work has also been carried out with rabbits and microorganisms and again MS analysis has revealed the formation of chlorohydroxybiphenyls.[16,33] In

TABLE 5

**Approximate Values for Mass Differences for Selected Classes of Pesticides and Some Natural Products (in m.m.u.)**

| Pesticide* group I | | Pesticide* group II | | Pesticide* group III | | Selected natural products | |
|---|---|---|---|---|---|---|---|
| Compound | Δ m.m.u. | Compound | Δ m.m.u. | Compound | Δ m.m.u. | Compound | Δ m.m.u. |
| Organochlorine | | Organometallic | | Triazine herbicides | | | |
| insecticides | | compounds | | Atrazine | +95 | Fatty acids | |
| DDT | −85 | Dimethylarsinic | | Ametryne | +120 | Sterols | >+200 |
| Aldrin | −125 | acid | −35 | | | Carotenoids | |
| BHC | −140 | Ethylmercuric | | Bipyridyl herbicides | | | |
| Endosulfan | −185 | chloride | −20 | Paraquat | +115 | Tryptamine | +100 |
| | | Methylmercuric | | | | Flavonoids | +40 |
| Polychlorinated** | | cyanoguanidine | +30 | Carbamates, ureas | | Nicotinic acid | +30 |
| biphenyls | | Dimethylmercury** | +18 | and amides | | | |
| 4 Chlorine/molecule | −75 | Triphenyltin | | Carbaryl | +80 | | |
| 6 Chlorine/molecule | −155 | hydroxide | +22 | Fenuron | +95 | | |
| | | | | Linuron | +12 | | |
| Organochlorine | | Organophosphorus | | Barban | · 0 | | |
| fungicides | | compounds | | Propanil | +5 | | |
| Pentachlorophenol | −155 | Methyl trithion | −85 | | | | |
| Hexachlorobenzene | −185 | Disulfoton | +30 | | | | |
| Dyrene | −42 | Azinophosmethyl | +5 | | | | |
| Captan | −65 | | | | | | |
| Halogenated | | | | | | | |
| aromatic herbicides | | | | | | | |
| Amiben | −30 | | | | | | |
| Bromoxynil | −145 | | | | | | |
| Ioxynil | −175 | | | | | | |
| 2,4-D | −30 | | | | | | |

*Groups selected on the basis of suitability for detection in crude extract: Group I (Δm.m.u. = −30 to −200), Group II (Δm.m.u. = −30 to +50), Group III (Δm.m.u. = >+50 with some exceptions).
**Not being used as pesticides; compounds included because of structural similarity to the groups listed.

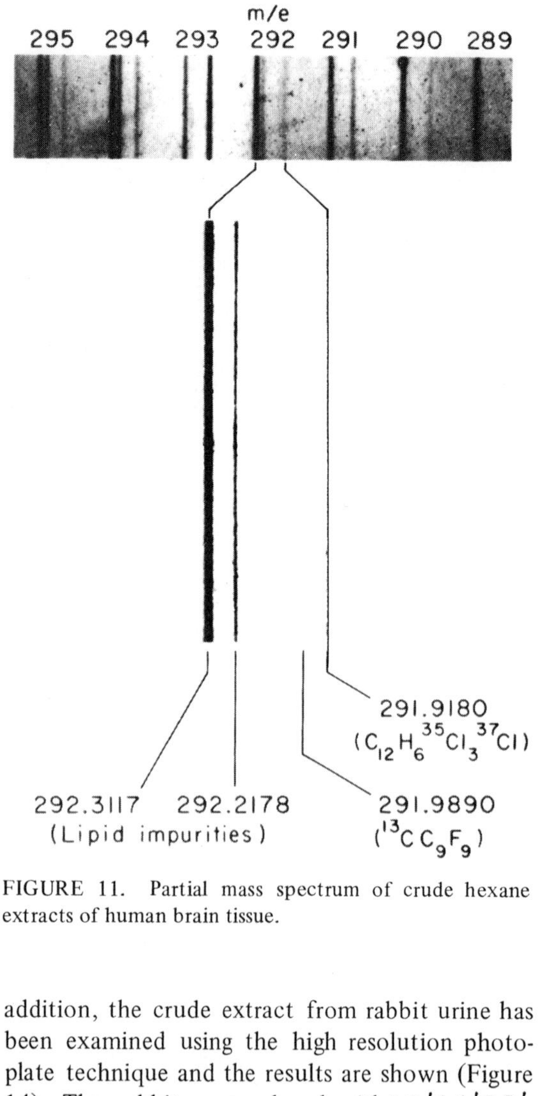


FIGURE 11. Partial mass spectrum of crude hexane extracts of human brain tissue.

addition, the crude extract from rabbit urine has been examined using the high resolution photoplate technique and the results are shown (Figure 14). The rabbits were dosed with 2,2,'4,4,'5,5,'-hexachlorobiphenyl and the results indicated the presence of a hexachlorohydroxybiphenyl, a pentachlorohydroxybiphenyl, and a pentachlorohydroxymethoxybiphenyl. The detection of the latter compound was only observed on the photoplate since the small quantities present were insufficient to detect it via the conventional electrical scans.

It has been shown that dansylation of hydroxybiphenyls and chlorohydroxybiphenyls enhances the chromatographic detection of these compounds as well as giving useful and characteristic mass spectra. Using this technique,[12] it was possible to detect trace quantities of a dichlorodihydroxybiphenyl metabolite from rat urine as the didansyl derivative and the mass spectrum of this derivative (Figure 15) confirmed the presence of the metabolite. Similarly, the dichlorohydroxybiphenyl metabolite from rat urine was also detected and confirmed using the dansyl method (Figure 16).

*Application of Chemical Ionization Mass Spectrometry to the Analysis of Chlorobiphenyls*

Chemical ionization (CI) mass spectrometry introduces a further simplification in PCB analysis. CI takes place at relatively high pressures (ca. 1 mm Hg) in which a reactant gas (e. g., methane) is first ionized and these species react with the sample molecules which are in turn ionized as a result of this ion molecule reaction. This method of ion formation is generally known as a soft-ionization process and usually results in the formation of abundant high molecular weight ionic species (e.g., molecular or quasi-molecular ions) and very few fragment ions.

This technique has been applied to the GC/MS analysis of Aroclor 1254.[32] A total ion chromatogram from the CI GC/MS analysis of the Aroclor mixture is shown (Figure 17a) and MS analysis of several of the peaks revealed two and three components were often present. The symmetrical peak between fractions 79 and 85 was examined in detail and spectrum number 82 is shown (Figure 17b) and clearly indicated the presence of both penta- and hexachlorobiphenyls. Because methane was used as the reactant gas, the $[M + 20]^+$ and $[M + 41]^+$ ions are present as well as the respective $[M + 1]^+$ due to the addition of $C_2H_5{}^+$ and $C_3H_5{}^+$ to the molecular species. It was also evident that the fragment ions of these biphenyl isomers were not evident in the low mass region of the spectrum.

*Mass Spectra of Chlorohydroxybiphenyls*

The mass spectra of chlorohydroxybiphenyls, the major PCB metabolites, have not been reported in the literature; however, the authors have examined the mass and IKE spectra of two representative compounds, 3,5-dichloro-4-hydroxybiphenyl and 4,4′-dichloro-3-hydroxybiphenyl. A summary of the results is shown in Table 7 and the data indicated some quantitative differences in the spectra. The table also indicates the fragmentation reactions which occur and typical loss of (H)Cl, (H)CO and $Cl_2$ are observed as well as elimination of ClCO from the m/e 202 ion. This latter reaction has also been noted in the fragmentation of isomeric dichlorophenols.[27]

TABLE 6

**GLC and Mass Spectral Data for Hexachlorobiphenyl Photolysis Products**

| | Retention time[a] (min) | $m/e$ of $M^+$ | Chlorine[b] content |
|---|---|---|---|
| TLC band I | | | |
| Peak 1 | 5.6 | —[c] | 0 |
| Peak 2 | 8.2 | —[c] | 0 |
| Peak 3 | 9.9 | 222 | 2 |
| Peak 4 | 13.5 | 256 | 3 |
| Peak 5 | 14.1 | 290 (256) | 4, ($3^{tr}$) |
| Peak 6 | 19.0 | 324 (290) | 5, ($4^{tr}$) |
| Peak 7 | 29.7 | 358 | 6 |
| Peak 8 | 56.9 | 304 | 2 |
| TLC band II | | | |
| Peak 1 | 19.6 | 290 | 4 |
| Peak 2 | 26.8 | 324 | 5 |
| Peak 3 | 35.2[d] | 358 | 6 |

[a]Conditions: instrument, Hewlett-Packard 5750; column, 6% QF-1 + 4% SE30 on "Chromosorb" (AW) 60–80 mesh 5′, ¼″ outside diameter, glass; injection port temperature, 220°C; column temperature, 175°C; flame ionization detector temperature and collection port temperature, 240°C; helium flow rate, 40 ml/min.
[b]From isotope peak distribution; tr, trace.
[c]Hydrocarbon peaks, $M^+$ not identified.
[d]Corresponding to starting material.

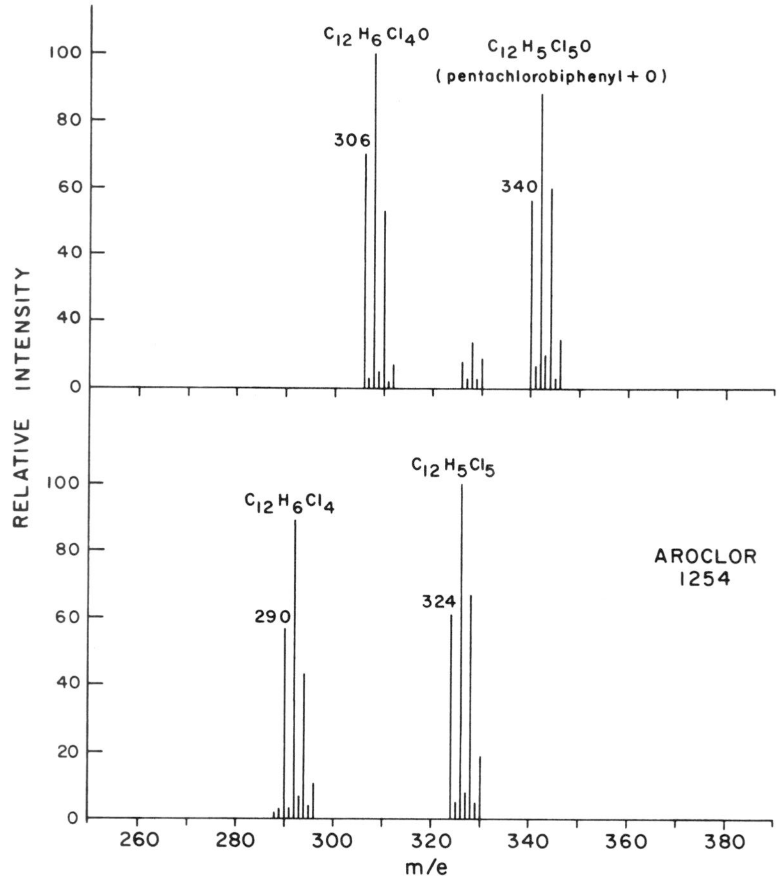


FIGURE 12. Top, partial mass spectrum of a polar fraction obtained from the photolysis of Aroclor 1254; bottom, partial mass spectrum of Aroclor 1254.

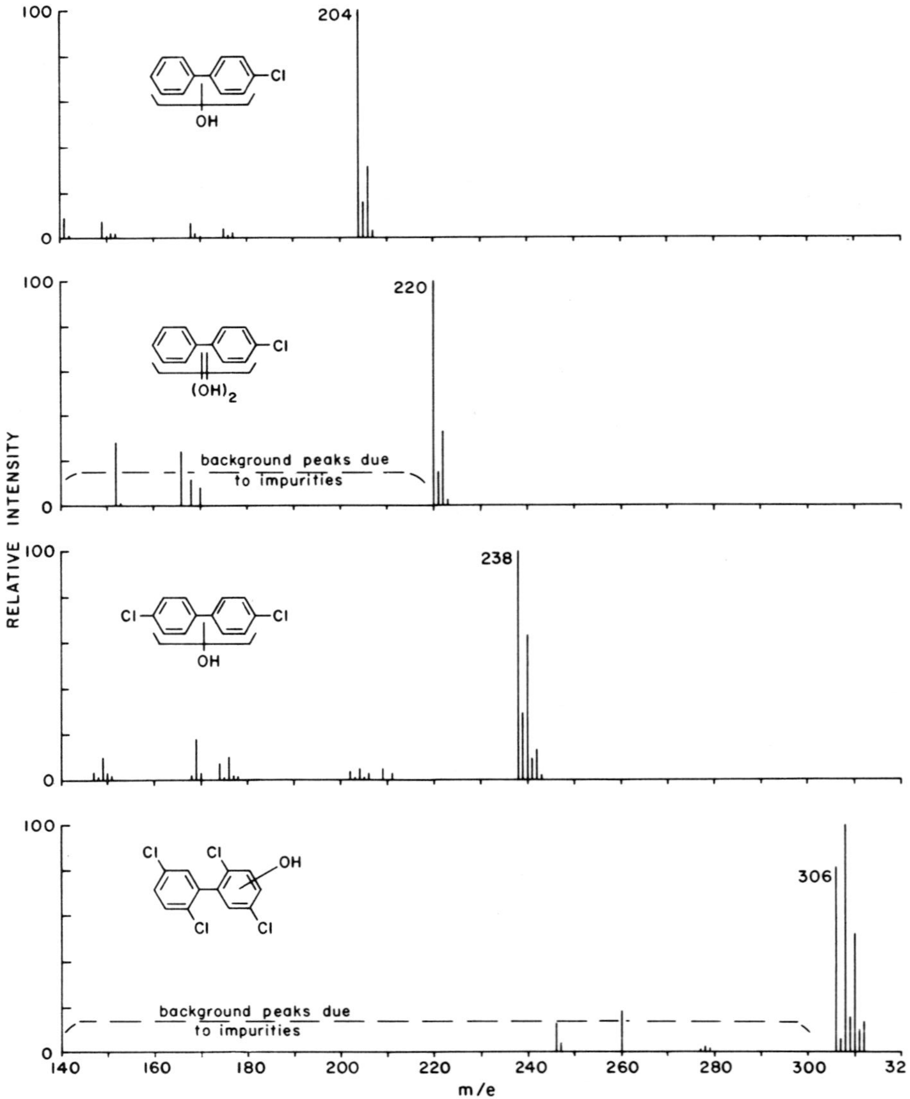


FIGURE 13. Mass spectra of hydroxy PCB metabolites.

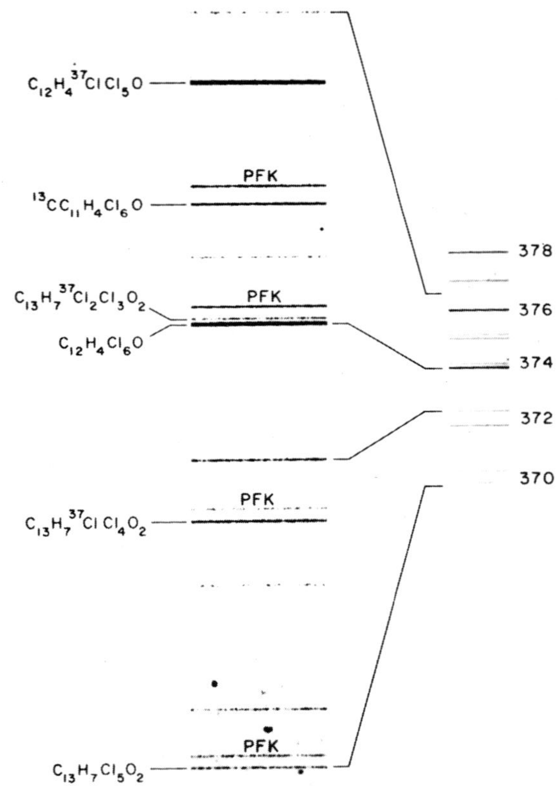


FIGURE 14. Partial mass spectrum of crude ether extract of urine from rabbit dosed with 2,2′,4,4′,5,5′-hexachlorobiphenyl.

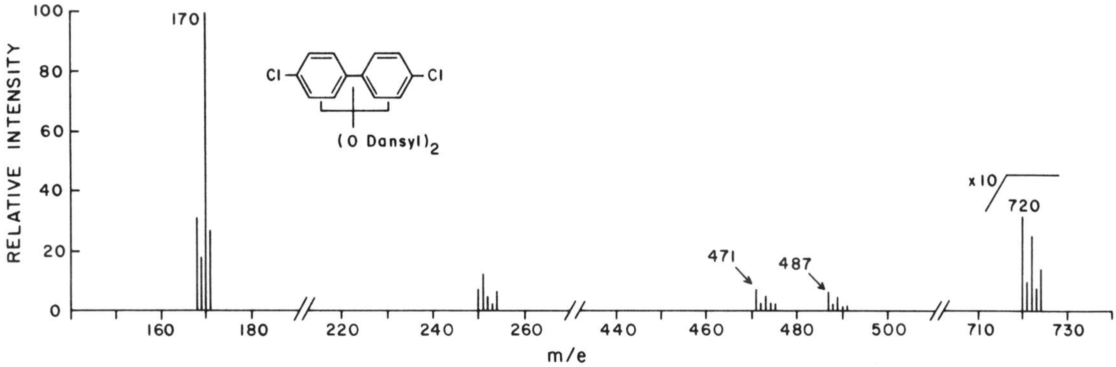


FIGURE 15. Mass spectrum of the dansyl derivative of a dihydroxy PCB metabolite.

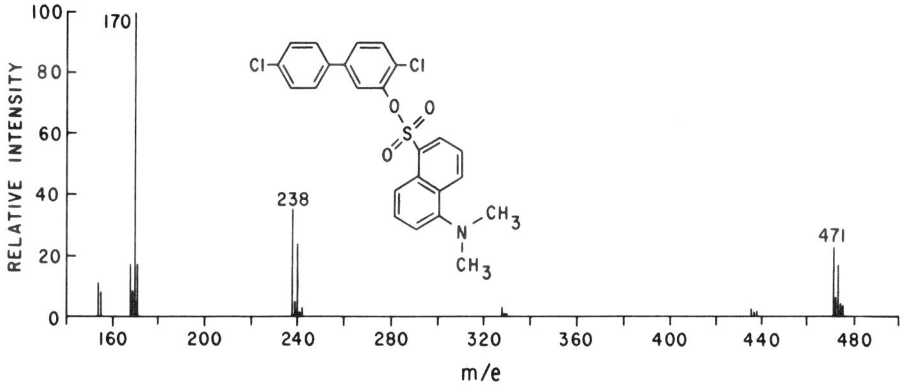


FIGURE 16. Mass spectrum of the dansyl derivative of the PCB metabolite 4,4-dichloro-3-hydroxybiphenyl.

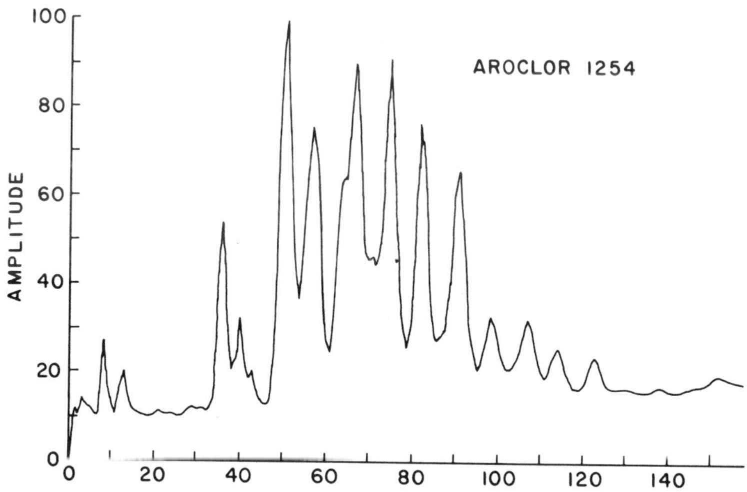


FIGURE 17a. Total ion chromatogram of Aroclor 1254.

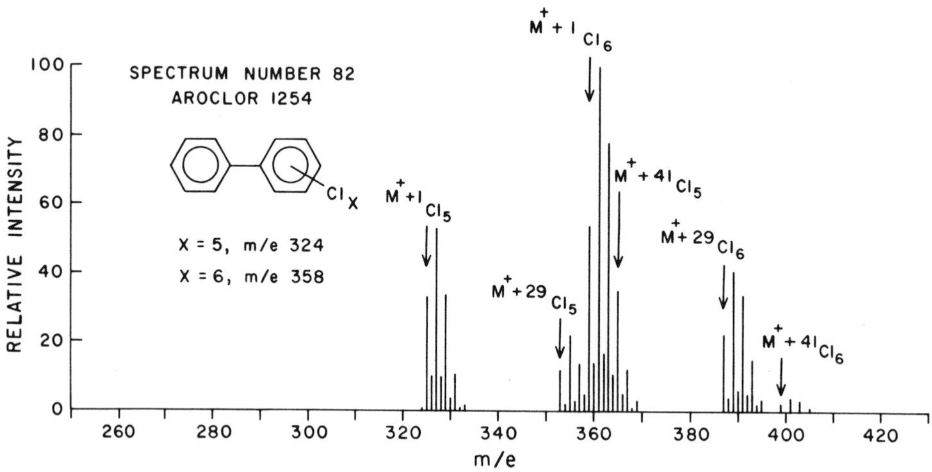


FIGURE 17b. Chemical ionization mass spectrum of fraction 82 (i.e., Figure 17a).

TABLE 7

**Mass and IKE Spectral Data for Dichlorohydroxybiphenyls**

| Compound | Ion Abundances M/E | | | | | | | |
|---|---|---|---|---|---|---|---|---|
| | 238 | 209 | 202 | 173 | 168 | 167 | 149 | 139 |
| 3,5-Dichloro-4-hydroxybiphenyl | 100 | 2 | 2 | 6 | 4 | 2 | 3 | 85 |
| 4,4′-Dichloro-3-hydroxybiphenyl | 100 | 2 | 2 | 3 | 7 | 0.5 | 3 | 26 |
| | Peak Abundances x $10^3$ (E) | | | | | | | |
| | 0.876 | 0.849 | 0.829 | 0.795 | 0.733 | 0.704 | 0.684 | 0.582 |
| 4,4′-Dichloro-3-hydroxybiphenyl | 1.1 | 7.9 | 8.5 | 6.8 | 6.7 | 5.5 | 2.7 | 6.0 |
| 3,5-Dichloro-4-hydroxybiphenyl | 1.5 | 8.9 | 6.8 | 7.0 | 3.3 | 1.8 | 6.1 | 10.3 |

238 → 209 (.876 E)
238 → 202 (.849 E)
209 → 273 (.829 E)
173 → 138 (.785 E)

238 → 173 (.733 E)
238 → 168 (.704 E)
202 → 139 (.684 E)
238 → 139 (.582 E)

*Plasma Chromatography of Chlorobiphenyls*

Plasma chromatography uses a $^{63}$Ni radioactive beta source to create ions which in turn react with trace constituents which are separated via gas chromatography. The resulting ion-molecules (positive and negative) which form are then separated and detected. The separated complexes appear as recorded plasmagrams of separated ion-molecule peaks and mass identification of these peaks has been achieved by mass spectrometric analysis. Composite negative and positive plasmagrams for several isomeric PCB's are shown (Figure 18).[18,19]

All the plasmagrams produced for the PCB compounds are quite simple. In cases in which a PCB-ion complex is formed, a single peak appears in the plasmagram. When a high reactivity in the positive mode is indicated by a strong plasmagram, little or nothing appears in the negative plasma-

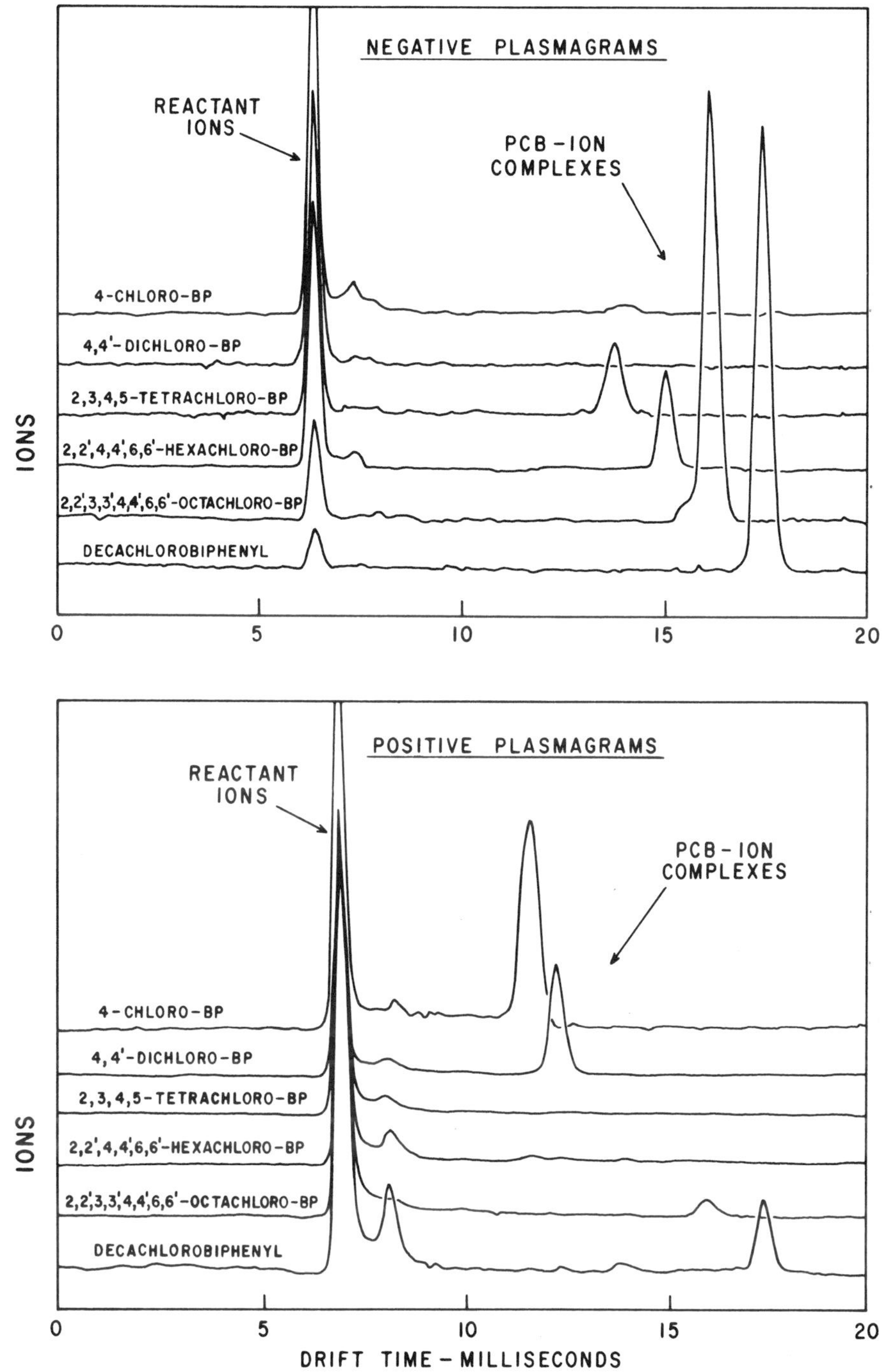


FIGURE 18. Top, negative plasmagrams for six PCB isomers; bottom, positive plasmagrams for six PCB isomers.

gram and this reactivity appears to be characteristic of the PCB compounds with lesser chlorine content. At the highest chlorine content of the PCB, the plasmagrams show exactly the opposite effect. That this is a consistent trend can be seen by the composite of the positive and the negative plasmagrams (Figure 18) for all six PCB compounds. Taken together, the positive and negative plasmagrams for a given compound comprise a unique set of "fingerprint" patterns for that compound and peaks occur in the plasmagram well resolved from each other.

## REFERENCES


1. **Bagley, G. E., Reichel, W. L., and Cromartie, E.,** *J. Assoc. Offic. Anal. Chem.,* 53, 251, 1970.
2. **Biros, F. J., Walker, A. C., and Medbery, A.,** *Bull. Environ. Contam. Toxicol.,* 5, 317, 1970.
3. **Beynon, J. H., Amy, J. W., and Baitinger, W. E.,** *Chem. Commun.,* 723, 1969.
4. **Bonelli, E. J.,** *Application Tips,* No. 16, Finnigan Corp., Sunnyvale, California, 1970.
5. **Bonelli, E. J.,** *Anal. Chem.,* 44, 603, 1972.
6. **Bonelli, E. J.,** *Am. Lab.,* 27, 1971.
7. **Chait, E. M. and Askew, W. B.,** *Org. Mass Spect.,* 5, 149, 1971.
8. **Hustert, K. and Korte, F.,** *Chemosphere,* 1, 7, 1972.
9. **Hutzinger, O., Jamieson, W. D., and Zitko, V.,** *Nature,* 225, 664, 1970.
10. **Hutzinger, O., Jamieson, W. D., Safe, S., MacNeil, J. D., and Zitko, A.,** A survey of the photochemical behavior of chlorobiphenyls in *PCB in the Environment,* Marcel Dekker, in press.
11. **Hutzinger, O., Nash, D. M., Safe, S., DeFreitas, A. S. W., Norstrom, R. J., Wildish, D. J., and Zitko, V.,** *Science,* 178, 312, 1972.
12. **Hutzinger, O., Safe, S., Ecobichon, D., DeFreitas, A. S. W., Norstrom, R. J., MacNeil, J. D., Wildish, D. J., and Zitko, V.,** A survey of the metabolism of chlorobiphenyl in *PCB in the Environment,* Marcel Dekker, in press.
13. **Hutzinger, O., Safe, S., and Zitko, V.,** *Environ. Health Perspect.,* 2, 95, 1972.
14. **Hutzinger, O., Safe, S., and Zitko, V.,** *Environ. Health Perspect.,* 1, 15, 1972.
15. **Hutzinger, O., Safe, S., and Zitko, V.,** *Bull Environ. Contam. Toxicol.,* 6, 209, 1971.
16. **Jamieson, W. D., Hutzinger, O., Safe, S., Crocker, J. F. S., and Zitko, V.,** The Use of High Resolution Photoplate Techniques in the Identification of Organochlorine Compounds and Their Metabolites in Biological Samples, Annual Conference of the American Society for Mass Spectrometry, San Francisco, May, 1973.
17. **Jensen, S.,** *New Scientist,* 32, 612, 1966.
18. **Karasek, F. W.,** *Anal. Chem.,* 43, 1982, 1971.
19. **Karasek, F. W., and Tatone, O. S.,** *Anal. Chem.,* 44, 1758, 1972.
20. **Koeman, J. H., ten Noever de Brauw, M. C., and de Vos, R. H.,** *Nature,* 221, 1126, 1969.
21. **Murata, T., Takahashi, S., and Tachikawa, R.,** *Shimadzu Rev.,* 28, 29, 1971.
22. **Rote, J. W., and Morris, W. J.,** *J. Assoc. Offic. Anal. Chem.,* 56, 188, 1973.
23. **Ruzo, L. R., Zabik, M. J., and Schuetz, R. D.,** *Bull. Environ. Contam. Toxicol.,* 8, 217, 1972.
24. **Safe, S., Jamieson, W. D., and Hutzinger, O.,** *Org. Mass Spect.,* 7, 169, 1973.
25. **Safe, S. and Hutzinger, O.,** *J. Chem. Soc.,* Perkin I, 686, 1972.
26. **Safe, S. and Hutzinger, O.,** *Chem. Commun.,* 446, 1971.
27. **Safe, S. and Hutzinger, O.,** *Mass spectrometry of pesticides and pollutants,* CRC Press, Inc., Cleveland, Ohio, 1973.
28. **Safe, S. and Hutzinger, O.,** *Nature,* 232, 641, 1971.
29. **Sissons, D. and Welti, D.,** *J. Chromatogr.,* 60, 15, 1971.
30. **Stalling, D. L. and Huckins, J. N.,** *J. Assoc. Offic. Anal. Chem.,* 54, 801, 1971.
31. **Stalling, D. L.,** 163rd National Meeting, American Chemical Society, New York, N.Y., Symposium on PCB, No. 22, August, 1972.
32. **Story, M. S. and Squires, R. B.,** *Finnigan Spectra,* Finnigan Corp. Sunnyvale, Cal., August 2, 1972.
33. **Wallnöfer, P. R., Engelhardt, G., Safe, S., and Hutzinger, O.,** *Chemosphere,* 69, 1973.
34. **Widmark, G.,** *J. Assoc. Offic. Anal. Chem.,* 50, 1069, 1972.
35. **Willis, D. E. and Addison, R. F.,** *J. Fish Res. Bd. Can.,* 29, 592, 1972.

Chapter 9

# NUCLEAR MAGNETIC RESONANCE SPECTROSCOPY OF CHLOROBIPHENYLS

Nuclear magnetic resonance (NMR) spectroscopy has been a useful technique for determining the precise structure of the components of the commercial Aroclors. NMR spectra have also been used in establishing the identity of PCB metabolites and, in addition, the spectra of several synthetic chlorohydroxybiphenyl isomers have also been reported. The latter group of compounds are important as possible PCB metabolites. Many of the above results have been obtained with 100 and 220 MHz NMR spectrometers which greatly simplifies interpretation of the spectra since second order H-H couplings are often minimized and in many cases eliminated. A further advantage of NMR analysis is that only relatively small samples are required since computer averaging techniques (CAT) can be used with microsamples.

**Proton Magnetic Resonance (PMR) Spectra of Individual Chlorobiphenyls**

Tarpley and Goldstein[17] and others[2-4,10,12-14,22] have reported the 60 MHz PMR of several chlorobiphenyl isomers and more recently Sissons and Welti[16,21] have published the 220 MHz spectra of a range of chlorobiphenyls many of which have been identified as components of commercial Aroclor samples. The proton chemical shifts and coupling constants for some of the major PCB components of Aroclor 1254 are given in Table 1 and the simulated 220 MHz NMR spectra are shown in Figures 1, 2, and 3. In addition Sissons and Welti[21] have also reported the spectra of 17 synthetic PCB standards and the chemical shifts and coupling constants for these compounds are given in Table 2.

The chemical shift values for the protons on a chlorinated biphenyl ring nucleus are influenced by a number of factors.

1. The position of the proton with respect to the chlorine groups present;

2. The numbers of chlorine atoms in both phenyl rings and particularly the number of chlorine atoms at the 2,2′,6 and 6′ positions; and

3. The electronic effect of the chlorine substituents on the biphenyl bridge.

The influence of chlorine atoms on the chemical shifts of aromatic protons is a well-studied effect. Introduction of chlorine groups into the benzenoid nucleus results in characteristic chemical shifts of the *ortho, meta,* and *para* protons. The magnitude of the shifts on the above protons is ca. −0.02, 0.06, and 0.04, respectively, with the negative value indicating a downfield shift (in ppm) with respect to the chemical shift (7.27 ppm) of benzene.

The freedom of rotation of the two phenyl rings is markedly affected by substitution at the 2,2′,6, and 6′ positions of the biphenyl nucleus. Increasing substitution at these positions *ortho* to the Ph-Ph bond markedly restricts the freedom of rotation and distorts the rings out of the coplanar conformation. The effects of chlorine substitution at these *ortho* positions on a 2 (or 6) proton are summarized in Table 3. The chemical shift values for the 2 (or 6) proton move upfield by ca. 0.15 ppm values for each Cl atom substituted at the 2′ and 6′ positions. The upfield shift of the 2-proton is diminished when there is a 6-chloro group present. In this case, with increasing 2′ and 6′ Cl substitution the upfield shift is only 0.05 ppm. It was also generally observed by Sissons and Welti that increasing Cl substitution in the other positions of both rings of the biphenyl nucleus tended to cause a shift to lower fields and a summary of some of these data is shown in Table 4. Presumably with increasing 4 and 3,4 substitution the coplanarity and cross conjugation between the two phenyl ring is enhanced with the chemical shift values of the remaining protons moving to lower fields

Recent work by Tas and co-workers[12] has established the structures of several major components of Phenochlor DP6 using, among other techniques, PMR analysis. The compounds identified in this fashion were 2,2′,4,4′,5,5′-hexachlorobiphenyl, 2,2′,3,4,4′,5′-hexachlorobiphenyl,

TABLE 1

**Chemical Shifts (δ ppm) and Approximate Coupling Constants (Hz) of Major PCB Components of Aroclor 1254**

| | Chemical shifts | | Coupling constants | | |
|---|---|---|---|---|---|
| Proton | $CDCl_3$ | $C_2Cl_4$ | Protons | $CDCl_3$ | $C_2Cl_4$ |
| 2,2′,5,5′-Tetrachlorobiphenyl (Peak 22) MS shows 4 Cl atoms | | | | | |
| 3 and 3′ | 7.383 | 7.333 | J(3,4) | ~8.5 | ~8.0 |
| 4 and 4′ | 7.305 | 7.250 | J(3,6) | unresolved | unresolved |
| 6 and 6′ | Coincident with $CHCl_3$ line (~7.250) | 7.210 | J(4,6) | ~2.5 | ~2.3 |
| 2,2′,3′,5,6′-Pentachlorobiphenyl (Peak 25) MS shows 5 Cl atoms | | | | | |
| 3 | 7.425 | 7.363 | J(3,4) | ~8.8 | ~9.3 |
| 4 | 7.342 | 7.281 | J(3,6) | unresolved | unresolved |
| 6 | 7.170 | 7.155 | J(4,6) | ~2.1 | ~2.2 |
| 4′ | 7.346 | 7.279 | J(4′,5′) | ~8.8 | ~8.7 |
| 5′ | 7.449 | 7.386 | | | |
| 2,2′,4′,5,5′-Pentachlorobiphenyl (Peak 39) MS shows 5 Cl atoms | | | | | |
| 3 | 7.394 | 7.342 | J(3,4) | ~8.6 | ~8.6 |
| 4 | 7.323 | 7.267 | J(3,6) | unresolved | unresolved |
| 6 | 7.217 | 7.206 | J(4,6) | ~2.3 | ~2.5 |
| 3′ | 7.579 | 7.541 | J(3′,6′) | unresolved | unresolved |
| 6′ | 7.339 | 7.311 | | | |
| 2,4-2′,4′,5′-Pentachlorobiphenyl (Peak 41) MS shows 5 Cl atoms | | | | | |
| 3 | 7.486 | 7.461 | J(3,5) | ~2.0 | ~2.0 |
| 5 | 7.303 | 7.250 | J(3,6) | unresolved | unresolved |
| 6 | 7.150 | 7.112 | J(5,6) | ~8.0 | ~8.0 |
| 3′ | 7.575 | 7.534 | J(3′,6′) | unresolved | unresolved |
| 6′ | 7.325 | 7.307 | | | |
| 2,5-2′,3′,4′-Pentachlorobiphenyl (Peak 43) MS shows 5 Cl atoms | | | | | |
| 3 | 7.391 | 7.335 | J(3,4) | ~8.5 | ~8.5 |
| 4 | 7.313 | 7.255 | J(3,6) | unresolved | unresolved |
| 6 | 7.201 | 7.191 | J(4,6) | ~2.5 | ~2.0 |
| 5′ | 7.432 | 7.362 | J(5′,6′) | ~8.3 | ~8.3 |
| 6′ | 7.080 | 7.022 | | | |
| 3,4-2′,3′,6′-Pentachlorobiphenyl (Peak 44) MS shows 5 Cl atoms | | | | | |
| 2 | 7.322 | 7.291 | J(2,5) | unresolved | unresolved |
| 5 | 7.533 | 7.498 | J(2,6) | ~2.0 | ~2 |
| 6 | 7.059 | 7.005 | J(5,6) | ~8.4 | ~8 |
| 4′ | 7.335 | 7.267 | J(4′,5′) | ~8.7 | ~8.5 |
| 5′ | 7.424 | 7.351 | | | |
| 2,3,6-2′,4′,5′-Hexachlorobiphenyl (Peak 48) MS shows 6 Cl atoms | | | | | |
| 4 | 7.347 | 7.278 | J(4,5) | ~8.3 | ~9 |
| 5 | 7.463 | 7.401 | J(3,6′) | unresolved | unresolved |
| 3′ | 7.614 | 7.566 | | | |
| 6′ | 7.280 | 7.243 | | | |
| 2,3,4-2′,3′,6′-Hexachlorobiphenyl (Peak 50) MS shows 6 Cl atoms | | | | | |
| 5 | 7.481 | | J(5,6) | ~8.5 | |
| 6 | 7.027 | | J(4,5) | ~8.3 | |
| 4′ | 7.350 | | | | |
| 5′ | 7.458 | | | | |

TABLE 1 (Continued)

**Chemical Shifts (δ ppm) and Approximate Coupling Constants (Hz) of Major PCB Components of Aroclor 1254**

| Chemical shifts | | | Coupling constants | | |
|---|---|---|---|---|---|
| Proton | $CDCl_3$ | $C_2Cl_4$ | Protons | $CDCl_3$ | $C_2Cl_4$ |
| 3,4-2′,4′,5′-Pentachlorobiphenyl (Peak 52) MS shows 5 Cl atoms | | | | | |
| 2 | 7.468 | 7.457 | *J*(2,5) | unresolved | unresolved |
| 5 | 7.490 | 7.433 | *J*(2,6) | ~2.4 | ~2.0 |
| 6 | 7.220 | 7.181 | *J*(5,6) | ~8.4 | ~8.5 |
| 3′ | 7.565 | 7.530 | *J*(3′,6′) | unresolved | unresolved |
| 6′ | 7.380 | 7.355 | | | |
| 3,4-2′,3′,4′-Pentachlorobiphenyl (Peak 55) MS shows 5 Cl atoms | | | | | |
| 2 | | 7.443 | *J*(2,5) | | unresolved |
| 5 | | 7.440 | *J*(2,6) | | ~2 |
| 6 | | 7.170 | *J*(5,6) | | ~8 |
| 5′ | | 7.365 | *J*(5′,6′) | | ~8 |
| 6′ | | 7.074 | | | |
| 2,4,5=2′,4′,5′-Hexachlorobiphenyl (Peak 56) MS shows 6 Cl atoms | | | | | |
| 3 and 3′ | 7.577 | 7.545 | *J*(3,6) | unresolved | unresolved |
| 6 and 6′ | 7.323 | 7.295 | | | |
| 2,3,4-2′,4′,5′-Hexachlorobiphenyl (Peak 59) MS shows 6 Cl atoms | | | | | |
| 5 | 7.449 | 7.377 | *J*(5,6) | ~8 | ~8.5 |
| 6 | 7.074 | 7.020 | *J*(3′,6′) | unresolved | unresolved |
| 3′ | 7.586 | 7.545 | | | |
| 6′ | 7.318 | 7.295 | | | |
| 2,3,4-2′,3′,4′-Hexachlorobiphenyl (Peak 62) MS shows 6 Cl atoms | | | | | |
| 5 and 5′ | | 7.011 | *J*(5,6) | | ~8 |
| 6 and 6′ | | 7.372 | | | |
| 3,4-2′,3′,4′,5′-Hexachlorobiphenyl (Peak 67) MS shows 6 Cl atoms | | | | | |
| 2 | | 7.442 | *J*(2,5) | | unresolved |
| 5 | | 7.450 | *J*(2,6) | | ~2.3 |
| 6 | | 7.165 | *J*(5,6) | | ~8.2 |
| 6′ | | 7.305 | | | |
| 2,3,4-2′,3′,4′,5′-Heptachlorobiphenyl (Peak 69) MS shows 7 Cl atoms | | | | | |
| 5 | | 7.395 | *J*(5,6) | | ~8.3 |
| 6 | | 7.014 | | | |
| 6′ | | 7.250 | | | |

2,2′,3,3′,6,6′-hexachlorobiphenyl, 2,2′,3,4′,5′,6-hexachlorobiphenyl, 2,2′,3,4,4′,5,5′-heptachlorobiphenyl, 2,2′,3,3′,4,4′,5′-heptachlorobiphenyl, and 2,2′,3,3′,4,5,6-heptachlorobiphenyl. Similarly Bartle[1] has also identified 2,2′,5,5′-tetrachlorobiphenyl and 2,2′,3,4,4′,5′-hexachlorobiphenyl from a commercial Clophen sample by PMR analysis.

**PMR Spectra of Commercial Aroclors**

The PMR spectra of four commercial Aroclor samples are shown in Figures 4 and 5 and although the spectra are significantly different the complexity of the samples precludes any chemical shift assignments.[11]

**PMR Spectra of Hydroxylated Chlorobiphenyls**

Recent work by Hutzinger and co-workers[6,7,20] has shown that hydroxylated chlorinated biphenyls are the most common PCB metabolites produced by a wide variety of biological systems (mammals, fish, birds, and microorganisms). In

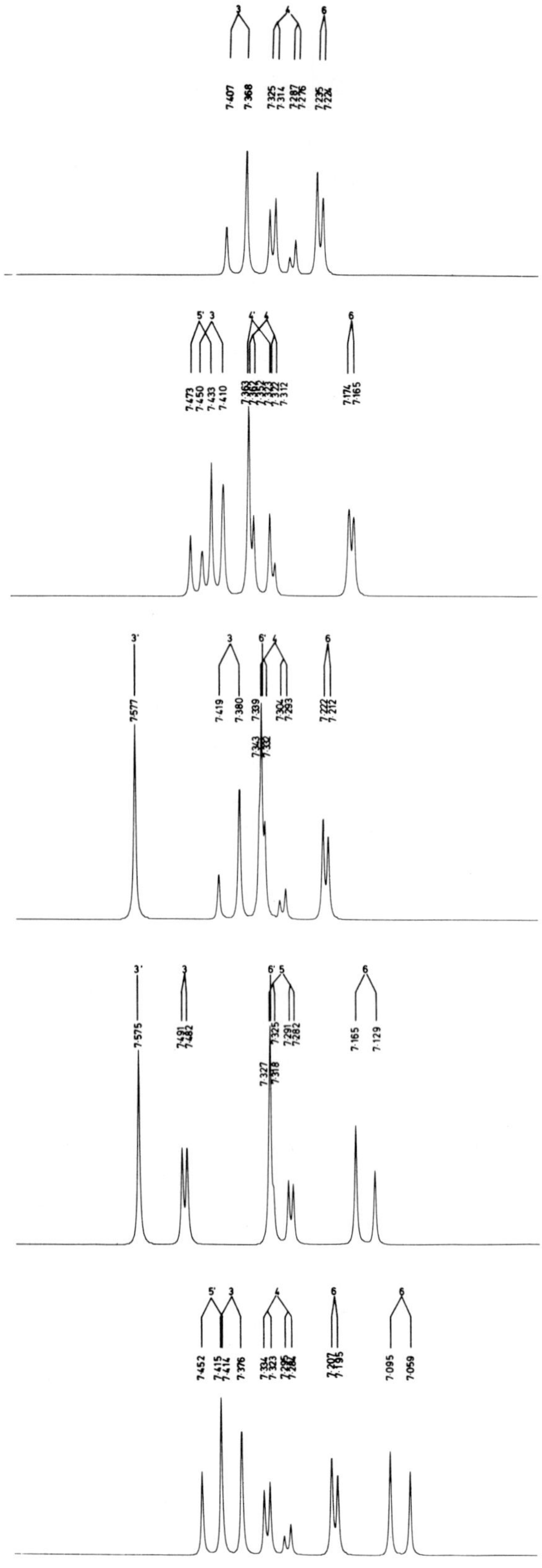

FIGURE 1. Simulated 220 MHz spectra of (a) 2,2′,5,5′-tetrachlorobiphenyl; (b) 2,2′,3′,5,6′-pentachlorobiphenyl; (c) 2,2′,4′,5,5′-pentachlorobiphenyl; (d) 2,2′,4,4′,5′-pentachlorobiphenyl; and (e) 2,2′,3′,4′,5-pentachlorobiphenyl.

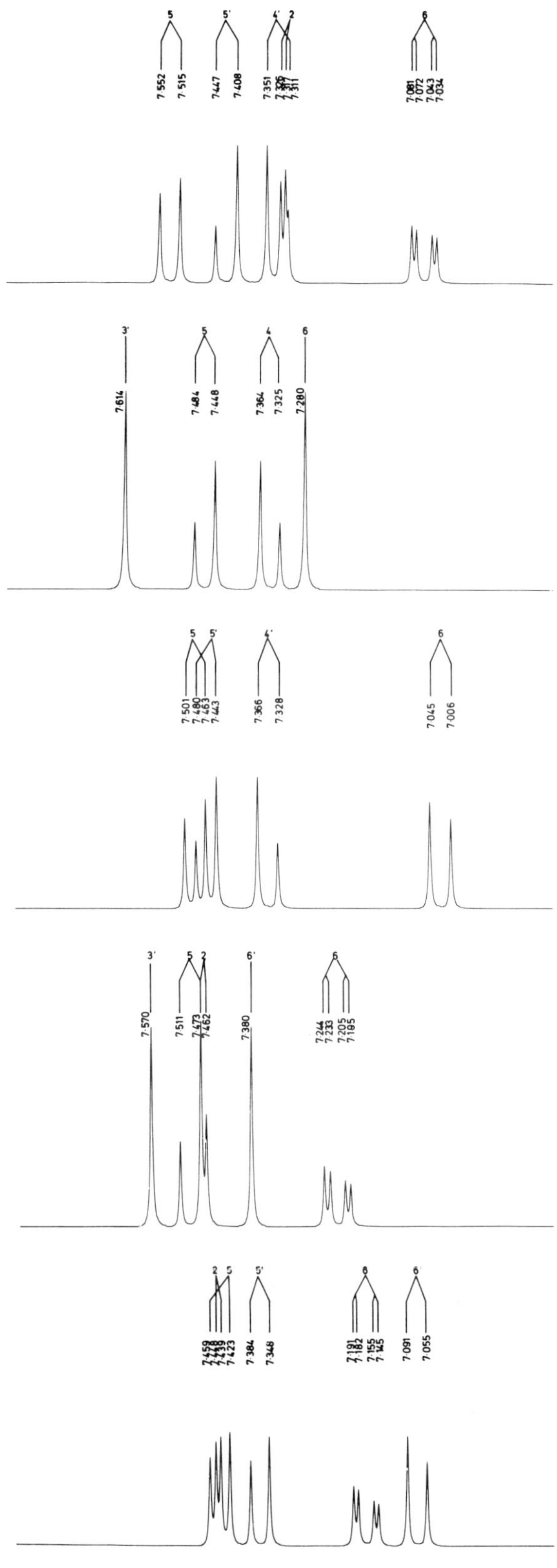

FIGURE 2. Simulated 220 MHz spectra of (f) 2′,3,3′,4,6′-pentachlorobiphenyl; (g) 2,2′,3,4′,5′,6-hexachlorobiphenyl; (h) 2,2′,3,3′,4,6′-hexachlorobiphenyl; (i) 2′,3,4,4′,5′-pentachlorobiphenyl; and (j) 2′,3,3′,4,4′-pentachlorobiphenyl.

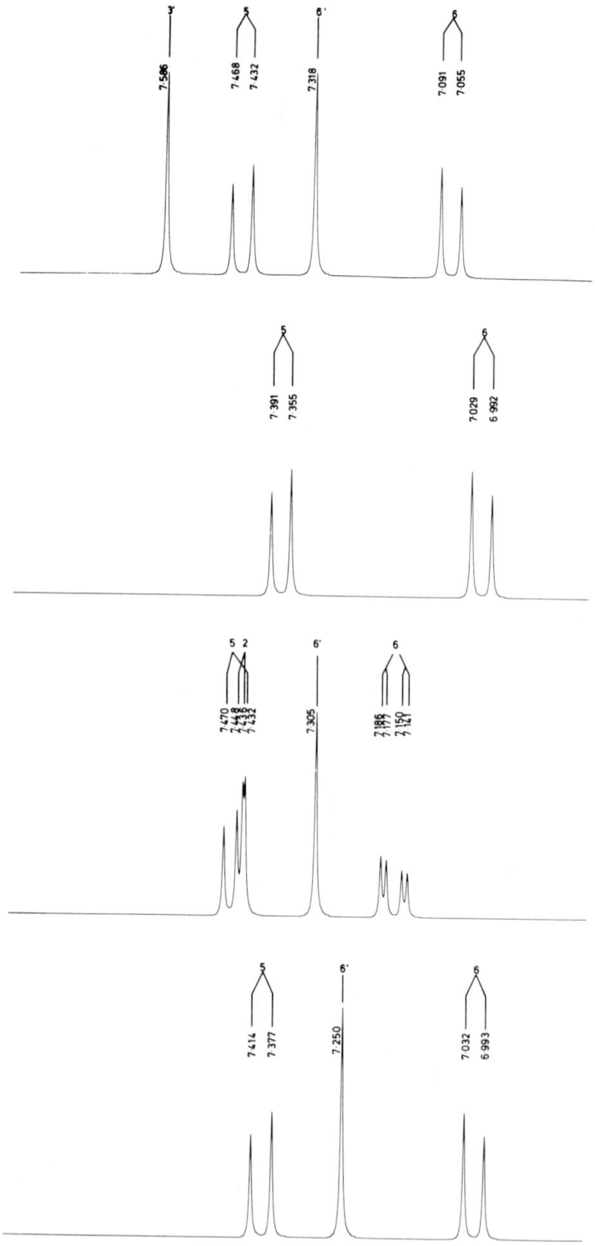


FIGURE 3. Simulated 220 MHz spectra of (k) 2,2′,3,4,4′,5,5′-hexachlorobiphenyl; (l) 2,2′,3,3′,4,4′-hexachlorobiphenyl; (m) 2′,3,3′,4,4′,5′-hexachlorobiphenyl; and (n) 2,2′,3,3′,4,4′,5-heptachlorobiphenyl.

addition, it has also been shown that irradiation of PCB isomers and Aroclor[5,8,15] samples all yield products whose mass spectra are consistent with the chlorohydroxybiphenyl structure. When rats are fed 4-chlorobiphenyl and 4,4′-dichlorobiphenyl, each isomer is converted into an hydroxy metabolite. The PMR spectra of the acetate derivatives of these two metabolites are shown (Figures 6 and 7) and the structures could be assigned as indicated (i.e., 4-chloro-4′-hydroxybiphenyl and 4,4′-dichloro-3-hydroxybiphenyl, respectively). Since the identification of hydroxylated PCB isomers is greatly facilitated by their respective PMR spectra, a number of these com-

TABLE 2

**Chemical Shifts and Coupling Constants for 17 Synthetic Standard PCB Isomers**

| Compound | Proton | Chemical shift (ppm) | Coupling constants (Hz) |
|---|---|---|---|
| 4-Chlorobiphenyl | 2,6 | 7.38 | $J_{2,3}=J_{5,6}\sim 8.2$ |
| | 3,5 | 7.49 | $J_{2,5}=J_{3,6}\sim 0.1$ |
| | 2′,6′ | 7.53 | $J_{2,6}=J_{3,5}\sim 2.5$ |
| | 3′,5′ | 7.42 | $J_{2,3}=J_{5',6}\sim 7.9$ |
| | 4′ | 7.33 | $J_{2',4'}=J_{4',6'}\sim 1.7$ |
| | | | $J_{2',5'}=J_{3',6'}\sim 0.5$ |
| | | | $J_{2',6'}\sim 2.0$ |
| | | | $J_{3',4'}=J_{4',5'}\sim 7.3$ |
| | | | $J_{3',5'}=1.8$ |
| 2,4-Dichlorobiphenyl | 3 | 7.43 | $J_{3,5}\sim 2.0$ |
| | 5 | 7.23 | $J_{3,6}\sim 0.5$ |
| | 6 | 7.21 | $J_{5,6}\sim 8.3$ |
| | 2′,3′,4′,5′,6′ | Not assigned | |
| 4,4′-Dichlorobiphenyl | 2,6 | 7.45 | $J_{2,3}=J_{5,6}\sim 8.3$ |
| | 3,5 | 7.39 | $J_{2,6}=J_{3,5}\sim 2.2$ |
| 2,6-Dichlorobiphenyl | 3,5 | 7.33 | $J_{3,4}=J_{4,5}\sim 7.8$ |
| | 4 | 7.14 | $J_{2',3'}=J_{5',6'}\sim 7.6$ |
| | 2′,6′ | 7.22 | $J_{2',4'}=J_{4',6'}\sim 1.0$ |
| | 3,5′ | 7.40 | $J_{2',5'}=J_{3',6'}\sim 0.5$ |
| | 4′ | 7.38 | $J_{2',6'}=J_{3',5'}\sim 1.2$ |
| | | | $J_{3',4'}=J_{4',5'}\sim 5.2$ |
| 2,2′-Dichlorobiphenyl | 3,3′ | 7.44 | $J_{3,4}=J_{3',4'}\sim 6.9$ |
| | 4,4′ | 7.30 | $J_{3',5'}=J_{3',5'}\sim 2.3$ |
| | 5,5′ | 7.28 | $J_{3,6}=J_{3',6'}\sim 0.3$ |
| | 6,6′ | 7.23 | $J_{4,5}=J_{4',5'}\sim 7.3$ |
| | | | $J_{4,6}=J_{4',6'}\sim 2.3$ |
| | | | $J_{5,6}=J_{5',6'}\sim 7.3$ |
| 3,3′-Dichlorobiphenyl | 2,2′ | 7.51 | $J_{2,4}=J_{2',4'}\sim 1.9$ |
| | 4,4′ | 7.32 | $J_{2,5}=J_{2',5'}\sim 0.5$ |
| | 5,5′ | 7.35 | $J_{2,6}=J_{2',6'}\sim 1.6$ |
| | 6,6′ | 7.40 | $J_{4,5}=J_{4',5'}\sim 8.1$ |
| | | | $J_{4,6}=J_{4',6'}\sim 1.9$ |
| | | | $J_{5,6}=J_{5',6'}\sim 7.4$ |
| 3,4-Dichlorobiphenyl | 2 | 7.61 | $J_{2,6}\sim 2.1$ |
| | 5 | 7.44 | $J_{5,6}\sim 8.2$ |
| | 6 | 7.35 | $J_{2',3'}=J_{5',6'}\sim 6.9$ |
| | 2′,6′ | 7.48 | $J_{2',4'}=J_{4',6'}\sim 1.4$ |
| | 3′,5′ | 7.40 | $J_{2',5'}=J_{3',6'}\sim 0.6$ |
| | 4′ | 7.34 | $J_{2',6'}\sim 2.0$ |
| | | | $J_{3',4'}=J_{4',5'}=7.2$ |
| | | | $J_{3',5'}=1.8$ |
| 2,4,4′-Trichlorobiphenyl | 3 | 7.46 | $J_{3,5}\sim 2.2$ |
| | 5 | 7.27 | $J_{5,6}\sim 8.0$ |
| | 6 | 7.21 | $J_{2,3'}=J_{5',6'}=8-9$ |
| | 2′,6′ | 7.31 | $J_{2',5'}=J_{3',6'}=0-0.5$ |
| | 3′,5′ | 7.38 | $J_{2',6'}=J_{3',5'}\sim 2$ |

TABLE 2 (Continued)

**Chemical Shifts and Coupling Constants for 17 Synthetic Standard PCB Isomers**

| Compound | Proton | Chemical shift (ppm) | Coupling constants (Hz) |
|---|---|---|---|
| 2,4,6-Trichlorobiphenyl | 3,5 | 7.41 | $J_{2',3'}=J_{5',6'}\sim8.1$ |
| | 2′,6′ | 7.21 | $J_{2',4'}=J_{4',6'}\sim1.4$ |
| | 3′,5′ | 7.45 | $J_{2,5'}=J_{3',6'}\sim0.2$ |
| | 4′ | 7.41 | $J_{3',4'}=J_{4',5'}\sim7.4$ |
| | | | $J_{2',6'}\sim2.1$ |
| | | | $J_{3',5'}\sim1.2$ |
| | | | $J_{3,5'}=0$ |
| 2,4,4′,6-Tetrachlorobiphenyl | 2′,6′ | 7.15 | $J_{2',3'}=J_{5',6'}\sim8.6$ |
| | 3′,5′ | 7.42 | $J_{2',5'}=J_{3',6'}\sim0.2$ |
| | 3,5 | 7.41 | $J_{2',6'}\sim2.0$ |
| | | | $J_{3',5'}\sim2.1$ |
| | | | $J_{3,5}=0$ |
| 2,3,4,5-Tetrachlorobiphenyl | 6 | 7.35 | $J_{2',3'}=J_{5',6'}\sim8.2$ |
| | 2′,6′ | 7.33 | $J_{2',4'}=J_{4',6'}\sim1.7$ |
| | 3′,5′ | 7.40 | $J_{2',5'}=J_{3',6'}\sim0.1$ |
| | 4′ | 7.38 | $J_{2',6'}\sim1.2$ |
| | | | $J_{3',4'}=J_{4',5'}\sim8.0$ |
| | | | $J_{3',5'}\sim2.2$ |
| 2,3,5,6-Tetrachlorobiphenyl | 4 | 7.60 | $J_{2',3'}=J_{5',6'}\sim8.1$ |
| | 2′,6′ | 7.17 | $J_{2',4'}=J_{4',6'}\sim1.1$ |
| | 3′,5′ | 7.45 | $J_{2',5'}=J_{3',6'}\sim0.3$ |
| | 4′ | 7.42 | $J_{2',6'}\sim1.8$ |
| | | | $J_{3',4'}=J_{4',5'}\sim7.3$ |
| | | | $J_{3',5'}\sim1.7$ |
| 2,2′,4,6-Tetrachlorobiphenyl | 3′ | 7.50 | $J_{3',4'}\sim8.5$ |
| | 4′ | 7.38 | $J_{3',5'}\sim0.8$ |
| | 5′ | 7.36 | $J_{3',6'}\sim0.1$ |
| | 6′ | 7.17 | $J_{4',5'}\sim7.5$ |
| | 3,5 | 7.43 | $J_{4',6'}\sim2.0$ |
| | | | $J_{5',6'}\sim7.3$ |
| | | | $J_{3,5}\sim0$ |
| 2,2′,6,6′-Tetrachlorobiphenyl | 3,3′ | 7.41 | $J_{3,4}=J_{3',4'}=J_{4,5}=J_{4',5}\sim8.0$ |
| | 4,4′ | 7.29 | $J_{3,5}=J_{3',5'}\sim0$ |
| | 5,5′ | 7.41 | |
| 2,2′,4,4′-Tetrachlorobiphenyl | 3,3′ | 7.47 | $J_{3,5}=J_{3',5'}\sim2.2$ |
| | 5,5′ | 7.28 | $J_{5,6}=J_{5',6'}\sim8.1$ |
| | 6,6′ | 7.14 | |
| 3,3′,4,4′-Tetrachlorobiphenyl | 2,2′ | 7.59 | $J_{2,6}=J_{2',6'}\sim2.1$ |
| | 5,5′ | 7.49 | $J_{5,6}=J_{5',6'}\sim8.4$ |
| | 6,6′ | 7.33 | |
| 2,2′,3,4′,5′-Pentachlorobiphenyl | 4 | 7.45 | $J_{4,5}\sim7.5$ |
| | 5 | 7.16 | $J_{4,6}\sim1.7$ |
| | 6 | 7.07 | $J_{5,6}\sim7.9$ |
| | 3′ | 7.53 | |
| | 6′ | 7.30 | |

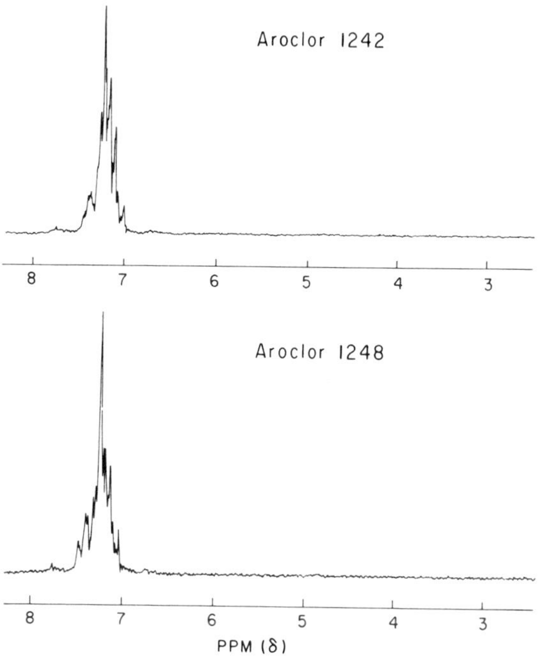


FIGURE 4. Top: NMR spectrum of Aroclor 1242; bottom: NMR spectrum of Aroclor 1248.

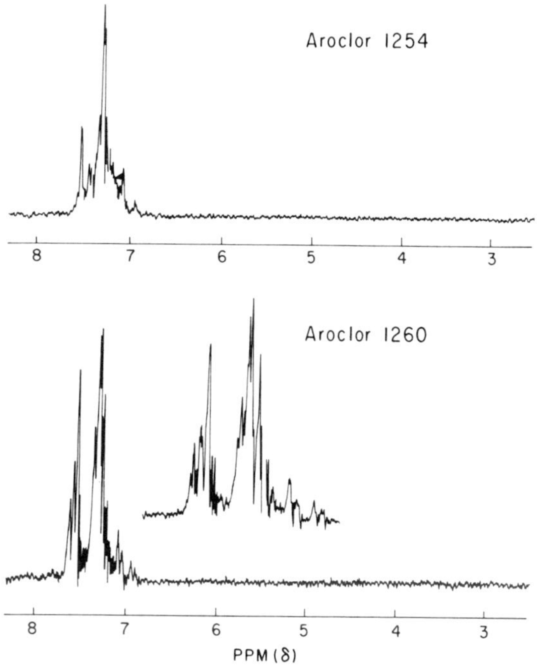


FIGURE 5. Top: NMR spectrum of Aroclor 1254; bottom: NMR spectrum of Aroclor 1260.

TABLE 3

**Relation of Chemical Shifts of *Ortho*-protons to Presence of *Ortho*-chlorine Atoms ($CDCl_3$ Solution)**

(a) No *ortho*-Cl in first ring

| | No. of *ortho*-Cl in second ring | | | | |
|---|---|---|---|---|---|
| First ring | 0 | 1 | 2 | Δ 0 to 1 | Δ 1 to 2 |
| Phenyl | 7.45–7.52 | 7.33 | 7.17–7.21 | 0.15 | 0.13 |
| 4-Chloro | 7.45–7.49 | 7.31–7.35 | 7.15 | 0.14 | 0.18 |
| 3,4-Dichloro | 7.59–7.61 | 7.47 | 7.32 | 0.13 | 0.15 |
| 4,5-Dichloro | 7.33–7.35 | 7.22 | 7.06 | 0.12 | 0.16 |
| | | | | Ave 0.13(5) | 0.15(5) |

(b) One *ortho*-Cl in first ring

| | No. of *ortho*-Cl in second ring | | | | |
|---|---|---|---|---|---|
| First ring | 0 | 1 | 2 | Δ 0 to 1 | Δ 1 to 2 |
| 2,5-Dichloro | – | 7.20–7.23 | 7.17 | | 0.05 |
| 2,4-Dichloro | 7.20 | 7.15 | | 0.05 | |
| 2,3,4-Trichloro | | 7.08 | 7.03 | 0.05 | |
| 2,4,5-Trichloro | 7.38 | 7.31–7.34 | 7.28 | 0.05 | 0.05 |
| | | | | Ave 0.05 | 0.05 |

pounds have been synthesized[9] and their spectra are shown in Figures 8 through 15. Assignments of chemical shifts and coupling constants await a more detailed analysis of these spectra.

TABLE 4

**Comparison of Proton Chemical Shifts (δ ppm) with Varying Substitution in the Second Ring**

| | *Ortho* positions in first ring | | | | | |
|---|---|---|---|---|---|---|
| | '2' | | | '6' | | |
| Substitution in first ring | $CDCl_3$ | $C_2Cl_4$ | Substitution in second ring | $CDCl_3$ | $C_2Cl_4$ | Substitution in second ring |
| None | 7.17 | | 2,3,5,6 | | | |
| | 7.21 | | 2,4,6 | | | |
| | 7.21 | | 2,6 | | | |
| | 7.33 | | 2,3,4,5 | | As for '2' | |
| | 7.45 | | – | | | |
| | 7.48 | | 3,4 | | | |
| | 7.52 | | 4 | | | |
| 2-Cl | | | | 7.17 | | 2,4,6 |
| | | | | 7.23 | | 2 |
| 3-Cl | 7.51 | | 3 | 7.40 | | 3 |
| 4-Cl | 7.15 | | 2,4,6 | | | |
| | 7.31 | | 2,4 | | | |
| | 7.35 | | 2 | | As for '2' | |
| | 7.45 | | 4 | | | |
| | 7.49 | | – | | | |
| 3,4-DiCl | 7.32 | 7.29 | 2,3,6 | 7.06 | 7.00 | 2,3,6 |
| | | 7.44 | 2,3,4,5 | | 7.16 | 2,3,4,5 |
| | | 7.45 | 2,3,4 | | 7.17 | 2,3,4 |
| | 7.47 | 7.46 | 2,4,5 | 7.22 | 7.18 | 2,4,5 |
| | | 7.48 | 2,5 | | 7.20 | 2,5 |
| | 7.59 | 7.57 | 3,4 | 7.33 | 7.28 | 3,4 |
| | 7.61 | | – | 7.35 | | – |
| 2,5-DiCl | | | | 7.17 | 7.15 | 2,3,6 |
| | | | | 7.20 | 7.19 | 2,3,4 |
| | | | | 7.21 | 7.20 | 2,4,5 |
| | | | | 7.23 | 7.21 | 2,5 |
| 2,3-DiCl | | | | 7.11 | 7.07 | 2,4,5 |
| 2,4-DiCl | | | | 7.15 | 7.11 | 2,4 |
| | | | | 7.15 | 7.11 | 2,4,5 |
| | | | | 7.20 | | – |
| | | | | 7.21 | | 4 |
| 2,6-DiCl | | | | | | |
| 3,4,5-TriCl | 7.41 | 7.40 | 2,4,5 | | As for '2' | |
| 2,3,4-TriCl | | | | 7.03 | | 2,3,6 |
| | | | | | 7.01 | 2,3,4 |
| | | | | | 7.01 | 2,3,4,5 |
| | | | | 7.07 | 7.02 | 2,4,5 |
| | | | | 7.08 | 7.02 | 2,5 |
| | | | | | 7.07 | 3,4 |
| 2,4,5-TriCl | | | | 7.28 | 7.25 | 2,3,6 |
| | | | | 7.31 | 7.30 | 2,3,4 |
| | | | | 7.32 | 7.30 | 2,4,5 |
| | | | | 7.32 | 7.31 | 2,4 |
| | | | | 7.33 | 7.30 | 2,3 |
| | | | | 7.34 | 7.31 | 2,5 |
| | | | | 7.38 | 7.35 | 3,4 |
| | | | | 7.38 | 7.37 | 3,4,5 |
| 2,3,4,5-TetraCl | | | | | 7.25 | 2,3,4 |
| | | | | | 7.30 | 3,4 |
| | | | | 7.35 | 7.35 | – |
| 2,3,6-TriCl | | | | | | |
| 2,4,6-TriCl | | | | | | |

TABLE 4 (Continued)

**Comparison of Proton Chemical Shifts (δ ppm) with Varying Substitution in the Second Ring**

| Substitution in first ring | *Meta* positions in first ring: '3' $CDCl_3$ | '3' $C_2Cl_4$ | '3' Substitution in second ring | '5' $CDCl_3$ | '5' $C_2Cl_4$ | '5' Substitution in second ring | *Para* position in first ring: '4' $CDCl_3$ | '4' $C_2Cl_4$ | '4' Substitution in second ring |
|---|---|---|---|---|---|---|---|---|---|
| None | 7.25 | | – | | | | 7.17 | | – |
| | 7.40 | | 3,4 | | | | 7.33 | | 4 |
| | 7.40 | | 2,6 | | | | 7.34 | | 3,4 |
| | 7.40 | | 2,3,4,5 | | As for '3' | | 7.38 | | 2,6 |
| | 7.41 | | 4 | | | | 7.38 | | 2,3,4,5 |
| | 7.45 | | 2,4,6 | | | | 7.41 | | 2,4,6 |
| | 7.45 | | 2,3,5,6 | | | | 7.42 | | 2,3,5,6 |
| 2-Cl | 7.44 | | 2 | 7.28 | | 2 | 7.30 | | 2 |
| | 7.50 | | 2,4,6 | 7.35 | | 2,4,6 | 7.37 | | 2,4,6 |
| 3-Cl | | | | 7.35 | | 3 | 7.32 | | 3 |
| 4-Cl | 7.38 | | – | | | | | | |
| | 7.38 | | 2 | | | | | | |
| | 7.38 | | 2,4 | | As for '3' | | | | |
| | 7.39 | | 4 | | | | | | |
| | 7.42 | | 2,4,6 | | | | | | |
| 3,4-DiCl | | | | 7.44 | | – | | | |
| | | | | 7.49 | 7.43 | 3,4 | | | |
| | | | | 7.49 | 7.43 | 2,4,5 | | | |
| | | | | | 7.43 | 2,5 | | | |
| | | | | | 7.44 | 2,3,4 | | | |
| | | | | | 7.45 | 2,3,4,5 | | | |
| | | | | 7.53 | 7.46 | 2,3,6 | | | |
| 2,5-DiCl | 7.38 | 7.33 | 2,5 | | | | 7.30 | 7.25 | 2,5 |
| | 7.39 | 7.34 | 2,3,4 | | | | 7.31 | 7.25 | 2,3,4 |
| | 7.40 | 7.34 | 2,4,5 | | | | 7.32 | 7.27 | 2,4,5 |
| | 7.43 | 7.36 | 2,3,6 | | | | 7.34 | 7.28 | 2,3,6 |
| 2,3-DiCl | | | | 7.25 | 7.16 | 2,4,5 | 7.50 | 7.45 | 2,4,5 |
| 2,4-DiCl | 7.44 | | – | 7.23 | | – | | | |
| | 7.46 | | 4 | 7.27 | | 4 | | | |
| | 7.47 | 7.45 | 2,4 | 7.28 | 7.23 | 2,4 | | | |
| | 7.49 | 7.46 | 2,4,5 | 7.30 | 7.25 | 2,4,5 | | | |
| 2,6-DiCl | 7.33 | | – | | As for '3' | | 7.14 | | – |
| | 7.41 | | 2,6 | | | | 7.30 | | 2,6 |
| 3,4,5-TriCl | | | | | | | | | |
| 2,3,4-TriCl | | | | | 7.36 | 3,4 | | | |
| | | | | 7.43 | 7.36 | 2,5 | | | |
| | | | | | 7.37 | 2,3,4 | | | |
| | | | | 7.45 | 7.38 | 2,4,5 | | | |
| | | | | | 7.40 | 2,3,4,5 | | | |
| | | | | 7.48 | | 2,3,6 | | | |
| 2,4,5-TriCl | | | | 7.56 | 7.53 | 3,4 | | | |
| | | | | 7.57 | 7.53 | 2,3 | | | |
| | | | | 7.57 | 7.54 | 2,4 | | | |
| | | | | 7.58 | 7.54 | 2,5 | | | |
| | | | | 7.58 | 7.55 | 2,4,5 | | | |
| | | | | 7.58 | 7.55 | 3,4,5 | | | |
| | | | | 7.59 | 7.55 | 2,3,4 | | | |
| | | | | 7.61 | 7.57 | 2,3,6 | | | |
| 2,3,4,5-TetraCl | | | | | | | | | |
| 2,3,6-TriCl | | | | 7.42 | 7.35 | 3,4 | 7.34 | 7.27 | 3,4 |
| | | | | 7.45 | 7.38 | 3,5 | 7.35 | 7.28 | 2,5 |
| | | | | 7.46 | | 2,3,4 | 7.35 | 7.28 | 2,4,5 |
| | | | | 7.46 | 7.40 | 2,4,5 | 7.35 | | 2,3,4 |
| 2,4,6-TriCl | 7.40 | | – | | As for '3' | | | | |
| | 7.40 | | 4 | | | | | | |

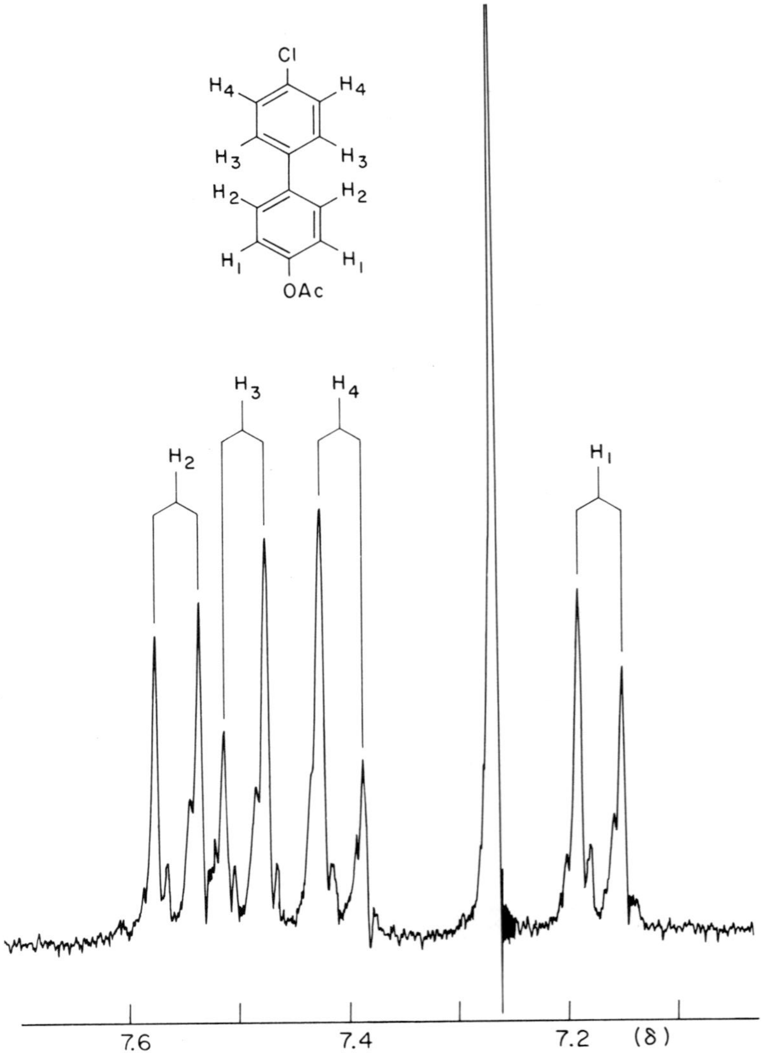


FIGURE 6. NMR spectrum of 4-acetoxy-4′-chlorobiphenyl.

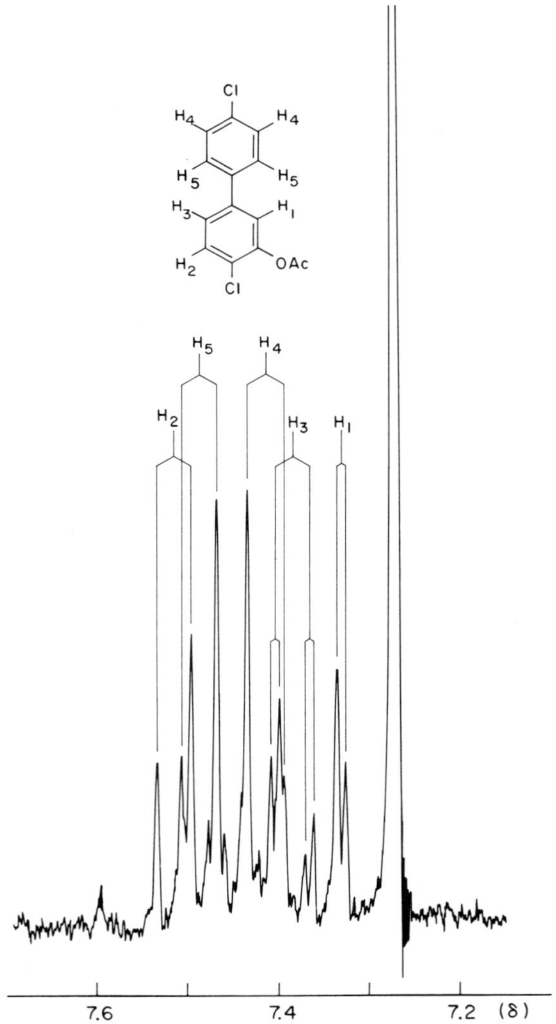


FIGURE 7. NMR spectrum of 3-acetoxy-4,4′-dichloro-biphenyl.

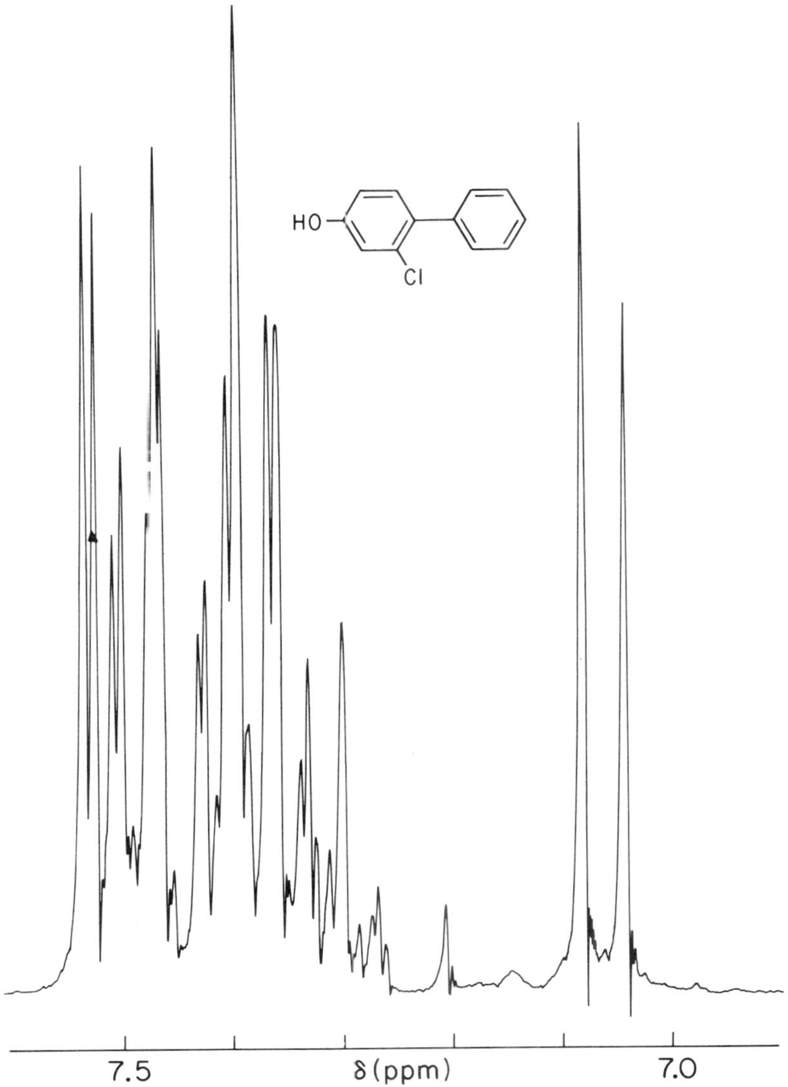

FIGURE 8. NMR spectrum of 2-chloro-4-hydroxybiphenyl.

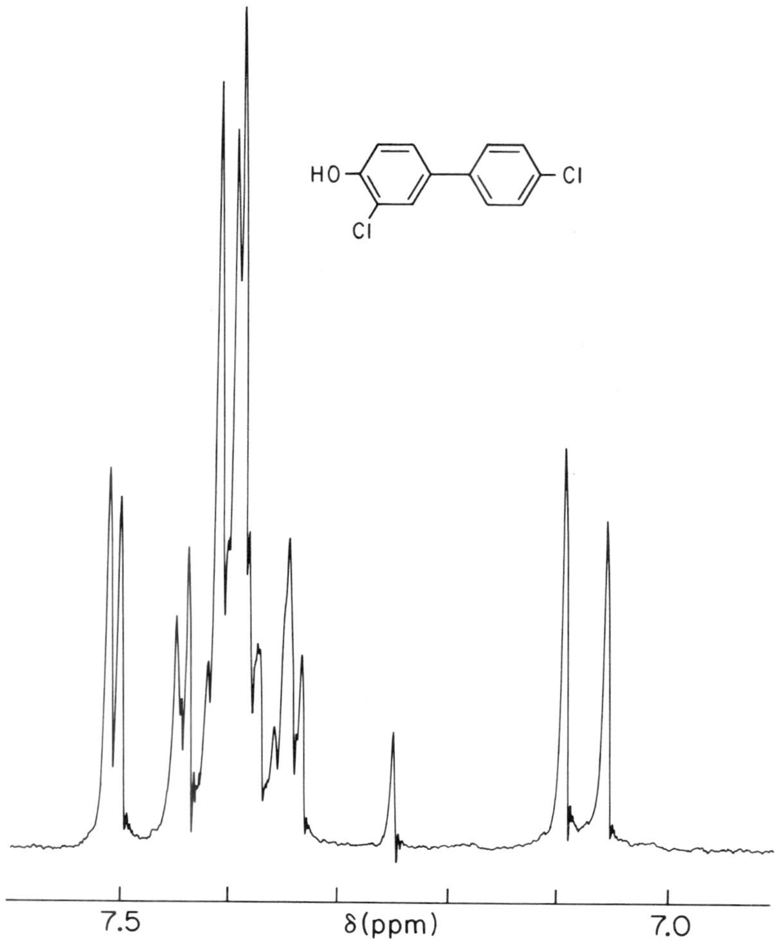

FIGURE 9. NMR spectrum of 3′,4-dichloro-4-hydroxybiphenyl.

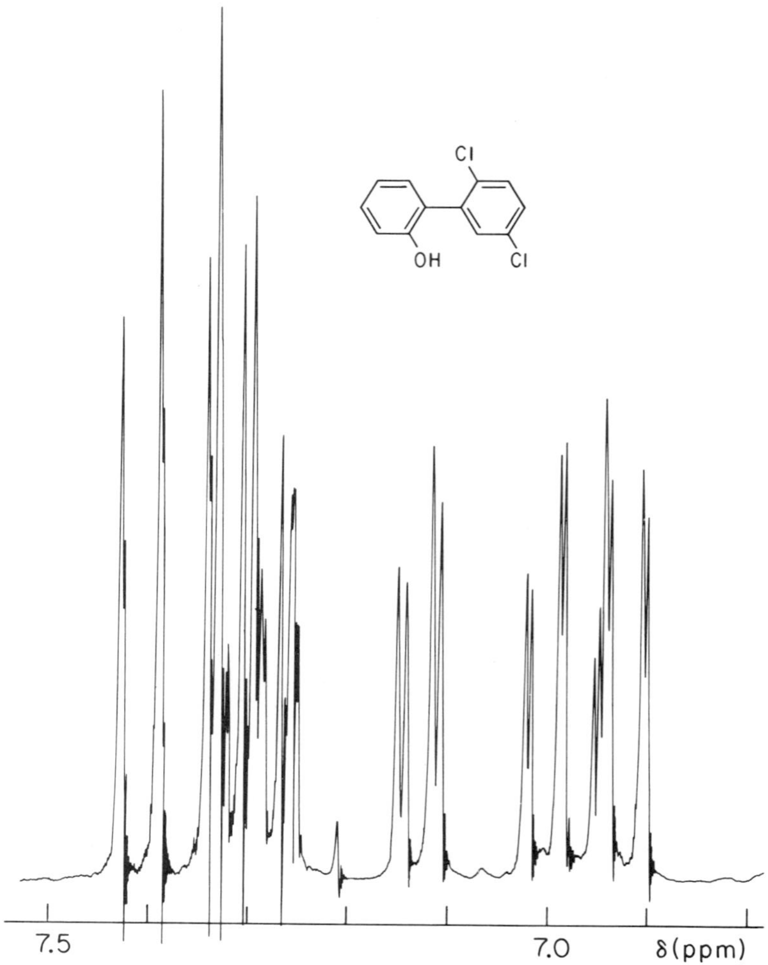


FIGURE 10. NMR spectrum of 2,5-dichloro-2′-hydroxybiphenyl.

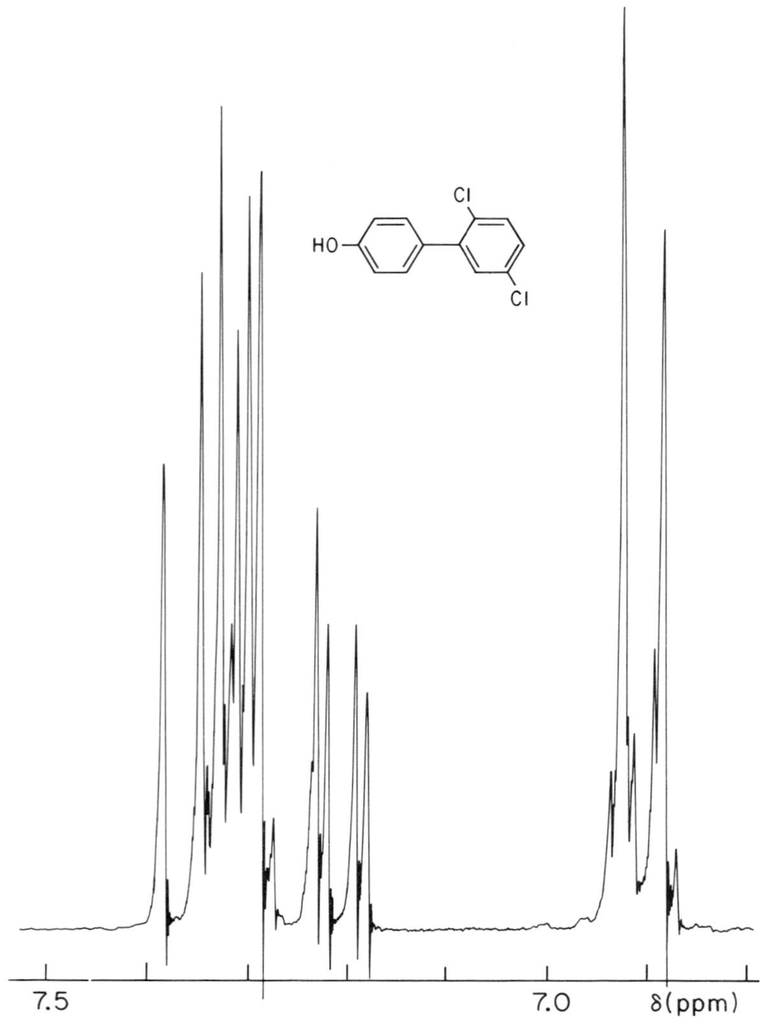


FIGURE 11. NMR spectrum of 2,5-dichloro-4′-hydroxybiphenyl.

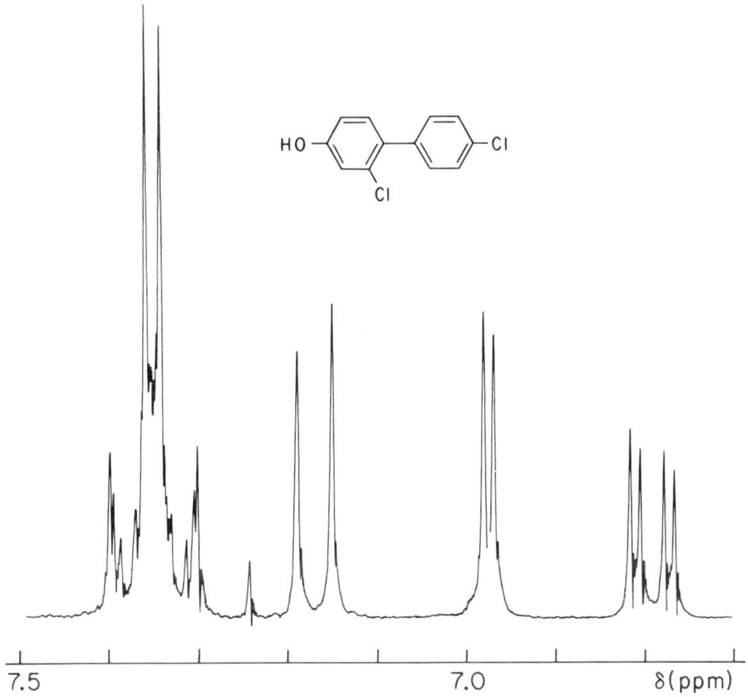


FIGURE 12. NMR spectrum of 2,4′-dichloro-4-hydroxybiphenyl.

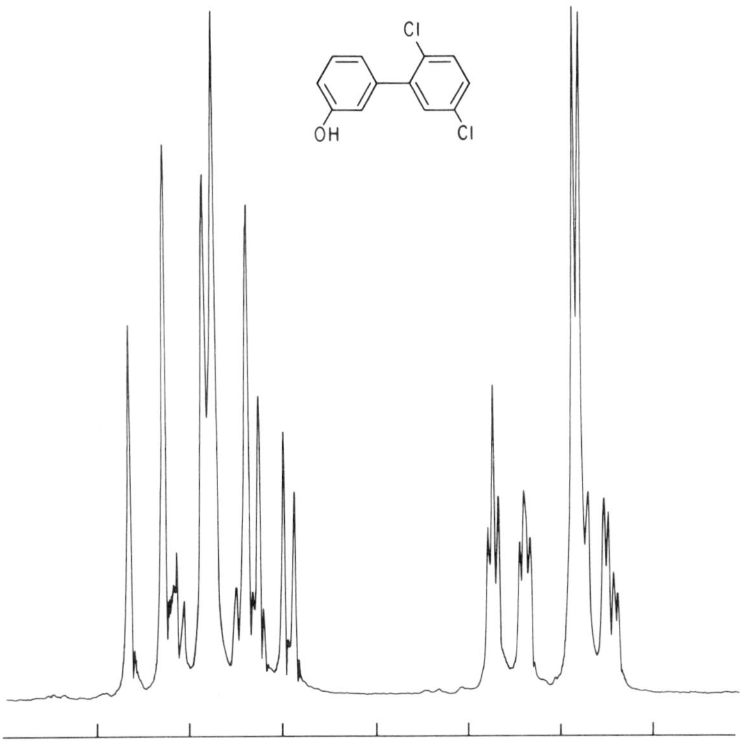


FIGURE 13. NMR spectrum of 2,5-dichloro-3′-hydroxybiphenyl.

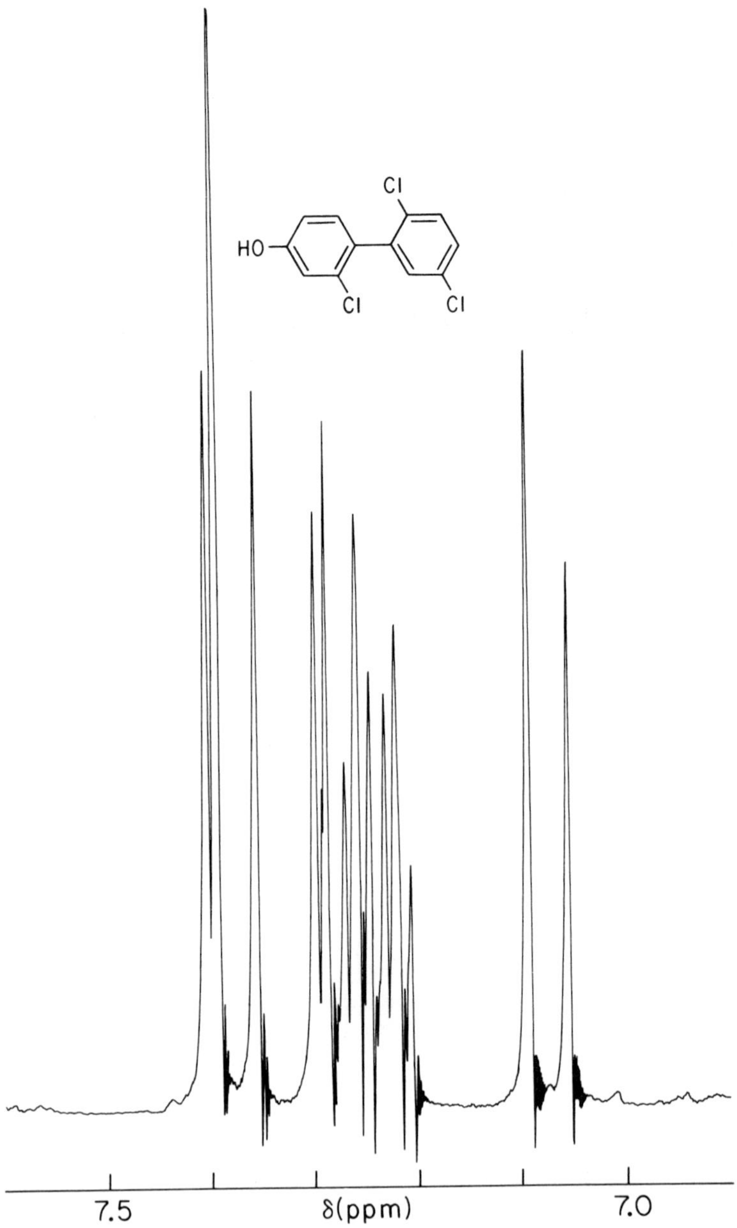


FIGURE 14. NMR spectrum of 2,2′,5′-trichloro-4-hydroxybiphenyl.

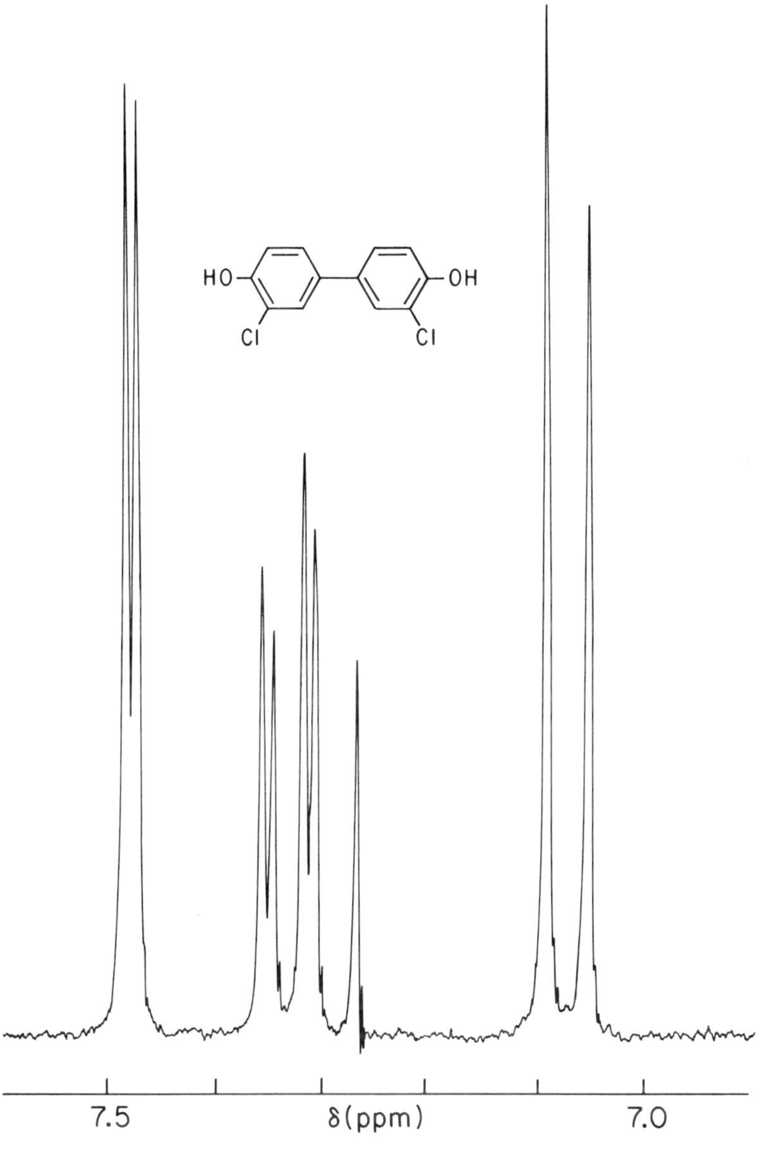


FIGURE 15. NMR spectrum of 3,3′-dichloro-4,4′-dihydroxybiphenyl.

# REFERENCES


1. **Bartle, K. D.,** *J. Assoc. Offic. Anal. Chem.,* 55, 1101, 1972.
2. **Brownstein, S.,** *J. Am. Chem. Soc.,* 80, 2300, 1958.
3. **Grant, D. N., Hirst, R. C., and Gutowsky, H. S.,** *J. Chem. Phys.,* 38, 470, 1963.
4. **Hoffman, R. A., Kinell, P. O., and Bergström, G.,** *Arch. Chem.,* 15, 533, 1960.
5. **Hutzinger, O., Jamieson, W. D., Safe, S., MacNeil, J. D., and Zitko, V.,** *Symposium on PCB, Still-Prevalent, Still-Persistent,* Proceedings 164th National ACS Meeting, New York, 1972, Marcel Dekker, New York, in preparation.
6. **Hutzinger, O., Nash, D. M., Safe, S., DeFreitas, A. S. W., Norstrom, R. J., Wildish, D. J., and Zitko, V.,** *Science,* 178, 312, 1972.
7. **Hutzinger, O., Safe, S., Ecobichon, D. J., DeFreitas, A. S. W., Norstrom, R. J., MacNeil, J. D., Wildish, D. J., and Zitko, V.,** *Symposium on PCB, Still-Prevalent, Still-Persistent,* Proceedings 164th National ACS Meeting, New York, 1972, Marcel Dekker, New York, in preparation.
8. **Hutzinger, O., Safe, S., and Zitko, V.,** *Environ. Health Persp.,* 1, 15, 1972.
9. **Hutzinger, O., Safe, S., and Zitko, V.,** unpublished results.
10. **Kurland, R. J. and Wise, W. B.,** *J. Am. Chem. Soc.,* 86, 1877, 1964.
11. Monsanto Methodology for Aroclor, Dec. 1970.
12. **Nakazaki, M. and Isoe, S.,** *Chem. Ind. (Lond.),* 468, 1965.
13. **Nakazaki, M. and Yamamoto, K.,** *Chem. Ind. (Lond.),* 224, 1965.
14. **Nomura, Y. and Takeuchi, Y.,** *J. Chem. Soc.,* (B), 956, 1970.
15. **Safe, S. and Hutzinger, O.,** *Nature,* 232, 641, 1971.
16. **Sissons, D. and Welti, D.,** *J. Chromatogr.,* 60, 15, 1971.
17. **Tarpley, A. R. and Goldstein, J. H.,** *J. Phys. Chem.,* 75, 421, 1971.
18. **Tas, A. C. and deVos,** *Environ. Sci. Toxicol.,* 5, 12, 1971.
19. **Tas, A. C. and Kleipool, R. J. C.,** *Bull. Environ. Contam. Toxicol.,* 8, 32, 1972.
20. **Wallnöfer, P. R., Engelhardt, G., Safe, S., and Hutzinger, O.,** *Chemosphere,* 69, 1973.
21. **Welti, D. and Sissons, D.,** *Org. Mag. Res.,* 4, 309, 1972.
22. **Yamamoto, K., Horikawa, T., and Nakazaki, M.,** *Tetrahedron Lett.,* 4551, 1969.

Chapter 10

# ULTRAVIOLET SPECTROSCOPY OF CHLOROBIPHENYLS

Due to the widespread occurrence of PCB in the environment, there has recently been considerable interest in the spectroscopic properties of these compounds.[8,10-12] The UV spectra of biphenyl and several substituted biphenyls have been the subject of a number of investigations.[2-5,9] A number of factors have been shown to affect the UV absorption bands and these can be summarized as follows:

1. The nature of the substituent,
2. The location of the substituent,
3. The number of substituents present in the biphenyl nucleus, and
4. Particularly, the degree of substitution at the positions *ortho* to the Ph-Ph bond (i.e., 2,2′,6 and 6′ positions).

The UV spectrum of biphenyl features two important absorption maxima: one at λmax 202 nm (ϵ, 44,000) and a second at λmax 242 nm (ϵ, 17,000). The former band is generally referred to as the "main band" and the latter is generally known as the κ band which is attributed to the conjugated biphenyl system with the contributions of both phenyl rings. The effects of chlorine substitution on the UV spectrum of biphenyl are summarized in the data shown in Tables 1 and 2.

## Chlorinated Biphenyls with Less than Two Chlorine Atoms *Ortho* to the Ph-Ph Bond

The results given in Tables 1 and 2 will be discussed according to degree of *ortho* Cl substitution since this factor markedly affects the UV spectral features. The data shown in Table 1 indicated that both the number and position of the Cl substituent caused some significant variations in the position and intensity of both the main absorption band and the κ band in the UV spectra.

For the monochlorobiphenyls, the main absorption band was only slightly changed in comparison to biphenyl. However, the 4- and 3-chloro groups induced a marked bathochromic shift of κ band with this shift greater for the *para* substituent (ca. 13 nm) than for the *meta*substituent (ca. 6 nm) (see Figure 1). The κ band for 2-chlorobiphenyl is shifted to slightly lower wave lengths with a

TABLE 1

**UV Spectra of Chlorobiphenyls (None or One *Ortho* Chlorine)**

| Chlorinated biphenyl Isomer | UV Maxima and Extinction Coefficients (x $10^{-3}$) "Main band" (nm) | κ band (nm) | Reference |
|---|---|---|---|
| 4 | 199 (43.3) | 253 (20.5) | 8 |
| 3 | 205 (42.8) | 248 (16.0) | 8 |
| 2 | 204 (39.2) | 240 (10.2) | 8 |
| 4,4′ | 200 (41.9) | 258 (22.9) | 8 |
| 3,3′ | | 248 (23.4) | 9 |
| 2,4 | 204 (42.2) | 255 (12.8) | 8 |
| 2,4,4′ | 205 (42.5) | 250 (14.8) | 8 |
| 2,3′,4 | 210 (44.9) | 246 (12.0) | 8 |
| 2,3′,4′,5 | 214 (42.0) | 248 (11.3) | 8 |
| 3,3′,4,4′ | | 260 (22.9) | 10 |
| 2,3′,4,4′ | | 253 (15.9) | 10 |
| 2,4,4′,5 | | 257 (15.1) | 10 |
| 2,3,4,4′ | | 250 (12.6) | 10 |
| 2,3′,4,4′,5 | | 253 (2.5) | 10 |
| 2,3,3′,4,4′ | | 253 (2.5) | 10 |
| 3,3′,4,4′,5,5′ | 222 (51.7) | 265 (27.7) | 8 |

TABLE 2

**Ultraviolet Spectra of Chlorobiphenyls (Two or More *Ortho* Chlorines)**

| Chlorinated biphenyl Isomer | UV Maxima and Extinction Coefficients (x $10^{-3}$) "Main band" (nm) | $\kappa$ band (nm) | "$\beta$ bands" (nm) | Reference |
|---|---|---|---|---|
| 2,2′ | 208 (36.0) | 230 (6.6) | 273 (.54)<br>266.5 (.74) | 3 |
| 2,2′,5 | 197 (62.5) | | 267 (1.10)<br>275 (1.17)<br>283 (0.82) | 8 |
| 2,2′,4,4′ | 207 (51.2) | 220 (29.4)* | 273 (1.49)<br>282 (0.83) | 8 |
| 2,2′,5,5′ | 204 (43.3) | 214 (34.9)* | 276 (1.32)<br>284 (1.25) | 8 |
| 2,2′,6,6′ | 197 (88.9) | | 272 (0.78)<br>280 (0.65) | 8 |
| 2,2′,4,4′,5,5′ | 211 (45.5) | | 282 (1.60)<br>290 (1.12) | 8 |
| 2,2′,4,4′,6,6′ | 202 (93.1) | | 267 (0.50)<br>275 (0.59)<br>288 (0.46) | 8 |
| 2,2′,3,4,5,5′,6 | 214 (100) | | 268 (1.37)<br>277 (1.76)<br>286 (1.82)<br>297 (0.63) | 8 |
| 2,2′,3,3′,4,4′,5,5′ | 210 (57.5) | | 285 (0.69)<br>294 (0.59) | 8 |
| 2,2′,3,3′,5,5′,6,6′ | 210 (91.6) | | 285 (2.04)<br>295 (2.31) | 8 |
| 2,2′,3,3′,4,4′,5,5′,6,6′ | 216 (108) | | 291.5 (1.10)<br>301.5 (1.22) | 8 |

*Shoulder

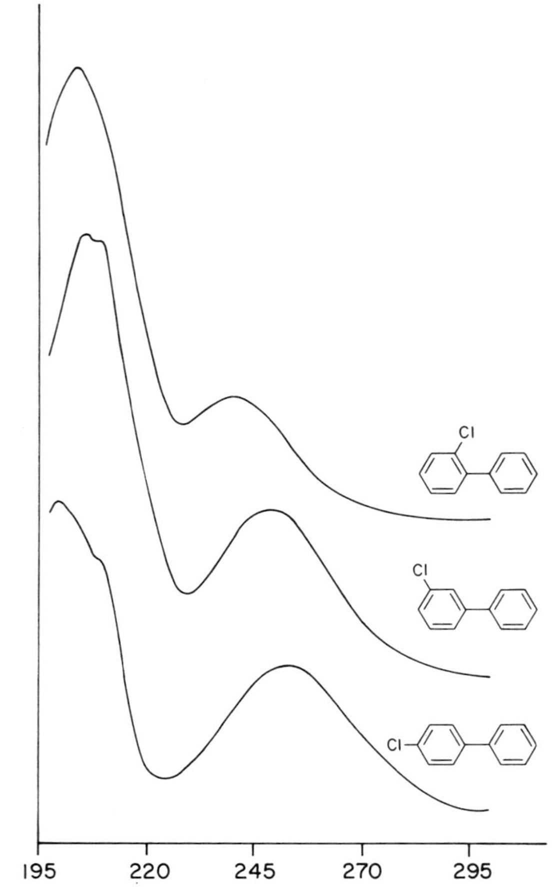


FIGURE 1. Ultraviolet spectra of the isomeric chlorobiphenyls.

diminished $\epsilon$ value. This effect was undoubtedly due to some steric inhibition of resonance between the two phenyl groups with the resultant hypsochromic shift and diminished intensity of the absorption maxima of the conjugation $\kappa$ band. This effect will be discussed in detail for the more highly hindered isomers given in Table 2.

Similarly, for the 4,4′- and 3,3′-isomers, the magnitude of the bathochromic shift for the $\kappa$ absorption maximum is greater for the *para* disubstituted derivative.

For the more highly substituted chlorinated biphenyl isomers both the main absorption band and the $\kappa$ band are shifted towards the visible with increasing Cl substitution (see Figure 2).

**Chlorinated Biphenyls with Two or More Chlorine Atoms *Ortho* to the Ph-Ph Bond**

The spectroscopic properties of the more highly hindered PCB isomers are shown in Table 2. It has

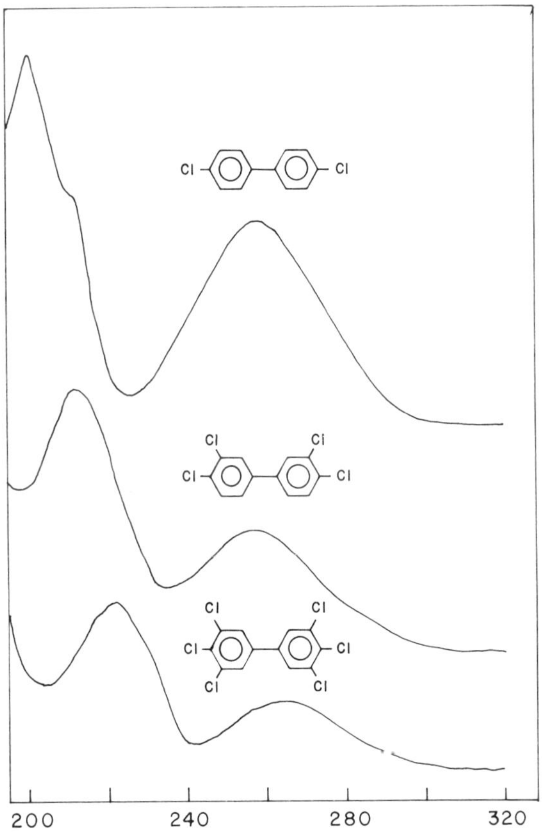


FIGURE 2. Ultraviolet spectra of 4,4′-dichlorobiphenyl, 3,3′,4,4′-tetrachlorobiphenyl and 3,3′,4,4′,5,5′-hexachlorobiphenyl.

been previously shown that introduction of substituents into the *ortho* positions (2,2′,6′, and 6′) of the biphenyl nucleus induces major changes in the UV absorption spectra.[2,3,7,9] It is generally accepted that in the highly hindered *ortho*-substituted biphenyls there is a considerable hindrance to free rotation which results in the loss of coplanarity between the two phenyl rings. This results in a subsequent marked diminution of the extinction coefficient of the conjugation band ($\kappa$ band) with a shift of the absorption maxima towards the ultraviolet. Inspection of the UV spectra of the more highly substituted chlorinated biphenyl isomers (Table 2) revealed a number of useful correlations.

Introduction of two or more *ortho* chloro substituents resulted in a marked diminution of the $\epsilon$ value of the $\kappa$ band which was also shifted to the ultraviolet. This is seen in the UV spectra of 2,2′-dichlorobiphenyl, 2,2′,4,4′-biphenyl, and 2,2′,5,5′-biphenyl where the absorption maxima appears as a shoulder on the more intense band in the 197 to 216 nm region of the spectra. For the remaining isomers, the $\kappa$ band was not observed and was completely masked by the above-mentioned absorption maxima. The effect of increasing *ortho*-chlorine substitution is illustrated in Figure 3 with a series of tetrachlorobiphenyls.

It was also observed that the intensity of the 197 to 216 nm absorption maxima was also affected by increasing substitution at the *ortho* position. The extinction coefficient of this band for the 2,2′,4,4′- and 2,2′,5,5′-tetrachlorobiphenyl isomers was 51,200 and 43,400, respectively, but the value for the 2,2′,6,6′-isomer was 88,900. This effect was also evident in comparing the 2,2′,4,4′,5,5′- and 2,2′,4,4′,6,6′-hexachlorobiphenyls and the 2,2′,3,3′,4,4′,5,5′- and 2,2′,3,3′,4,4′,6,6′-octachlorobiphenyls.

These highly *ortho*-substituted PCB isomers also exhibited a series of weak absorption maxima between 268 and 302 nm (e.g., Figure 4). These bands were reminiscent of the fine-structured B bands observed in the UV spectra of benzene and substituted benzenes. The signals are attributed to forbidden transitions to an excited state with increased contributions from homopolar structures.[6] Thus the maxima in the 268 to 302 nm region can be attributed to the individual contribution of the phenyl rings of the biphenyl nucleus. It was previously noted by Pickett and co-workers[9]

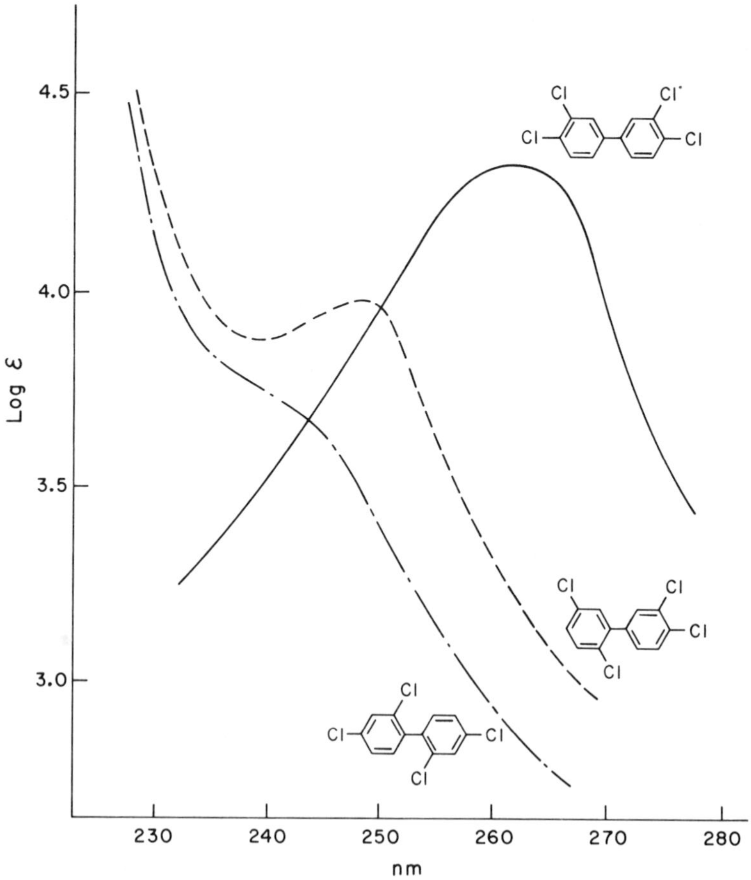


FIGURE 3. Ultraviolet spectra of some isomeric tetrachlorobiphenyls.

that the "B band" absorption maxima obtained for 2,2′,4,4′,6,6′-hexachlorobiphenyl were remarkably coincident with the B band maxima obtained for 1,3,5-trichlorobenzene. Moreover the extinction coefficients of the band maxima for the biphenyl derivative were two to three times greater than those observed for the trichlorobenzene isomer indicating that the absorption is nearly additive for the two benzene rings of the biphenyl derivative. Similar results have been obtained with highly hindered alkyl biphenyls. Ballester and Castaner[1] have reported that the B bands for pentachlorobenzene were at λmax 289 (ε, 390) and λmax 298 (ε, 370). The position of these maxima was similar to that obtained for decachlorobiphenyl (Table 2) with the extinction coefficients ca. 2 to 2.5 times greater in the latter compound.

Thus, it is clear that the UV spectra of PCB isomers are highly diagnostic particularly with respect to the degree of substitution at the 2,2′,6 and 6′ positions. The UV data have been used in the identification of some of the main isomers of Kanechlor[10] and the UV spectra of several commercial Aroclors have also been reported[11,12] (Figure 5). The Aroclor spectra are all different and quite clearly reflect their respective chlorine contents. In addition the spectrum of the highly chlorinated Aroclor 1260 sample also exhibited the characteristic long wave length "B bands" which indicated a relatively high degree of *ortho*-chlorosubstituted isomers in this mixture.

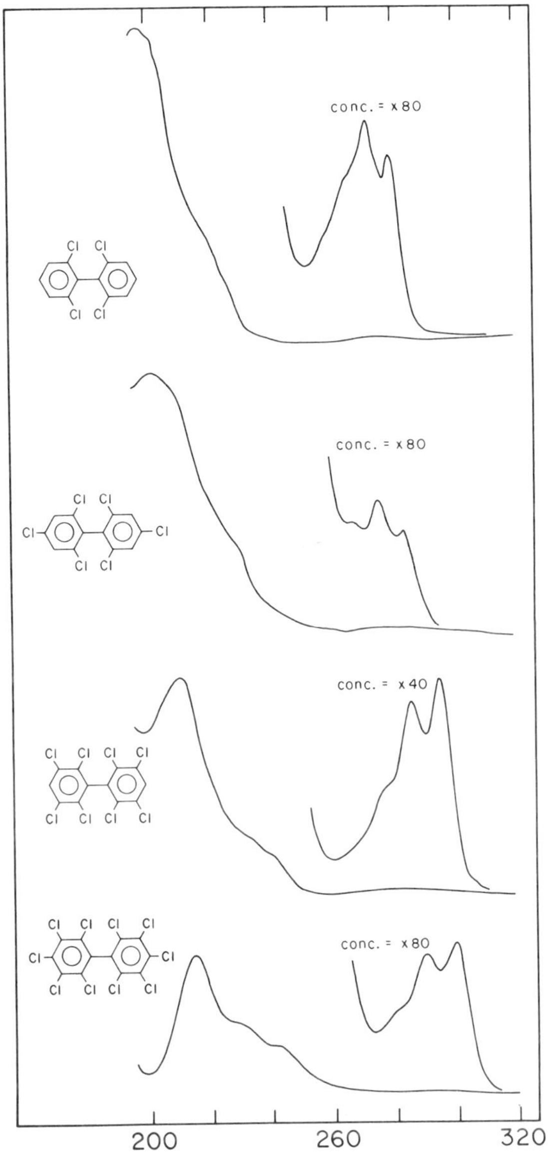


FIGURE 4. Ultraviolet spectra of some substituted 2,2′,6,6′-tetrachlorobiphenyls.

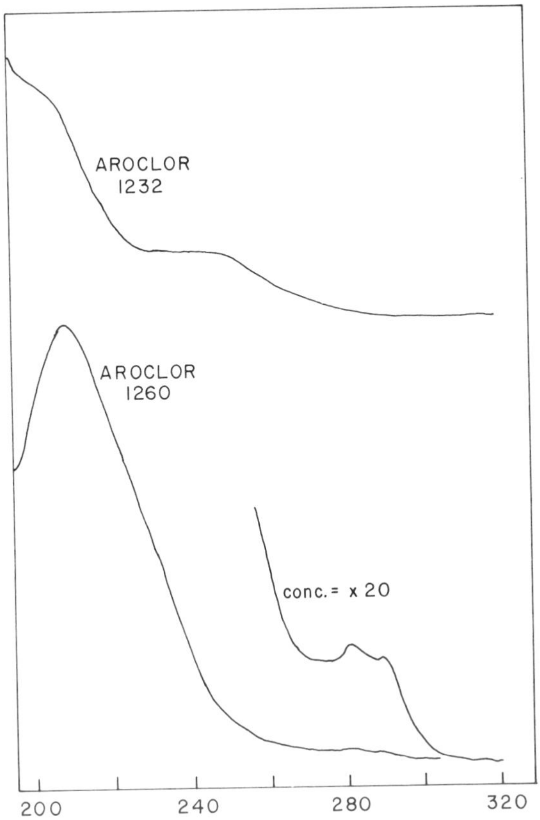


FIGURE 5. Ultraviolet spectra of Aroclor 1232 and 1260.

## REFERENCES


1. **Ballester, M. and Castaner, J.,** *J. Am. Chem. Soc.,* 82, 4259, 1960.
2. **Beaven, G. H.,** in *Steric Effects of Conjugated Systems,* Grey, G. W., Ed., *Chem. Soc. Lond.,* Butterworths, 1958.
3. **Beaven, G. H. and Hall, M. D.,** *J. Chem. Soc.,* 4637, 1956.
4. **Brocklehurst, P., Burawoy, A., and Thompson, A. R.,** *Tetrahedron,* 10, 102, 1960.
5. **Burawoy, A. and Thompson, A. R.,** *J. Chem. Soc.,* 4319, 1956.
6. **Gillam, A. E. and Stern, E. S.,** in *An Introduction to Electronic Absorption Spectroscopy in Organic Chemistry,* 2nd ed., Edward Arnold, London, 1957.
7. **Hall, D. M. and Minhay, F.,** *J. Chem. Soc.,* 4584, 1957.
8. **MacNeil, J. D., Safe, S., and Hutzinger, O.,** unpublished results.
9. **Pickett, L. W., Walter, G. F., and France, H.,** *J. Am. Chem. Soc.,* 58, 2296, 1936.
10. **Saeki, S., Tsutsui, A., Oguri, K., Yoshimura, H., and Hamana, M.,** *Fukuoka Med. J.,* 62, 20, 1971.
11. **Zitko, V.,** Manuscript Report Series, No. 1083, Fisheries Research Board of Canada, 1970.
12. **Zitko, V.,** *Bull. Environ. Contam. Toxicol.,* 5, 279. 1970.

Chapter 11

# INFRARED SPECTROMETRY OF CHLOROBIPHENYLS

Several groups have reported the infrared (ir) and Raman spectra of biphenyl and some of its deuterated analogs.[2,3,7,8,11] These studies were not only concerned with the molecular conformations of the molecule but with the identification of characteristic ir bands which are useful in analytical and structural studies. Sandroni and Geiss[7] have discussed in detail the main ir bands of biphenyl which occur in the three different regions of the spectrum:

1. 4,000 to 2,000 $cm^{-1}$, (C-H stretching vibrations)
2. 2,000 to 1,250 $cm^{-1}$ (C=C stretching vibrations)
3. 1,250 to 250 $cm^{-1}$ (bending vibrations and deformations)

The results obtained for biphenyl are shown in Table 1. Using specifically deuterated moieties, it was also possible to assign the specific contributions of the individual hydrogen atoms of the biphenyl nucleus. Biphenyl appears to be planar in the solid state but the configuration of biphenyl in solution has not unequivocally been ascertained.

The ir spectra of 4,4′-dichlorobiphenyl and several other disubstituted biphenyls have also been investigated. In all cases, the solution spectra of dichlorobiphenyl indicated that the molecule favors a planar conformation.[1,4-6]

Webb and McCall[12] have recently used infrared spectroscopy in the analysis of commercial Aroclor 1221, 1242, and 1254 samples. In the analysis of Aroclor 1242, the fractions with retention times of 29 to 29.5 (relative to *p,p*′-DDE) consisted of two phases. The solid phase was recovered (retention time 29.5) and shown to be identical to 4,4′-dichlorobiphenyl. The ir spectrum of the liquid phase is shown (Figure 1) and repeated gas chromatographic analysis indicated the presence of two isomers with retention times of 29 and 29.2, respectively. The ir spectra of 2,5,2,′- and 2,4,2′-trichlorobiphenyl and a mixture of two isomers (see Figure 1) clearly indicated that these two isomers were the components of the above liquid phase. The infrared spectrum of an Aroclor 1254 fraction which contained two peaks, a major component (relative retention time 48.0) and a minor component (relative retention time 49.5) is shown in Figure 2. Comparison of the ir spectra of 2,2′,5,5′- and 2,2′,4,5′-tetrachlorobiphenyl with the Aroclor fraction clearly indicates that the 2,2′,5,5′ isomer is the major component with the minor constituent being the 2,2′,4,5′ isomer. Thus, the characteristic ir spectra of PCB isomers particularly in the 1,200 to 300 $cm^{-1}$ region which denotes the aromatic C-H bending vibrations and deformations as well as the C-Cl stretching vibrations are a useful analytical tool for the identification of the components of complex commercial

TABLE 1

**Ir Frequencies for Biphenyl***

| 4,000–2,000 $cm^{-1}$ | 2,000–1,250 $cm^{-1}$ | 1,250 → 300 |
|---|---|---|
| 3,140 vw | 1,962 m | 1,238 w |
| 3,126 sh | 1,945 m | 1,217 vw |
| 3,115 sh | 1,900 w | 1,176 w |
| 3,106 w | 1,883 m | 1,156 m |
| 3,082 sh | 1,869 m | 1,108 m |
| 3,060 vs | 1,823 w | 1,074 m |
| 3,030 vs | 1,803 m | 1,043 m |
| 3,020 s-sh | 1,782 vw | 1,008 m |
| 2,980 w | 1,758 w | 992 w |
| 2,930 w | 1,745 w | 983 vw |
| 2,875 w | 1,687 w | 965 vw |
| 2,620 vw | 1,675 w | 919 m |
| 2,600 vw | 1,610 w | 903 s |
| 2,315 w | 1,597 s | 839 m |
| | 1,572 m | 780 s |
| | 1,482 vs | 735 vs |
| | 1,457 m | 698 vs |
| | 1,432 vs | 669 m |
| | 1,383 m | 666 w |
| | 1,340 w | 654 vw |
| | 1,317 w | 626 vw |
| | 1,303 w | 609 m |
| | 1,283 w | 546 m |
| | 1,269 m | 488 m |
| | | 404 w |
| | | 383 w |

*vs = very strong; s = strong; m = medium; w = weak; vw = very weak; br = broad; sh = shoulder; imp = band due to impurity.

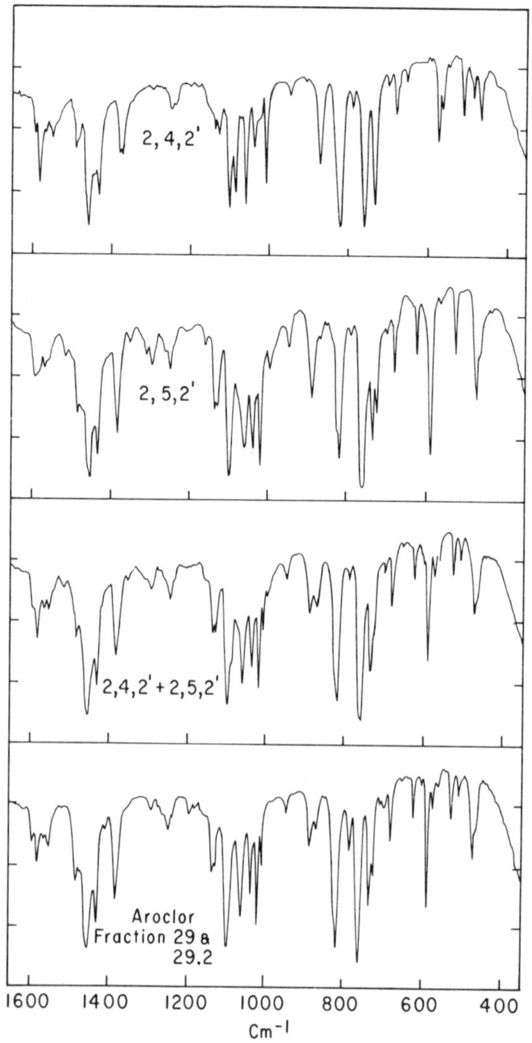


FIGURE 1. Infrared spectra of authentic 2,2′,4-trichlorobiphenyl, 2,2′,5-trichlorobiphenyl, a mixture of 2,2′,4- and 2,2′,5-trichlorobiphenyl and Aroclor fraction 29 and 29.2 respectively.

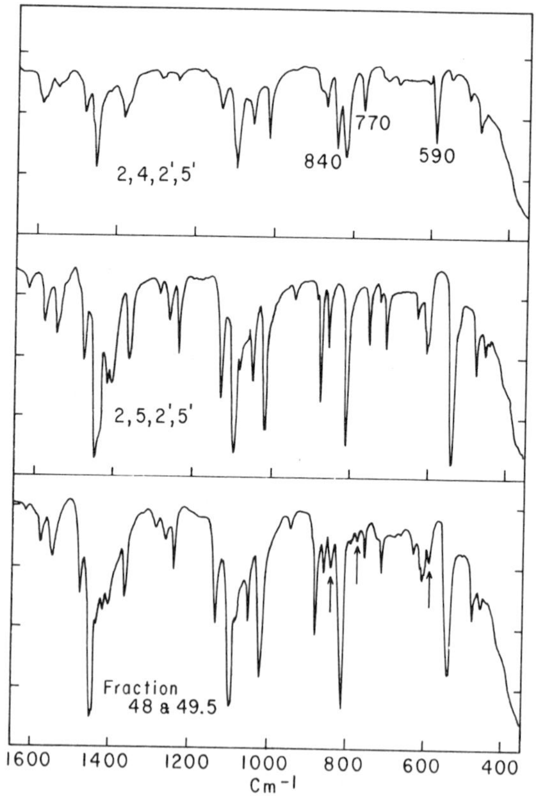


FIGURE 2. Infrared spectra of 2,2′,4,5′-tetrachlorobiphenyl, 2,2′,5,5′-tetrachlorobiphenyl and Aroclor fractions 48 and 49.5 respectively.

PCB formulations. This technique will be of more importance in the future with the use of the highly sensitive Fourier transform techniques and the availability of many pure PCB standards.

Recent analyses of PCB mixtures have also used ir spectrometry as an aid in establishing the identity of several pure components of these mixtures.[9,10]

## REFERENCES


1. **Barrett, R. M. and Steele, D.,** *J. Mol. Struct.,* 11, 105, 1972.
2. **Katon, J. E. and Lippincott, E. R.,** *Spectrochim. Acta,* 15, 627, 1959.
3. **Kouner, M. A.,** *Opt. Spektrosk.,* 1, 742, 1956.
4. **Michelsen, H., Klaeboe, P., Hagen, G., and Stroger-Hansen, T.,** *Acta Chem. Scand.,* 26, 1576, 1972.
5. **Nanni, P., Viani, F., and Lorenzelli, V.,** *J. Mol. Struct.,* 6, 133, 1970.
6. **Nanni, P., Viani, F., and Pontoglio, A.,** *Ann. Chim.,* 60, 132, 1970.
7. **Sandroni, S. and Geiss, F.,** *Spectrochim. Acta,* 22, 235, 1966.
8. **Steele, D. and Lippincott, E. R.,** *J. Mol. Spectrosc.,* 6, 238, 1961.
9. **Tas, A. C. and deVos, R. H.,** *Environ. Sci. Technol.,* 5, 1216, 1971.
10. **Tas, A. C. and Kleipool, R. J. C.,** *Bull. Environ. Contam. Toxicol.,* 8, 32, 1972.
11. **Tomoko, O. and Oikawa, H.,** *Nippon Kagaku Zasshi,* 86, 366, 1965.
12. **Webb, R. G. and McCall, A. C.,** *J. Assoc. Offic. Anal. Chem.,* 15, 746, 1972.

Chapter 12

# DETERMINATION OF PCB'S

Many technical materials may contain or be contaminated by PCB's and great care must be exercised to avoid the contamination of samples from this source during sampling and analysis. Otherwise, the determination of PCB's in environmental samples differs little from that of chlorinated hydrocarbon pesticides. The individual stages including the sampling itself, extraction, cleanup, separation of PCB's from interfering chlorinated hydrocarbon pesticides and other compounds, and quantitation of PCB's are discussed below. A brief review of current environmental levels of PCB's is also included.

## Sampling

The techniques of obtaining representative samples are discussed in monographs on statistical methods. There are a few additional points which should be emphasized in the case of water and wildlife samples.

It is difficult to distinguish between PCB's dissolved in water and PCB's adsorbed on particulate matter. Filtration of samples prior to extraction removes the particulate fraction, but dissolved PCB's may also be adsorbed on the filter, or, inversely adsorbed PCB's may be desorbed in the filtration process. High-speed centrifugation of the samples may be better than filtration but sufficient data on this are not yet available. In sampling of plankton, large volumes of water pass through a fine net and the already accumulated plankton. Adsorption and desorption of PCB's as in the case of water samples is a distinct possibility. Recently, Harvey and Teal showed that nylon netting adsorbs PCB's and hydrocarbons and that these are again desorbed if exposed to water containing PCB's and hydrocarbons in lower than equilibrium concentration.[46] Since the collection of plankton is essentially a filtration, non-plankton particulate material may also be collected. This may lead to erroneous PCB levels since the foreign material may either contain high concentration of PCB's or in case of low PCB concentration, it may dilute the plankton sample. Polystyrene particles (0.1 to 2 mm in diameter) recovered from the ocean contained 5 μg/g of Aroclor 1254.[27] The concentration of PCB's in "tar balls" found frequently in the ocean is not known.

The vessel itself may be an important source of contamination in sampling of water, plankton, and marine organisms. Grice et al.[44] list the following possible sources of contamination:

1. Pumping of the bilge,
2. Discharge of the septic tank,
3. Dumping of garbage, debris, or engine room waste,
4. Cleaning of boilers, and
5. Chipping of paint

Jensen et al.[60] demonstrated that plankton samples collected from the wake of the boat were contaminated by PCB's from the ship's antifouling paint and contained 4 to 14 times more PCB's than samples taken 7 ft abeam of the boat. A sample collected in front of the boat had the lowest concentration of PCB's.

Obviously, great care must be exercised in taking plankton samples to avoid contamination from the boat, from non-planktonic material, and to correct for the PCB adsorption or desorption due to the large volume of water passing through the bed of already collected plankton.

Contamination problems are less severe in the case of larger aquatic organisms. Grice et al.[44] recommend:

1. To keep the fishing gear as clean as possible,
2. To wash hands, tools, sorting trays, etc., with pharmaceutical-grade 95% ethyl alcohol,
3. To clean the surface of organisms with freshly dipped sea water followed by 95% ethyl alcohol, and
4. To preserve the specimens at −10 to −20°C in glass containers or in ethanol-washed aluminum foil.

In our experience, polyethylene bags are suitable containers for deep-freeze storage of whole fish. However, each batch of bags should be checked for PCB contamination. We have not encountered contamination from this source.

Aqueous 4% formaldehyde is also a suitable preservative for whole fish and seal blubber samples. The latter usually contain very high concentration of PCB's and, consequently, the danger of contamination is less acute. It is a good practice to keep control samples of the preservative.

The levels of chlorinated hydrocarbon pesticides in wildlife vary a great deal and the variation between individuals of a given population is very high.[75] The distribution is often skewed, tailing towards the higher concentrations. The distribution of PCB's in two species of fish followed the same pattern in spite of the fact that the fish were of the same sex and age.[118]

In order to determine meaningful levels of PCB's in a given species, biologically well-defined samples (sex, age, spawning state, etc.) must be collected, and at least 20 to 25 organisms must be individually analyzed. Unfortunately, this is an expensive and time-consuming proposition.

**Precautions Against Contamination in the Laboratory**

Water, glassware, chemicals, and the laboratory itself are potential sources of PCB contamination. It is imperative that blanks be run frequently to detect and eliminate in time any source of contamination.

Nanograde petroleum ether doubly distilled through a 30-cm glass column did not show presence of electron-capturing impurities after 200-fold concentration. However, interfering peaks were apparent after 1,000-fold concentration.[42] Water should be triply distilled in the presence of 0.1 to 0.2 g of potassium permanganate per 3 l of water.[42] An all-glass system, consisting of a 3-l flask and a Friedrichs condenser precleaned with dichromate in sulfuric acid and kept at 200° for 16 hr produces water of satisfactory quality.[25] Problems of contamination in water analysis for pesticide residues were studied by Bevenue et al.[24] and their results are equally applicable to the determination of PCB's. Glassware washing sequence (ethanol-acetone-hexane-air dry) did not completely remove the contaminants and the authors recommend any of the three following sequences: (1) dichromate + sulfuric acid (16 hr-soak) – tapwater – distilled water – acetone – air dry; (2) acetone-air dry-heat (16 hr, 200°); (3) dichromate + sulfuric acid (16 hr-soak) – tapwater – distilled water – acetone – air dry-heat (16 hr, 200°).

Giam and Wong removed contaminants from glassware by washing with detergent and tapwater, rinsing with distilled water and acetone, and heating at 300° overnight. Waring blender was rinsed with 100-ml portions of acetone followed by 100-ml portions of the solvent used for extraction. Florisil, sodium sulfate, sodium chloride, glass wool, and aluminum foil were either heated at 300 to 350° overnight, or washed with petroleum ether.[42]

Bevenue et al. cleaned silica gel by heating at 300° for 16 hr and reported that this treatment did not change the TLC properties.[24] It would be interesting to see whether the treatment is also applicable to silica gel used in the column chromatography of PCB's. Interfering peaks from glass wool can be eliminated by heating at 600° for 2 hr.[66]

In our laboratory, glassware is rinsed with acetone and pesticide-grade hexane and only occasionally washed with a detergent in tapwater and rinsed with distilled water before the solvent treatment. It is expedient to use only the minimum necessary amount of glassware so that each piece is washed several times a day and to have clearly designated areas for dirty, acetone- and hexane-rinsed glassware on the bench. Extraction thimbles are pre-extracted with hexane and silica gel is washed with hexane in a glass column (47 mm i.d. x 600 mm) until a 50-ml portion of the effluent concentrated to 0.2 ml shows no interfering peaks on gas chromatography. Sodium sulfate is treated in the same way. The activation of alumina (800°, 4 hr) efficiently removes all interfering peaks. Lamont and Cromartie recommend pre-extracting extraction thimbles with methylene chloride. A pinhole is made in the bottom of the thimble and the thimble is placed inverted into the Soxhlet extractor.[64]

Cleanliness of the laboratory is of utmost importance. The benches should be vacuum-cleaned and washed once a week. This is an easy task when only the necessary pieces of equipment are on the bench.

The sources of PCB contamination may be quite unexpected. A batch of floor wax used in our laboratory contained Aroclor 1254 at a level of 50 μg/g, and the dust generated by polishing the floor had a PCB concentration of 45 μg/g.

## Extraction of PCB's

### *Air*

PCB's were extracted from air using sintered glass absorbers or impingers filled with 75 ml amyl acetate or secondary butyl alcohol. The methods were developed before the advent of gas chromatography and PCB's were quantitated by determining chlorine after combustion over heated platinum or in a special burner (sulfur lamp).[37,59] The analysis could easily be carried out by gas chromatography, but to the best of our knowledge, this has not been reported in the literature.

### *Water*

Hexane is the solvent of choice for the extraction of PCB's from water. A simple batch extraction may be used when PCB concentration is in the 0.1 μg/l range.[104] In the ng/l range, the batch extraction is increasingly cumbersome and it is of advantage to preconcentrate PCB's.

Ahling and Jensen used reversed liquid-liquid partition between water and Chromosorb® W coated with *n*-undecane and Carbowax® 4000 (30 g undecane and 10 g Carbowax 4000 on 100 g Chromosorb W). This material (3g) was placed in a 10 i.d. x 300 mm glass column, fitted with a glass filter disc (G1, 100 to 120 μ).[2] Water flow was 65 to 130 ml/min. Waters with a high particle content were treated with aluminum sulfate (300 mg/l) and filtered before passing them through the Chromosorb W column. From 10 to 176 l of water were passed through the column and chlorinated hydrocarbons were extracted from the column with 10 ml of petroleum ether. The aluminum sulfate precipitate was air dried and extracted in a small column (10 x 300 mm) with 6 ml of acetone, followed by 5 ml of petroleum ether. Elemental sulfur is also adsorbed on the Chromosorb W column and can be removed from the petroleum ether extract by treating it with potassium hydroxide in methanol.

Gesser et al.[40] described the use of polyurethane foam culture stoppers (diSPo Stoppers, Canlab) to quantitatively extract PCB's from water. The plugs were washed by a 1:1 hexane-acetone solution, pushed into a 20-mm, i.d. glass column, which was then washed with 25 ml of ethyl alcohol and 200 ml of distilled water. The sampled water was passed through the column at a flow rate of 250 ml/min. Water was squeezed from the plugs and PCB's were extracted with 20 ml of acetone followed by 100 ml of hexane. The recovery of PCB's from 1 l of water by a column consisting of two plugs was usually better than 90%. In the original paper, the method was tested on 1-l samples, spiked to contain 2 and 20 μg/l of an unspecified PCB preparation.

A scaled-up procedure was recently published.[41] Two polyurethane foam plugs (8.3 cm diameter, 5 cm length), in a 76 mm i.d. glass column were used, the flow rate was 5 to 10 l/hr, and 1,400 to 1,800 l of water was extracted. The recovery on this scale is not known. The detected level of PCB's is from 5 to 10 ng/l. The type of PCB's is not specified in the paper, but from the published GLC tracings it may be Aroclor 1242.

Polyurethane foams coated with different liquid phases (DC-200, OV series, SE-30, QF-1, and DEGS) were tested for extraction of chlorinated hydrocarbon pesticides from water. The best recovery was obtained with a DC-200 coated (10 to 20 mg per plug) polyurethane foam. The behavior of PCB's on this adsorbent was not studied.[103]

Aue et al.[11,12] showed the applicability of surface-bonded silicones (octadecyltrichlorosilane-treated Chromosorb G) to adsorb PCB's from water. Up to 10 l of water, prefiltered if necessary, was aspirated at a rate of 50 to 55 ml/min through a 10 mm i.d. x 350 mm column containing 7 g of the surface-bonded silicone material. The column was then dried with nitrogen and the adsorbed organochlorine compounds were extracted with pentane. From the published chromatogram, the recovery of Aroclor 1254 appears to be practically quantitative. The application of surface-bonded silicones to the extraction of PCB's from water deserves further attention.

### *Soil, Sediment, Sludge, and Paper*

More polar solvent mixtures are usually used to extract PCB's from this type of material. Wakimoto et al.[105] refluxed 10 to 20 g of air-dried top- or subsoil with 50 ml of 1 *N* KOH in ethyl alcohol, transferred the mixture with 10 to 20 ml of 1.1 hexane-ethanol into a separatory funnel, added 20 to 25 ml water, separated the hexane layer, and extracted the aqueous layer with additional hexane (4 x 50 ml).

Acetone-hexane 6:4 was used to extract sand samples.[18] If an interphase between the sample and the solvent was formed (too much water in the sample), the amount of acetone was increased.

Holden used a mixture of hexane and isopropanol to extract PCB's from sewage sludge.[49] The recovery and other experimental details of the procedure have not been published. Schmidt et al.[93] extracted PCB's from sewage effluent with 15% ether in hexane. Grab samples (1 l) were taken in 1-gal glass containers; 75 ml of the extraction solvent were immediately added and the container was vigorously shaken for 2 min. In the laboratory, the organic phase was separated and the extraction was repeated once more with new solvent. Recovery of PCB's from spiked samples using this procedure was not reported.

PCB's from non-carbon copy paper were extracted with acetone in a Soxhlet apparatus for 3 days.[63] Several other studies of the PCB content in paper have been published, but experimental details of the extraction were not given.[62,102]

*Biological Samples*

Many different solvents or solvent mixtures have been used to extract PCB's from biological samples. Except for a few plankton and diatom samples, all of the samples analyzed were of animal origin. The solvents used range from nonpolar petroleum ether, pentane, or hexane through solvent mixtures, such as ethyl ether-hexane, or acetone-hexane to chloroform and chloroform-methanol. The samples are either dehydrated using anhydrous sodium sulfate or, at times, calcium chloride, or extracted as such, in which case enough of a water-miscible solvent is usually present to form a homogeneous liquid phase. The extraction may be carried out in a blender or a high-speed mixer, in a column, or in a Soxhlet extractor.

The recoveries of PCB's from spiked samples are generally better than 90% for all the solvents and experimental conditions. However, one has to keep in mind that PCB's, added to the sample, may be more easily extractable than PCB's incorporated into the sample through normal biological processes. This problem was studied with labeled dieldrin translocated from the soil to the plant and it was shown that dieldrin was completely extracted only in a 12-hr Soxhlet extraction with 1:1 chloroform-methanol.[77] As far as we know, no similar study was reported with PCB's.

Apart from the recovery, the possibility of contamination during the extraction must be taken into consideration. From this point of view, an ideal extraction would use a few ingredients, equipment which can be cleaned easily and would have a capacity of 10 to 15 samples per man-day. A smaller amount of ingredients mean less chance of introducing contaminants and less time spent in cleaning contaminated ingredients. Most of the commercial blenders are very difficult to clean. A simple chromatographic column is easy to clean, but on the other hand, it may require more solvent per unit sample weight. A Soxhlet extractor and a corresponding condenser can be cleaned fairly easily. The solvent is recycled through the sample many times, but since the extraction is carried out at an elevated temperature, losses of the more volatile components may occur. In fact, Gesser et al. reported losses of some lower chlorinated PCB's on Soxhlet extraction, but have not provided any quantitative data.[40]

From the above discussion, it can be seen that there is no clearly preferable extraction technique. The analyst must use his own judgment and choose the most convenient and efficient procedure, depending on the type of samples and equipment available. A few examples of typical extraction techniques are given below.

***Soxhlet Extraction with Hexane***

Sample of tissue (6 g) is ground with anhydrous sodium sulfate (30 g) using a mortar and a pestle; the resulting, free-flowing powder is quantitatively transferred into an extraction thimble and extracted in a Soxhlet extractor with 100 ml hexane for 1 hr.[114] Others homogenized the sample similarly, but extracted PCB's with petroleum ether in a Soxhlet for 7 hr.[14]

***Extraction of Adipose Tissue***

The tissue (5 to 6 g) is homogenized in a mortar with sea sand (1:3), the mixture is transferred into a 18 mm i.d. x 150 mm glass column, and the column is percolated with 50 ml of *n*-pentane, 25 ml of ethanol, and 50 ml of pentane. The combined eluates are washed with a 2% solution of sodium sulfate (2 x 50 ml) in the presence of approximately 100 mg of powdered cellulose to improve the separation of the phases, and the pentane phase is filtered through a 3-cm layer of anhydrous sodium sulfate.[1]

***Extraction of Fish Lipids***

Whole fish is ground with dry ice in a blender

and 20 g of the powder resulting after the evaporation of dry ice is either blended with chloroform-methanol (2:1, 3 x 100 ml) or mixed with 80 g of anhydrous sodium sulfate. In the latter case the mixture is packed into a chromatographic column, and extracted by percolating 250 ml of cyclohexane through the column.[98]

### *Chloroform Extraction and Saponification*

The tissue sample (1 g) is homogenized with anhydrous sodium sulfate and the powder is shaken with chloroform (2 x 50 ml). The extract is filtered off and evaporated to dryness in a rotatory evaporator. An aliquot of the residue (50 mg) is saponified by refluxing with 10% potassium hydroxide in ethanol. Ethanol is distilled off, the residue is taken up in 3 ml of water, and the aqueous phase is extracted with 6 ml of heptane.[20]

### *Extraction after Saponification*

The sample (5 to 10 g) in a 200-ml Erlenmeyer flask is refluxed with 50 ml of 1 *N* potassium hydroxide in ethanol for 1 hr. After cooling to 50°, 50 ml of hexane is added and the solution is stirred and cooled to room temperature. The solution is then transferred into a 300-ml separatory funnel using 10 to 20 ml of hexane-ethanol (1:1). Water (20 to 25 ml) is added and the funnel is shaken for several seconds. The extraction is repeated three times, each time with fresh 50 ml of hexane. The hexane extract is filtered through anhydrous calcium chloride and concentrated in a Kuderna-Danish evaporator.[105]

### *Extraction with Acetonitrile-acetone*

The sample (5 g) is homogenized in an explosion-proof Waring Blender with acetonitrile, acetone, and sodium sulfate (45 ml, 15 ml, and 5 g, respectively). The homogenate is filtered through a medium porosity fritted glass filter and the residue is washed with acetonitrile (2 x 10 ml).[92]

### *Extraction of PCB's with Ethyl Ether-petroleum Ether*

This is a procedure developed for milk. Potassium oxalate (1 g) and ethanol (100 ml) are added to the sample (100 ml) in a 500-ml centrifuge bottle which is shaken for 10 min. Ethyl ether (50 ml) and petroleum ether (50 ml) are added and the bottle is shaken vigorously after each addition. The mixture is centrifuged until the phases separate and the aqueous phase is reextracted twice with 50 ml of 1:1 petroleum ether-ethyl ether. The combined extracts are washed with water (100 ml).[80]

### *Extraction with Acetonitrile-methylene Chloride*

This method was used to extract PCB's from fecal matter. The sample (10 g) in a 250-ml centrifuge bottle is shaken for 1 min with water and acetonitrile (25 and 150 ml, respectively) and centrifuged for 20 min at 1,500 rpm. The supernatant is decanted and the extraction is repeated with 100 ml of acetonitrile. The supernatants, filtered through glass wool, are combined, acetonitrile is evaporated in vacuum, and the aqueous residue is extracted four times with equal volumes of methylene chloride. The extracts are filtered through a column of anhydrous sodium sulfate and concentrated in vacuum.[85]

### *Other Extraction Procedures*

The procedures given above were chosen to illustrate different techniques of extraction of PCB's from biological samples. The reader may consult *Official Methods of Analysis* and analytical manuals dealing with the determination of pesticides for additional extraction procedures and recoveries of pesticides.[5,53]

## Cleanup of Extracts

A number of compounds other than PCB's, particularly lipids and other chlorinated hydrocarbons, are coextracted under the conditions described in the previous section. Lipids would interfere with the determination of PCB's by contaminating the GLC column and detector. Other chlorinated hydrocarbons give peaks which might interfere with the quantitation of PCB's by gas chromatography. The removal of lipids from the extracts is generally referred to as the cleanup. The separation of PCB's from other chlorinated hydrocarbons is discussed in the next section.

The cleanup procedures may be divided into three groups:

1. Methods based on chemical decomposition or physical removal of lipids,
2. Methods based on solvent partition, and
3. Methods based on column or thin-layer chromatography.

PCB's are stable in the presence of concentrated sulfuric acid and reasonably stable in alkaline media, whereas lipids are destroyed under these conditions. Cleanup based on the chemical decomposition of lipids is fast, simple, and suitable for routine analysis. The main disadvantage of this procedure is the decomposition of some chlorinated hydrocarbon pesticides. Hence, this cleanup cannot be used if the analyst is also interested in the pesticide residues, but it may be valuable for confirmatory purposes.

PCB's are preferentially extracted from hexane into DMF or acetonitrile and very likely also into DMSO, whereas lipids remain in the hexane phase. Lipids may be very effectively removed by this procedure. However, it has been reported that the partitioning may change the relative magnitude of the PCB peaks, i.e., the distribution coefficients of individual biphenyls differ. The effect of the changed PCB pattern on the quantitation may be eliminated by partitioning the appropriate standards under identical conditions. The distribution coefficients may also be affected by the concentration of lipids in the original hexane solution and it may be necessary to keep the lipid concentration within a certain range. The removal of water-soluble solvents from the extract is not really a cleanup since lipids remain together with PCB's in the nonpolar solvent phase.

Alumina and Florisil® are most often used to remove lipids by column chromatography. The chromatography on Florisil also separates PCB's from some chlorinated hydrocarbon pesticides and this aspect is described in the next section. Similarly, silica gel, which can handle a limited amount of lipid, but is used primarily to separate PCB's and chlorinated hydrocarbon pesticides, is discussed in the next section. A somewhat unique approach to the separation of lipids from PCB's is gel permeation chromatography.[98] This cleanup uses the same column over and over again and may be easily automated. However, the technique has not been tested as yet in many laboratories.

Again as in the case of PCB extraction, the analyst must use his own judgment and choose the most suitable cleanup depending on the type and number of samples, type of analysis (PCB's only or PCB's and chlorinated hydrocarbon pesticides), and equipment available. Typical examples of cleanup procedures are given below.

*Sulfuric Acid Cleanup*

The extract (5 to 10 ml) is shaken with 1 ml of concentrated sulfuric acid for 30 sec. If necessary, the procedure is repeated with fresh acid. Dieldrin, malathion, and parathion are destroyed by this treatment; the recovery of PCB's (Aroclor 1254) and the common chlorinated hydrocarbon pesticides, including endrin and heptachlor epoxide, is practically quantitative.[78]

*Microscale Alkali Treatment*

This procedure is useful for additional cleanup and confirmation of alkali-stable compounds. Precleaned extract in a concentration suitable for GLC analysis (2 ml) is placed into a 10-ml Mills tube. Ethanolic potassium hydroxide (2%, 1 ml) and a few carborundum chips are added and the contents are gently boiled down to 0.2 ml on a boiling water bath. Any precipitate formed is dissolved by adding a few drops of ethanolic potassium hydroxide. The contents are diluted with aqueous ethanol (1:1, 2 ml) and extracted with 1 to 2 ml of hexane. The recovery of PCB's is quantitative; the recoveries of aldrin, dieldrin, and endrin range from 70 to 90%. The recovery of heptachlor, its epoxide, and of mirex is 30 to 50%. BHC isomers and endosulfan are completely eliminated and DDD and DDT isomers are converted to the corresponding olefins. Sulfur is completely eliminated by this treatment.[111]

*Low-temperature Precipitation of Lipids*

Lipids congeal when cooled to −75° and can be removed by filtration through a layer of Solka Floc. More than 98% of the lipids may be removed by this procedure but some samples may require an additional cleanup.[70]

*Acetonitrile Partitioning*

The following is a procedure given in the *Official Methods of Analysis*[53] (reprinted with permission):

"Weigh ≤ 3 g fat into 125 ml separator, and add petroleum ether so that total volume of fat and petroleum ether in separator is 15 ml. Add 30 ml of acetonitrile saturated with petroleum ether, shake vigorously 1 min, let layers separate, and drain acetonitrile into 1 l separator containing 650 ml water, 40 ml sodium chloride solution, and 100 ml petroleum ether. Extract petroleum ether solution in 125 ml separator with 3 additional 30 ml portions of acetonitrile saturated with petroleum ether, shaking vigorously 1 min each time. Combine all extracts in a 1 l separator. Hold separator in horizontal position and mix thoroughly 30–45 sec. Let layers separate and drain aqueous layer into second 1 l separator. Add 100 ml petroleum ether to second separator, shake vigorously 15 sec, and let layers separate. Discard aqueous layer, combine with petroleum ether in original separator, and wash with two 100 ml portions of water. Discard washings and draw off petroleum ether layer through 25 o.d. x 50 mm column of anhydrous sodium sulfate into 500 ml Kuderna-Danish concentrator. Rinse separator and the column with three ca 10 ml portions of petroleum ether. Evaporate combined extracts and rinses to ca 10 ml in Kuderna-Danish concentrator for transfer to Florisil column.

"Purify technical acetonitrile as follows: To 4 l of acetonitrile add 1 ml of phosphoric acid, 30 g phosphorus pentoxide and boiling chips, and distill in all-glass apparatus at 81–82°. Do not exceed 82°. Saturate acetonitrile with redistilled petroleum ether.

"Some lots of reagent grade acetonitrile are impure and require distillation. Generally vapors from such lots will turn moistened red litmus paper blue when held over mouth of storage container. Pronounced amine odor is detectable."

Further cleanup on Florisil usually follows the acetonitrile partitioning.

*Chromatography on Alumina*

Holden and Marsden used alumina activated at 800° for 4 hr and deactivated by the addition of 5% water. The chromatography was carried out in a 5 mm i.d. x 450 mm column, containing a small glass wool plug at the lower end and charged with 2 g of alumina.[51] In our experience, this column removes about 50 mg of lipid from the extract; however, the extract may be applied only from a hexane solution. PCB's and common chlorinated hydrocarbon pesticides are eluted in 20 ml of hexane. Even a small concentration of a polar solvent would elute an excessive amount of lipid.[114]

McLure recommends alumina activated at 400° for a minimum of 4 hr and then deactivated by 1% methanol in benzene. The adsorbent can be stored under the deactivating solvent for several months without changes in activity. The separation of chlorinated hydrocarbons from fish and plankton lipids is achieved on 5 mm i.d. x 75 mm column containing 1.7±0.1 g of alumina. The column can handle up to 45 mg of lipid, but the recommended amount is 15 mg. PCB's (type not stated) and chlorinated hydrocarbon pesticides were eluted in the first 3.5 ml of hexane.[69]

*Chromatography on Florisil*

The following is the technique according to the *Official Methods of Analysis*[53] (reprinted with permission):

"Prepare 22 mm i.d. Florisil column (with Teflon stopcocks and coarse fritted plate or glass wool plug; 22 mm i.d. x 300 mm) containing 4″, after settling, of activated Florisil (60/000 PR grade, activated at 650°C available from Floridin Co.) When activated Florisil is obtained in bulk, transfer immediately after opening to ca 1 pt glass jars, or bottles, with glass-stoppered or foil-lined, screw-top lids, and store in dark. Heat ≥5 h at 130° before use. Store at 130° in glass-stoppered bottles or in desiccator at room temperature and reheat at 130° after 2 days), topped with ca 0.5″ anhydrous sodium sulfate. Prewet column with 40–50 ml petroleum ether. Place Kuderna-Danish concentrator with volumetric or graduated collection flask under column to receive eluate. Transfer petroleum ether extract or concentrate to column, letting it pass through at <5 ml/min. Rinse containers and sodium sulfate with two ca 5 ml portions petroleum ether, pour rinsings onto column, rinse walls of tube with additional small portions of petroleum ether, and elute at ca 5 ml/min with 200 ml 6% eluting solvent. (Dilute 60 ml ethyl ether to 1 l with redistilled petroleum ether. Ethyl ether redistilled at 34–35°, and stored under N. Must be peroxide-free. Add 2% alcohol. Petroleum ether is reagent grade, redistilled in all glass apparatus at 30–60°.) Change receivers and elute with 200 ml 15% eluting solvent (150 ml ethyl ether to 1 l with redistilled petroleum ether) at ca 5 ml/min. Change receivers and elute with 200 ml 50% eluting solvent (500 ml of ethyl ether to 1 l with redistilled petroleum ether) at ca 5 ml/min.

"Concentrate each eluate to suitable definite volume in Kuderna-Danish evaporator apparatus. When volume <5 ml is needed, use 2-ball micro-Snyder or micro-Vigreaux column.

"First eluate (6%) contains PCB's, chlorinated pesticides (aldrin, BHC, DDE, DDD (TDE), *o,p′*- and *p,p′*-DDT, heptachlor, heptachlor epoxide, lindane, and methoxychlor) and phosphated pesticides (carbophenothion, ethion, and ronnel) and is usually suitable for GLC directly. If further cleanup is necessary, repeat Florisil cleanup, using new column. Second eluate (15%) contains chlorinated pesticides (dieldrin and endrin) and phosphated pesticides (Diazinon, Me parathion, and parathion)."

The adsorptive capacity of Florisil may vary significantly from lot to lot and a method based on the adsorption of lauric acid was developed to characterize and standardize individual batches.[45,72]

Mills et al. described another solvent system for the Florisil chromatography.[73] Eluant A is 20% (v/v) methylene chloride in hexane, eluant B is 50% methylene chloride, 0.35% acetonitrile, 49.65% hexane (v/v/v), and eluant C is 50% methylene chloride, 1.5% acetonitrile, 48.5% hexane (v/v/v). The column is eluted successively with 200 ml of each eluant. PCB's, DDT and metabolites, hexachlorobenzene, and some other chlorinated pesticides are eluted with the eluant A; more polar organochlorine pesticides such as dieldrin, endrin, and endosulfan, and some organophosphates are eluted with the eluant B, and mostly organophosphates appear in the fraction C.

Large amounts of lipid can be removed by mixing it with unactivated Florisil (8 g lipid per 25 g Florisil), placing the mixture in a glass column (22 mm i.d. x 250 mm) and percolating the column with 150 ml of acetonitrile, containing 10% water. PCB's and organochlorine pesticides are recovered from the eluate by partitioning into petroleum ether (100 ml) from water-diluted acetonitrile (10 ml saturated sodium chloride and 600 ml of water).[83] Final cleanup is then accomplished by column chromatography on activated Florisil (130°C) according to Mills et al.[73] The recovery of Aroclor 1254 from corn oil spiked with a concentration of 0.1 μg/g was 85%.[83]

### Separation of PCB's from Chlorinated Hydrocarbon Pesticides and Other Chlorinated Compounds

The cleanup procedures mentioned in the previous section do not, or in the case of Florisil, only partially separate PCB's from chlorinated hydrocarbon pesticides. Additional separation may be required before either PCB's, or pesticides, or both can be quantitated. The necessity of additional separation depends very much on the relative ratio of PCB's and other chlorinated compounds and on the purpose of the analysis. If, for example, samples from a feeding experiment with a given commercial PCB preparation are analyzed, and the levels of PCB's are high, it is hardly worthwhile to separate trace amounts of DDT and metabolites since they would not significantly interfere with the quantitation of PCB's. Besides, it is almost always very difficult to determine a small amount of one compound in the presence of a large excess of another compound with similar properties.

The chromatography on silica is most frequently used to separate PCB's from the DDT group of compounds. The ease of this separation decreases in the order *p,p'*-DDD > p,p'-DDT > p,p'-DDE, and to separate the last compound is very difficult.

The activity of the adsorbent and the purity of the solvents are of crucial importance and it is not uncommon that a procedure developed in one cannot be reproduced in another laboratory. It may often be necessary to modify the conditions (most frequently the volumes of effluent collected) to achieve the desired separation.

Chromatography on silicic acid-Celite, described by Armour and Burke, is often used to separate PCB's from DDT and metabolites including *p,p'*-DDE.[8] In this procedure (see below for details), PCB's are eluted with petroleum ether, DDT and metabolites with a mixture of acetonitrile, hexane, and methylene chloride. The margin of separation between PCB's and *p,p'*-DDE is very small and some carry-over of PCB's into the pesticide fraction was observed when more than 400 mg of lipid was applied to the column.[19] Silicic acid (Mallinckrodt No. 2847, 100 mesh) is activated from 7 to 24 hr at 130°, deactivated by the addition of 3% water, and mixed with Celite 545 (5 g Celite and 20 g silicic acid). The mixture is then slurried with 80 ml petroleum ether, transferred into a chromatographic column (22 mm i.d. x 400 mm, with a coarse fritted plate and a Teflon stopcock), stirred to remove air bubbles, and petroleum ether is drained leaving only a 3-mm layer above the adsorbent. An aliquot of a cleaned-up extract (max 5 ml) in petroleum ether is applied to the column and washed into the column by additional small portions of petroleum ether. PCB's are eluted in 250 ml of petroleum ether, *p,p'*-DDE, and other pesticides in 200 ml of acetonitrile-hexane-methylene chloride 1:19:80 (v/v/v).

Masumoto studied in detail factors affecting the separation of PCB's and *p,p'*-DDE by silicic acid chromatography.[68] Complete separation of Aroclor 1254 from PCB's could not be achieved. Variability between batches of silicic acid activated at 130°C and deactivated by 3% water was caused by variable activation time. No further water loss was observed after 24 hr at 130°, whereas the weight loss between 7 and 24 hr was about 0.5%. Variations between separation patterns obtained on columns prepared from the same batch of silicic acid were shown to be due to Celite, which is added to improve the flow rate. However, columns without Celite behaved reproducibly and the effects of aging on the chromatographic patterns were not studied.

Aroclors 1260 and 1254 could be reasonably well separated from *p,p'*-DDE. However, Aroclor 1221 was eluted in the *p,p'*-DDE fraction and

Aroclors 1232 and 1242 were distributed between the PCB and *p,p'*-DDE fractions.[68]

Retention volumes were 540, 490, 660, 630, 370, and 890 ml for 4-chloro-, 3-chloro-, 2-chloro-, 2,2'-dichloro-, 4,4'-dichloro-biphenyl, and biphenyl, respectively.[68]

Two other procedures for the separation of PCB's from *p,p'*-DDE by chromatography on silicic acid have been described. Snyder and Reinert used silica gel (Davison Chemical Division, W. R. Grace, Baltimore; grade 950 activated desiccant, 60-200 mesh), received from freshly opened cans, or reactivated at 200°C for 8 hr and kept under pentane. Glass columns (10 mm i.d. x 200 mm) were filled to a length of 7.7 cm (about 3 g dry weight) and PCB's were eluted in 38 ml of pentane, DDT and metabolites in 36 ml of benzene. The method was tested on PCB-spiked hexane extracts of fish but the published data do not indicate the amount of lipid actually applied to the column.[95]

According to McClure silica gel (Matheson Coleman & Bell Sx-144-6 activated at 180°, deactivated with 0.5% ethyl ether in benzene, which in turn was washed out with hexane 5 ml of hexane per 1 g of adsorbent), separated PCB's from *p,p'*-DDT and metabolites. PCB's were eluted in 10 ml of hexane, *p,p'*-DDT, and metabolites in 5 ml of 0.5% ether in hexane. After rinsing with hexane, the column (4 mm i.d. x 230 mm) was ready for reuse.[69]

Other authors limited themselves to the separation of PCB's from *p,p'*-DDT and *p,p'*-DDD. Thus, in the procedure of Holden and Marsden, *p,p'*-DDE is eluted together with PCB's from the silica column by hexane.[51] Using this method, we have noticed that the elution volume for PCB's and pesticides varied with different batches of hexane. The source of these variations was traced to the varying amounts of benzene in commercial pesticide-grade hexanes.[113] It is practical to adjust the concentration of benzene to 5 ml/l hexane. Under these conditions, PCB's and *p,p'*-DDE are eluted in 15 ml of hexane and *p,p'*-DDD, *p,p'*-DDT, and dieldrin are eluted in 10 ml of 10% diethyl ether in hexane. A 5 mm i.d. x 45 mm glass column, charged with 2 g of silica (Silicar, Mallinckrodt 100-200 mesh, activated at 130° overnight and deactivated with 3% of water) was used. This procedure is fast and suitable for routine analyses. A correction can be made for the PCB's interference in the quantitation of *p,p'*-DDE whereas peaks with retention times different from that of *p,p'*-DDE are used to quantitate PCB's (see section on quantitation).

Collins et al. used silica gel (BDH Limited, 60-100 mesh), activated at 500° for 4 hr, and deactivated by the addition of 2.5 ml water per 100 g of silica gel, to separate PCB's from *p,p'*-DDT and *p,p'*-DDD.[28] An 8 mm i.d. x 300 mm glass column, containing 5 g of silica gel, was used. PCB's were eluted in 42 ml of hexane (flow rate 0.7 ml/min). This fraction also contained hexachlorobenzene, aldrin, *o,p'*-DDT, and *p,p'*-DDE. Other organochlorine pesticides were eluted with 50 ml of 10% ether in hexane. Collins et al.[28] quantitated *p,p'*-DDE as *p,p'*-dichlorobenzophenone after oxidation with chromium trioxide. The hexane fraction from the silica gel column was concentrated to 2 ml, 2 ml of acetic acid was added, hexane was evaporated using a micro-Snyder column, 100 mg of chromium trioxide were added, and the mixture was placed in a boiling-water bath for 15 to 20 min. Hexane (2 ml) was added, the tube was shaken vigorously, the acid was neutralized with 6 to 7 ml of 5 *N* sodium hydroxide and the contents were shaken again. Miles dehydrochlorinated DDT and metabolites by 1,5-diazobicyclo[5.4.0]undec-5-ene and oxidized the corresponding olefins by chromic acid to dichlorobenzophenone.[71]

Leoni studied the chromatography of PCB's and pesticides on silica gel (grade 950, 60-200 mesh, Davison Code 950-08-08-226, Grace Davison Chemical, Baltimore, Md.), activated at 130° for 2 hr and deactivated by 5% of water. Under these conditions, PCB's (Fenclor with 40, 52, 60, and 65% chlorine) were eluted with hexane together with *p,p'*-DDT and metabolites. The paper presents data on the chromatographic behavior of many chlorinated hydrocarbon and organophosphate pesticides.[65]

Florisil was also used to separate PCB's from the DDT compounds. The former were eluted with hexane, the latter with a mixture of ethyl ether and hexane.[89] Edwards reported satisfactory separations using this procedure,[34] but Bevenue and Ogata were not able to obtain a good separation.[23] Hexanes with different benzene content may be responsible for this discrepancy. According to a later report by Edwards, a separation of *p,p'*-DDE from PCB's is not possible.[35] Berg et al. separated PCB's from DDT and metabolites by column chromatography on charcoal.[21]

Only one type of charcoal (Fisher Scientific 5-690 charcoal, activated, 50-200 mesh) was suitable. Before use, the charcoal was repeatedly boiled with acetone, washed with cold acetone, air-dried, and stored at 135°. A 6 mm i.d. x 140 mm glass column, containing a layer of sand (25 mm) to retain the charcoal fines and a layer of charcoal (90 mm, 1.1 g dry weight), was used. The charcoal was applied to the column as an acetone slurry. DDT and metabolites were eluted with 90 ml of 25% acetone in ethyl ether, PCB's with 60 ml of benzene.

The problems in the separation of the DDT compounds from PCB's are obvious from the preceding data. Each laboratory must adjust the conditions and frequently check the performance of the procedure.

Two other approaches to the elimination of the separation difficulties have been described. Asai et al. dechlorinated PCB's (Aroclor 1260) to a mixture of biphenyl, cyclohexylbenzene, and dicyclohexyl.[10] These compounds can be easily determined by gas chromatography, very likely without much interference from the dechlorination products of the DDT compounds. Unfortunately, with the loss of chlorine the high sensitivity of the electron-capture detector is also lost and the method is not suitable for the detection of low levels of PCB's. At times, it may be useful for confirmatory purposes. The other possibility is the perchlorination of PCB's to yield decachlorobiphenyl which is easily separated by GLC from the perchlorinated products of *p,p'*-DDE. The perchlorination techniques will be discussed in more detail in the section on PCB quantitation.

PCB's interfere with the determination of chlorinated dibenzodioxins and dibenzofurans, particularly those containing from 2 to 5 chlorine atoms in the molecule.

Porter and Burke separated 2,3-di-, 2,3,7-tri-, and 2,3,7,8-tetrachlorodibenzo-*p*-dioxin from Aroclors 1254, 1260, and 1262 by using a 25-mm layer of alumina in a 4 mm i.d. x 100 mm disposable pipet.[82] Alumina (for chromatographic adsorption, Merck) was heated overnight to 130°C and cooled in a desiccator. PCB's were eluted with 6 ml of 1% methylene chloride in hexane, chlorodibenzo-*p*-dioxins with 6 ml of 20% methylene chloride in hexane. Using alumina from another supplier (Fisher Scientific A-540, 2 g in a 5 mm i.d. x 450 mm column), we had to use 2% methylene chloride in hexane to elute PCB's (Aroclor 1254 and PCB's from wildlife samples, resembling it) in 20 ml of effluent. Chlorodibenzo-*p*-dioxins (2,3,7,8-tetra, two hexa- and octachloro) were eluted with 10 ml of 20% methylene chloride in hexane. Under the same conditions, di-, tri-, tetra-, and octachlorodibenzofurans can be separated from PCB's of the Aroclor 1254 and 1260 type.[115] Some peaks, probably chlorobiphenyls, appear in the 20% methylene chloride fraction when large amounts of Aroclors 1221, 1232, 1242, and 1248 (800 μg) are applied to the column. These peaks are not removed in the 2% methylene chloride fraction even after repeated chromatography. The peaks could be mistaken for chlorinated dibenzo-*p*-dioxins or dibenzofurans and a confirmation by perchlorination or by mass spectrometry is absolutely necessary.

Chlorinated naphthalenes accompany PCB's on silica chromatography.[9] No method for a chromatographic separation of PCB's from chlorinated naphthalenes has been reported. The latter have not been found in environmental samples; however, they may be present in some and masked by or mistaken for PCB's. A sufficiently high concentration of chlorinated naphthalenes could be confirmed by UV spectrophotometry (UV maximum at 306 nm, $A^{1\%}_{1\,cm}$ = 329).[114] Holmes and Wallen reported that chlorinated naphthalenes are relatively easily oxidized by chromium trioxide and this procedure was suggested for the removal of chlorinated naphthalenes from mixtures with PCB's.[52]

**Thin-layer Chromatography of PCB's**

Several authors studied the thin-layer chromatography of commercial PCB preparations. Alumina G, sometimes with incorporated silver nitrate or silica gel G, was the adsorbent; hexane, heptane, 5% benzene in hexane, or 2% acetone in heptane were the developing solvents. Layers not containing silver nitrate were sprayed with a silver nitrate solution, i.e., 1.7 g silver nitrate in 200 ml of ethyl alcohol. Layers with incorporated silver nitrate were sprayed with a solution of 2-phenoxyethanol in aqueous acetone (5 ml of water, 10 ml of 2-phenoxyethanol, 200 ml acetone, and a few drops of 3% hydrogen peroxide). PCB's and other chlorinated compounds were then visualized under UV light. One-dimensional TLC did not separate PCB's from *p,p'*-DDE. PCB's were not resolved into individual components and migrated as a single spot with an $R_f$ value of approximately 0.9. PCB's were, however, clearly separated from dichlorodibenzophenone ($R_f$ approximately 0.3).[7,17,28,76,88,107]

Fehringer and Westfall separated PCB's from *p,p'*-DDE using two-dimensional TLC on silica gel (MN-Kieselgel G-HR, 30 g and 50 ml of a freshly prepared 0.2% silver nitrate solution, sufficient for five 20 x 20 cm plates at 250 μ), activated at 80° for 20 min. The plates were developed in one direction with *n*-heptane, in the other direction with *n*-heptane-acetone 98:2. Even PCB's were resolved into several components.[38]

A much better separation of PCB components may be achieved by reverse phase thin-layer chromatography (RPTLC). This technique was first used by De Vos and Peet.[31] Plates coated with silica gel G were dipped into an 8% solution of liquid paraffin in petroleum ether and air-dried. The developing solvent was a mixture of acetonitrile-acetone-methanol-water (40:18:40:2). The technique was developed further by Stalling and Huckins.[97] Silica gel G was again coated with liquid paraffin. Best separation of PCB components was achieved by either of the following solvent mixtures: water-acetonitrile-methanol (15:40:45); and water-acetone-acetonitrile-methanol (5:15:40:40). The resolved components were extracted from the plates and analyzed by gas chromatography and mass spectrometry. Most of the individual TLC spots yielded single peaks on gas chromatography (Table 1). $R_f$ values generally increased with decreasing number of chlorine atoms in the chlorobiphenyl molecule. However (as can be seen from Table 1), it would be difficult to separate *p,p'*-DDE from some chlorobiphenyls. The detection limits for the major components of PCB preparations were 15 μg for Aroclor 1232, 8 μg for Aroclors 1248 and 1254, and 2 μg for Aroclor 1260.

RPTLC is a useful technique for the study of individual PCB components and deserves further attention.

Thin-layer chromatography on alumina was used to separate Aroclor 1221, 1254, and 1260, and several chlorobiphenyls from 2,3,7,8-tetra- and octachlorodibenzo-*p*-dioxin.[109] The adsorbent was activated by heating at 120° for 2 to 3 hr and developed (7 cm) with tetrachloroethylene-acetone (1 + 4) and then in the same direction (10 to 11 cm) with tetrachloroethylene-methanol-water (5 + 45 + 1). In both solvents, PCB's moved close to the solvent front. The two chlorodibenzo-*p*-dioxins did not move in the second solvent.

**Confirmation of PCB's**

PCB's are most often detected and quantitated by gas chromatography. The GC patterns characteristic for different commercial preparations were presented already in Chapter 2. The GC pattern together with the elution of PCB's in the PCB fraction during the separations of PCB's from chlorinated hydrocarbon pesticides leaves little doubt about the correct identification of these compounds and an additional confirmation may not be required. In some cases, the pattern may be changed beyond recognition and a confirmation is necessary. Mass spectrometry coupled with gas chromatography is frequently used to confirm the identity of PCB's (see Chapter 8). The instrumentation is rather expensive and not available to all laboratories. For this reason, and also in an attempt to develop a better quantitation technique, several authors studied the perchlorination of PCB's and related compounds. The GC pattern of Aroclor 1254 before and after perchlorination is illustrated in Figure 1.

Berg et al. used antimony pentachloride to perchlorinate PCB's.[21] The reaction was carried out at 150 to 170° on a preparative scale (0.5 g, 3 ml of antimony pentachloride) in Teflon-lined, screw-cap vials or on an analytical scale (1 μg, 0.2 ml of antimony pentachloride) in sealed glass tubes. From 5 to 8 hr was necessary for complete conversion of Aroclor 1254 to decachlorobiphenyl. The perchlorination mixtures were then treated with 20% hydrochloric acid and extracted with benzene. The yield of decachlorobiphenyl was about 90% on the preparative scale and 85% on the analytical scale. Several other laboratories reported that the yield of decachlorobiphenyl depended on the chlorine content of the starting PCB preparation, lower chlorinated preparations giving lower recoveries. Mizutani and Matsumoto perchlorinated 0.1 μg of PCB's with 0.5 ml of antimony pentachloride in a sealed tube at 220° for 2 hr.[74] Huckins et al. somewhat modified the perchlorination with antimony pentachloride and obtained decachlorobiphenyl in a yield of 85 and 104% from Aroclors 1016 and 1254, respectively.[54] The sample in petroleum ether was placed into a Mini-Aktor® vial (Applied Science Laboratories No. 12993), petroleum ether was

TABLE 1

**Chromatographic Characterization of Polychlorinated Biphenyls (Aroclors): Correlation of Reverse-phase Thin-layer Chromatographic (RPTLC) Spots and GLC peaks[96]**

| RPTLC | | | GLC[b] | | GLC-MS | |
|---|---|---|---|---|---|---|
| Solvent system | Spot No. | $R_f$[a] | Peak No. | $R_t$ relative to *p,p'*-DDE | No. of chlorines | Molecular weight |
| | | | Aroclor 1232 | | | |
| Water- | 1 | 0.35 | 13 | 1.46 | 6 | 358 |
| Acetonitrile- | 2 | 0.40 | 11 | 1.07 | 5 | 324 |
| Methanol | 3 | 0.46 | 12 | 1.29 | 6 | 358 |
| 15 + 40 + 45 | 4 | 0.55 | 8 | 0.64 | 5 | 324 |
| | 5 | 0.60 | 1 | 0.07 | 1 | 188 |
| | 6 | 0.66 | 10 | 0.97 | 5 | 324 |
| | 7 | 0.72 | 9, 6 | 0.76, 0.40 | 5, 4 | 324, 290 |
| | 8 | 0.75 | 5, 7B | 0.30, 0.52 | 3, 4 | 256, 290 |
| | 9 | 0.78 | 7A | 0.48 | 4 | 290 |
| | 10 | 0.84 | 7, 5 | 0.48, 0.30 | 4, 3 | 290, 256 |
| | 11 | 0.87 | 3 | 0.21 | 2 | 222 |
| | 12 | 0.89 | 2 | 0.15 | 2 | 222 |
| | | | Aroclor 1248 | | | |
| | 1 | 0.37 | 11 | 1.06 | 5 | 324 |
| | 2 | 0.42 | 12, 9 | 1.28, 0.76 | 6, 5 | 358, 324 |
| | 3 | 0.47 | 8 | 0.63 | 5 | 324 |
| | 4 | 0.51 | 10 | 0.99 | 5 | 324 |
| | 5 | 0.57 | 9 | 0.76 | 5 | 324 |
| | 6 | 0.63 | 5, 6 | 0.30, 0.40 | 3, 4 | 256, 290 |
| | 7 | 0.68 | 7A & B[c] | 0.47, 0.51 | 4 | 290 |
| | 8 | 0.76 | 3, 2 | 0.21, 0.15 | 2 | 222 |
| | 9 | 0.79 | 4 | 0.24 | 3 | 256 |
| | | | Aroclor 1254 | | | |
| | 1 | 0.21 | 16 | 2.36 | 6 | 358 |
| | 2 | 0.27 | 13 | 1.55 | 6 | 358 |
| | 3 | 0.35 | 12, 15 | 1.28, 1.95 | 6 | 358 |
| | 4 | 0.40 | 14 | 1.61 | 6 | 358 |
| | 5 | 0.45 | 9 | 0.79 | 5 | 324 |
| | 6 | 0.55 | 11 | 1.06 | 5 | 324 |
| | 7 | 0.69 | 8, 6 | 0.66, 0.40 | 5, 4 | 324, 290 |
| | 8 | 0.76 | 7A | 0.48 | 4 | 290 |
| | | | Aroclor 1260 | | | |
| Water- | 1 | 0.50 | 18 | 5.23 | 7 | 392 |
| Acetone- | 2 | 0.60 | 16 | 3.10 | 7 | 392 |
| Acetonitrile- | 3 | 0.68 | 12A, 14, 17 | 1.51, 2.12, | | |
| Methanol | | | | 3.96 | 6, 7 | 358, 392 |
| 5 + 15 + 40 + 40 | 4 | 0.76 | 12B, 13, 15 | 1.67, 1.89, | | |
| | | | | 2.58 | 6, 7 | 358, 392 |
| | 5 | 0.87 | 9, 11 | 0.82, 1.28 | 5, 6 | 324, 358 |
| | 6 | 0.92 | 10, 12A | 1.22, 1.51 | 6 | 358 |
| | 7 | 0.96 | 8, 10 | 0.72, 1.12 | 5, 6 | 324, 358 |

[a]$R_f$ values of *p,p'*-DDE were 0.60 and 0.92, respectively.

[b]A column temperature of 160°C was used, except for GLC of Aroclor 1260 which was resolved at 190°C, glass column, 7 ft x 2 mm, packed with 0.3% OV-7 on glass beads, carrier gas nitrogen 40 ml/min.

[c]Two components under the same peak.

Courtesy of *J. Assoc. Offic. Anal. Chem.*

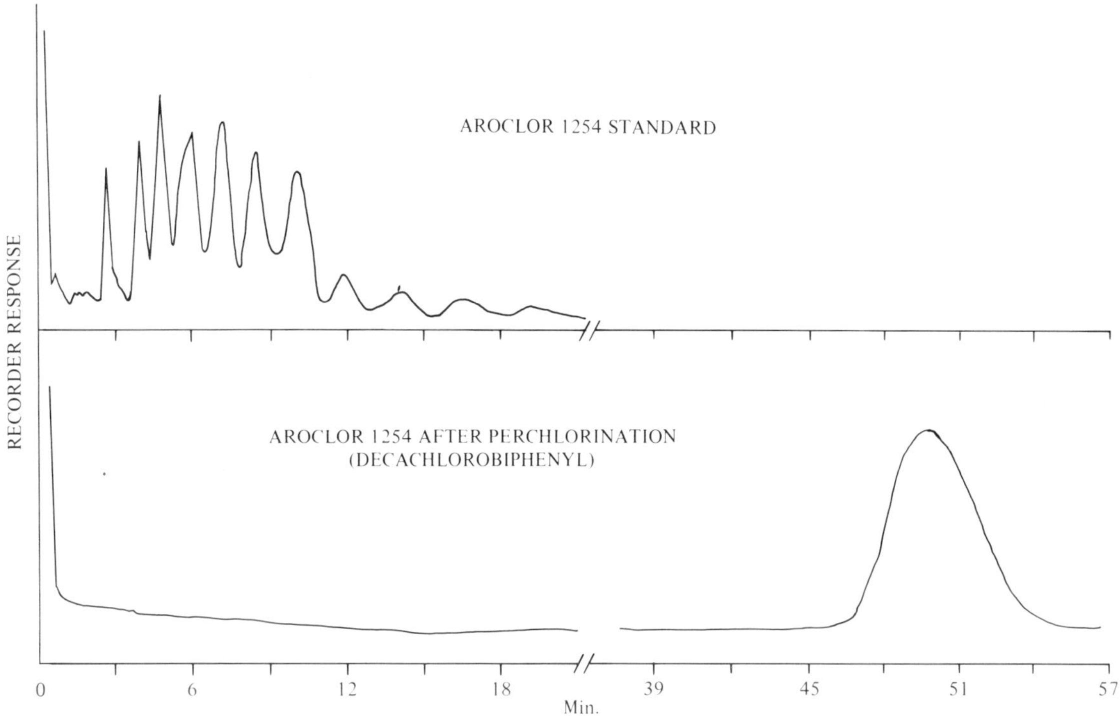


FIGURE 1. GC pattern of Aroclor 1254 before and after perchlorination (200°, 6ft x 4 mm glass column containing 4% SE-30 on acid-washed Chromosorb W 60-80 mesh, carrier gas nitrogen, 60 ml/min).

evaporated at 35° under nitrogen, 0.2 ml of antimony pentachloride was added, and the vial was capped and heated for 3 hr at 165° in a sand bath. After the mixture was allowed to cool, hydrochloric acid (20%, 0.5 ml) was added and decachlorobiphenyl was extracted with warm benzene (5 x 1 ml) in a supermixer (Arthur H. Thomas Co. No. 8929-W). The extracts were filtered through separate 9 mm i.d. glass columns containing 2 g of sodium sulfate. Each column was rinsed with 5 ml of benzene.

Huckins et al. noticed that some batches of antimony pentachloride were contaminated with bromine (antimony pentabromide ?) and consequently yielded bromononachlorobiphenyl eluted after decachlorobiphenyl on gas chromatography.[54]

Hutzinger et al.[55] investigated several chlorination reagents. Best results were obtained with a mixture of sulfuryl chloride, sulfur monochloride and anhydrous aluminum chloride (BMC-mixture),[15,39] and with a mixture of antimony pentachloride and iodine.

Prior to use, the BMC reagent was prepared by mixing equal volumes of 1% solution of sulfur monochloride in sulfuryl chloride and a 0.5% solution of anhydrous aluminum chloride in the same solvent. The sample was dissolved in the reagent (100 mg, 10 ml) and the solution was heated for 2 hr in a loosely capped flask at about 69°. Approximately 75% of sulfuryl chloride was evaporated during this time. Hydrochloric acid (25%, 30 ml) was added to the residue and the mixture was heated (60 to 80°, 10 to 60 min) to yield the perchlorinated compound as a precipitate.

The antimony pentachloride-iodine reagent was a solution of these compounds in sulfuryl chloride (0.5 g, 10 ml, and 90 ml respectively). The sample was dissolved in the reagent (100 mg, 10 ml) and the temperature was gradually raised to 120 to 130° and maintained for 2 hr. The residue was then heated for 10 min with hydrochloric acid (25%, 10 ml).

A third reagent, trichlorosulfur tetrachloroaluminate, transformed Aroclor 1254 into decachlorobiphenyl under relatively mild conditions.[33,56] This reagent was prepared by mixing sulfur dichloride and liquid chlorine (5.3 g and 3.2 g, respectively) at −80° and by carefully adding powdered anhydrous aluminum chloride at the same temperature. The mixture was dissolved in sulfuryl chloride (50 mg in 5 ml) and heated with Aroclor 1254 (1mg) for 3 hr at 70°. Hydrochloric acid (20%, 5 ml was added) and decachlorobiphenyl was extracted with benzene.

The properties of decachlorobiphenyl and of perchloro-derivatives of some other compounds which occur or may occur in environmental samples are summarized in Table 2. The perchlorinated compounds have a very long retention time on the GC columns usually employed in the determination of PCB's. A glass column (6 ft x 4 mm) containing 3% OV-210 (Pierce Chemical Company) on Chromosorb WAW 60/80 at 200° was used to obtain the GC data in Table 2. Berg et al. used a short (2 ft x 1/8") stainless steel column with 5% SE-30 on Chromosorb W to detect decachlorobiphenyl (elution time 9 min at 215°).[21] Huckins et al. determined decachlorobiphenyl with a 4 ft x 2 mm glass column, filled with 0.3% of OV-7 glass beads, and operated at 220°.[54]

It can be seen from Table 2 that perchlorination would not distinguish between chlorinated naphthalenes and PCB's. On the other hand, *p,p'*-DDE, which occurs together with PCB's in most biological samples, gives a number of peaks (possibly due to thermal decomposition on the column, since the purified perchlorinated product gave the expected mass spectrum of dodecachloro-DDE). Fortunately, none of these coincides with the peak of decachlorobiphenyl under the given GC conditions. It is also important to note that peaks of the dodecachloro-DDE compounds would not be mistaken for those of octachlorodibenzo-*p*-dioxin and octachlorodibenzofuran. However, these two compounds are not clearly separated. The three peaks of tetradecachloroterphenyl correspond in order of increasing retention time to the *o*-, *p*-, and *m*-isomer, respectively.

**Quantitation of PCB's**

Gas chromatography is the method of choice for the detection of PCB's. With the exception of model experiments with individual chlorobi-

TABLE 2

**Properties of Perchloro-derivatives[55]**

| Compound | (lit.m.p.) | Analyses[a] C | Analyses[a] Cl | Retention times[b] | Method of preparation |
|---|---|---|---|---|---|
| Octachlorodibenzofuran | 256–7[c] | 32.43<br>32.69 | 63.96<br>63.41 | 2.55[d] | BMC |
| Octachlorodibenzodioxin | 328<br>(330) | –[e] | – | 2.60[d] | BMC |
| Octachloronaphthalene | 196–7<br>(From 197 to 202) | 29.70<br>30.00 | 70.30<br>70.23 | 1 | $SO_2Cl_2$-$SbCl_5$[f] |
| Decachloro-1,4-dihydronaphthalene | 208<br>(208) | 25.26<br>25.26 | 74.74<br>74.20 | 1 | BMC |
| Decachlorobiphenyl | 304<br>(305) | –[e] | – | 1 | BMC and $SbCl_5$-$I_2$ |
| Dodecachloro-DDE | >360° | 28.28<br>28.27 | 71.72<br>70.87 | 1.9, 2, 2.9<br>3.6, 4.7, 5.9 | $SbCl_5$-$I_2$ |
| Tetradecachloroterphenyl | (From 290 to >360°) | 30.29<br>29.91 | 69.71<br>69.15 | 11.9, 19.2<br>22.2[g] | $SbCl_5$-$I_2$ |

[a]Calculated over found. Values for chlorine are consistently low for most of these perchloro-derivatives, possibly due to incomplete decomposition.
[b]Retention time relative to decachlorobiphenyl.
[c]Identical to a sample provided by Dr. Pohland.
[d]Identical to samples provided by Dr. Firestone.
[e]Samples were identical to those prepared by the methods described in the literature.
[f]Halowax or naphthalene was heated, under reflux, in a mixture of sulfuryl chloride and antimony pentachloride 9:1 (v/v) for 1 hr and the octachloronaphthalene was recovered as described for the other perchloro-compounds. Heating with neat $SbCl_5$ or $SO_2Cl_2$-$SbCl_5$-$I_2$ leads to extensive decomposition.
[g]This product changes m.p. on crystallization. It consists of a mixture of *o*-, *m*- and *p*-perchloroterphenyl.

Courtesy of *Int. J. Environ. Anal. Chem.*

phenyls, multicomponent mixtures of chlorobiphenyls are always encountered. These more or less resemble commercial PCB preparations which are then used as the standards for quantitation. The quantitation cannot be accurate since it is not based on a complete resolution of all chlorobiphenyls, and quantitative detector responses to all the chlorobiphenyls are not known. However, the quantitation can be precise and the degree of precision, tested in a multilaboratory exchange-sample program, is about the same as the precision and accuracy of the determination of the common chlorinated hydrocarbon pesticides (standard deviation ±20%).[50]

Many different procedures are used to quantitate GC tracings of PCB's. Some authors use peak area and others use peak height of one, two, three or more peaks. PCB patterns in wildlife samples usually resemble and are quantitated as Aroclor 1254 or 1260. At times the patterns of Aroclor 1232 or 1242 may also be recognizable. A technique based on computer-matching of the observed PCB patterns in the sample with patterns obtained by mixing various Aroclors in different ratios was described.[96]

In our laboratory, dealing mostly with aquatic wildlife samples containing PCB's resembling Aroclor 1254, the total height of five out of the six major Aroclor 1254 peaks is used for quantitation. The sixth peak (third in order of increasing retention time) is not used since it overlaps with that of *p,p'*-DDE (chromatography on silica not separating PCB's from *p,p'*-DDE is routinely carried out during the analyses). A correction is applied to the height of the *p,p'*-DDE peak. The peak height of the overlapping PCB peak per unit weight of Aroclor 1254 is about 10% of the peak height of *p,p'*-DDE.

The described quantitation procedures are suitable for monitoring purposes. However, the precision decreases if the PCB pattern in the sample differs from that of the appropriate commercial PCB preparation. To eliminate this decrease in precision, it was suggested to quantitate PCB's after perchlorination in terms of decachlorobiphenyl.[21] This approach may increase the precision of the quantitation but has no effect on its accuracy since the proportions of individual chlorobiphenyls in the original mixture remain unknown. Additional manipulations required for the perchlorination and variable yields of decachlorobiphenyl, discussed in the preceding section, may well eliminate the increase in precision expected from the perchlorination technique. Rote and Murphy obtained a semilogarithmic relationship between the EC detector total peak area (y) and the average number of chlorine atoms (x) in Aroclors 1232, 1242, 1248, 1254, 1260, and 1262:

$$\log y = 0.1512x - 0.7428,$$

and calculated from this relation the amount of chlorinated biphenyl in each peak of the analyzed sample using published mass spectrometry data to identify the number of chlorine atoms in the individual peaks.[90]

Again, this approach has not much effect on the accuracy of the quantitation of PCB's since, as will be shown later in this section, the electron-capture detector response to chlorobiphenyls with the same number of chlorine atoms may vary significantly with the substitution patterns.

GC columns with a high resolving capability have been described in Chapter 2. Many chlorobiphenyls present in commercial PCB preparations have been identified or tentatively identified and sooner or later, known detector responses to the identified chlorobiphenyls will make an accurate quantitation of PCB's possible. For the analyst engaged in routine monitoring work this approach may not be practical because of the increased complexity of the quantitation procedure. The situation may change drastically when, based for example on toxicological data, there will be a special interest in an accurate determination of some particular chlorobiphenyls.

A few observations about the quantitative response of the flame-ionization and electron-capture detectors to individual chlorobiphenyls can be made.

Relative molar flame ionization responses (biphenyl = 100) of several chlorobiphenyls were determined by Albro and Fishbein and also calculated using the following method: A response value of 100 was assigned to biphenyl. Three response units were subtracted for each chlorine substitution in positions 2,2',6,6', and 6 additional units were subtracted for each additional substitution after the first. Eleven units were subtracted for substitutions in positions 3,3',5,5', and 16 units for substitution in positions 4 and 4'. If both rings contained chlorine, 10 units were added for each chlorine in the least substituted ring. Vicinal substitution raised the response by 1.5 units in positions 2,3, and 5,6, and by 9.5 units in the positions 3,4 or 4,5.[3]

Flame-ionization detector is seldom used in the determination of PCB's due to its relatively low sensitivity. However, in the case of mono- and dichlorobiphenyls, its sensitivity may be comparable to the sensitivity of the electron-capture detector.

Electron-capture detector response to mono- and dichlorobiphenyls was reported by Gregory and we have determined the response to some mono-, di-, tri-, tetra-, penta-, hexa-, and octachlorobiphenyls.[43,117] The detector response increases sharply with increasing numbers of chlorine atoms. For example, the response to decachlorobiphenyl is 500 times stronger than the response to 4-chlorobiphenyl. Most of the increase occurs in the mono- to trichlorobiphenyl range and the response increases only by a factor of 2 to 3 between tetra- and decachlorobiphenyl.

As in the case of the flame-ionization detector, the chlorine substitution patterns affect the response of the electron-capture detector. Not enough data are available for generalization of these effects, but as a rule, chlorine substitution in positions 2 and 6 decreases and vicinal substitution, particularly in the positions 3 and 4, increases the response.

The responses of flame-ionization and electron-capture detectors to some chlorobiphenyls are summarized in Table 3.

In addition to the two detectors mentioned above, total-ion currect of GC-MS systems, microcoulometric, and electrolytic conductivity detectors sometimes are used to quantitate PCB's. The quantitation procedures are identical with those already described. These detectors are somewhat less sensitive than the electron-capture detector. However, the last two are more selective.

A selective determination of chlorinated hydrocarbon pesticides in the presence of PCB's was achieved by manipulating the operating parameters of a Coulson electrolytic conductivity detector. PCB's did not interfere with the determination of chlorinated hydrocarbon pesticides when the detector furnace was kept at 600°C. The detector response to PCB's was enhanced at temperatures above 820°.[32]

UV spectrophotometry was occasionally used to quantitatively determine PCB's. We have used this technique to monitor concentration of Aroclors 1221 and 1254 in water during toxicity

TABLE 3

**Relative Molar Responses of Electron-capture and Flame-ionization Detectors to Some Chlorobiphenyls**

| Chlorobiphenyl | Relative molar response | |
|---|---|---|
| | Electron capture | Flame ionization |
| 2- | 1.00 | 1.00 |
| 3- | 0.20 | 0.92 |
| 4- | 1.10 | 0.87 |
| 2,2′-di | 5.16 | 0.99 |
| 2,4-di | 17.7 | 0.86 |
| 2,6-di | 32.0 | 0.91 |
| 3,3′-di | 6.10 | 0.94 |
| 3,4-di | 15.2 | 0.86 |
| 4,4′-di | 5.97 | 0.81 |
| 2,4,4′-tri | 135 | 0.78 |
| 2,2′,4,4′-tetra | 106 | 0.87 |
| 2,2′,6,6′-tetra | 20.6 | 0.90 |
| 3,3′,4,4′-tetra | 396 | 0.87 |
| 3,3′,5,5′-tetra | 320 | 0.85 |
| 2,3,4,5-tetra | 367 | 0.87 |
| 2,3,5,6-tetra | 259 | 0.71 |
| 2,2′,4,4′,6,6′-hexa | 347 | |
| 3,3′,4,4′,5,5′-hexa | 726 | |
| 2,2′,3,3′,4,4′,6,6′-octa | 1,180 | |
| 2,2′,3,3′,5,5′,6,6′-octa | 1,150 | |
| deca | 1,410 | |

tests with fish. PCB's were solubilized in water by nonionic surfactants yielding optically homogeneous solutions and the detection limit was approximately 0.5 to 1.0 mg/l. The method is much faster than the extraction and gas chromatography, but also much less sensitive.[112]

Benthe et al. used UV spectrophotometry to quantitate PCB's (Pydraul A200, Monsanto, chlorine content 42%) in organs of rats, exposed to PCB aerosols. The levels of PCB's in the tissues were very high (40 to 250 μg/g).[20]

UV spectrofluorimetry was also used to determine Aroclors 1221 and 1254 in water, but again, the sensitivity, although better than that of UV spectrophotometry, was not very high (0.025 to 0.2 mg/l).[112]

The determination of a chlorobiphenyl in rat adipose tissue by neutron activation and measurement of $^{38}Cl$ was reported.[67] It is not likely that this technique will be widely used.

**Determination of Chlorobiphenylols**

Little is known about the analytical properties of chlorobiphenylols, a group of metabolites of chlorobiphenyls (Chapter 7). On the other hand, the determination of biphenylols was studied in detail in connection with the metabolism of the parent compound, and the methods for the determination of mono- and dichlorobiphenylols may not require many modifications. At the same time, pentachlorophenol may be chosen for an approximation of the analytical behavior of the polychlorobiphenylols. Some of the typical procedures are described below to stimulate the development of analytical techniques for the determination of chlorobiphenylols.

In a model experiment Bache and Lisk extracted 2,2′,6,6′-tetrachloro-4,4′-biphenyldiol with 2% bicarbonate from a hexane solution containing Aroclor 1254 and the above compound. The bicarbonate solution was acidified and the chlorobiphenyldiol was extracted with diethyl ether and quantitated by gas chromatography of the corresponding dimethyl ether.[13]

The commercial availability makes the use of 2,2′,6,6′-tetrachloro-4,4′-biphenyldiol attractive for the development of analytical techniques. In our experience, however, this compound is rather unstable.

Pillion separated 4- and 6-chloro-2-biphenylol and 2-chloro-1-biphenylol by GC on a 10% XE-60 on Diaport® S, 60-80 mesh and on 10% SE-30 on Chromosorb W, both columns at 210°. As in the case of chlorobiphenyls, substitution in 2 and 6 positions decreased the retention time.[81] Hutzinger et al. examined the GC behavior of a number of chlorobiphenylols and chlorobiphenyldiols and the results are summarized in Table 4. The data are not sufficient to correlate the retention times and detector responses with the substitution patterns of the chlorobiphenylols. It can be seen that at a given chlorine substitution, the retention times increase from 2- to 4-biphenylols.

Tomori separated 3,3′-dichloro-, 3,5′-dichloro-, and 5,5′-dichloro-2,2′-dihydroxybisphenol (increasing order of retention time) in the form of the corresponding trimethylsilyl (TMS) ethers on a polyethylene glycol adipate column. TMS ethers were prepared using hexamethyldisilazane and trimethylchlorosilane in pyridine (60 mg of compound, 0.2, 0.1, and 1 ml, respectively).[101]

The isolation of the metabolites of biphenyl from rat urine was described by West et al. in 1956.[106] Free biphenylols were recovered by extraction with diethyl ether and further fractionated by solvent extraction. Benzene extracted 4-biphenylol from the crude mixture precipitated with hydrochloric acid and carbon disulfide was used to separate 3,4-biphenyldiol (soluble) from 4,4′-biphenyldiol (insoluble). Conjugated biphenylols precipitated from the urine after dilution with concentrated hydrochloric acid (1/10 of urine volume) on standing for several days in the cold. The precipitate was extracted with chloroform to yield the chloroform-soluble biphenylmercapturic acid and the chloroform insoluble *p*-biphenyl-*β*-*D*-glucuronide.

Raig and Ammon determined 2- and 4-biphenylol, 3,4- and 4,4′-biphenyldiol by GC of their TMS ethers in urine of rabbits fed biphenyl.[86] The urine was treated with concentrated hydrochloric acid (1/10 of urine volume) and extracted with ether in a continuous extractor for 30 hr. The ether phase was then extracted in succession with saturated sodium bicarbonate, 10% sodium carbonate and 2 *N* sodium hydroxide. The solutions were acidified, extracted with ether, trimethylsilylated, and examined by gas chromatography. The trimethylsilylation was carried out using bis-trimethylsilylacetamide (BSA, 50-80 x excess, chloroform, 1 hr at 50°).

Columns with either 5% SE-52 on silica 60-100 mesh or 4% XE-60 or Chromosorb® G, 80-100

TABLE 4

**Relative Retention Times and EC Detector Responses of Some Chlorobiphenyls and Chlorobiphenylol Acetates***

| Compound | Retention time | Response |
|---|---|---|
| 2,2′,5,5′-tetrachlorobiphenyl | 1.00 | 1.00 |
| 2-chloro-4-biphenylol | 0.94 | 0.13 |
| 3-chloro-4′-biphenylol | 0.90 | 0.10 |
| 3,5-dichloro-2-biphenylol | 0.84 | 0.62 |
| 3,5-dichloro-4-biphenylol | 1.51 | 0.68 |
| 4,4′-dichloro-3-biphenylol | 1.91 | 0.74 |
| 2,4′-dichloro-4-biphenylol | 1.63 | 0.54 |
| 3,4′-dichloro-4-biphenylol | 2.02 | 0.34 |
| 2,5-dichloro-2′-biphenylol | 0.68 | 3.23 |
| 2,5-dichloro-3′-biphenylol | 1.48 | 0.39 |
| 2,5-dichloro-4′-biphenylol | 1.64 | 0.36 |
| 2,2′,5′-trichloro-4-biphenylol | 2.74 | 0.80 |
| 3,4,5-trichloro-4′-biphenylol | 3.83 | 1.89 |
| 3,3′,5,5′-tetrachloro-2,2′-biphenyldiol | 3.72 | 3.30 |
| 3,3′,5,5′-tetrachloro-4,4′-biphenyldiol | 14.7 | 1.79 |
| 3,3′,4,4′-tetrachlorobiphenyl | 2.41 | 7.52 |
| 2,2′,4,4′,6,6′-hexachlorobiphenyl | 1.81 | 3.91 |
| 2,2′,4,4′,5,5′-hexachlorobiphenyl | 3.62 | 10.3 |

*Varian 600D gas chromatograph with a glass column 5 ft x 1/8 in., containing 4% SE-30 on acid-washed Chromosorb W 100-120 mesh, at 184°C; carrier gas nitrogen, 70 ml/min).

mesh were used isothermally at 150° and then with a linear increase of 7.5°/min to 230°.[86,87]

Rudling extracted pentachlorophenol with a mixture of isopropanol and hexane (1 + 5 v/v, 5 ml) from an acidified aqueous homogenate of the sample (1 g sample, 5 ml water, 1 ml 6 *M* sulfuric acid, 10 min hydrolysis). The hexane layer was extracted with 0.1 *M* borax (2 x 2 ml) and hexane (0.5 ml) and the acetylation reagent (acetic anhydride-pyridine 0.4 + 1 v/v, 40 μl) were added. The mixture was shaken for 1 min and pentachlorophenol acetate was determined by gas chromatography of the hexane phase. The recoveries of pentachlorophenol from spiked samples were quantitative.[91] According to our preliminary experiments, 0.1 *M* borax does not extract chlorobiphenylols under these conditions and 0.1 *N* sodium hydroxide must be used. Chlorobiphenylols are recovered with hexane after acidification of the aqueous phase and acetylated in hexane using the above described reagent.

Pentachlorophenol may be also determined directly by GC on a column containing 3% diethylene glycol succinate and 2% sirupy phosphoric acid on Chromosorb G. This phase bleeds appreciably and a $^{63}$Ni detector at higher temperature should be used.[16]

The use of various pentachlorophenol alkyl ethers ranging from methyl to amyl was described for the determination of pentachlorophenol in urine and water.[29] The ethers were prepared by using the appropriate diazoalkanes. Concentrated sulfuric acid cleanup may be used to isolate pentachlorophenol from lipids.

Fat (5 g) is dissolved in 10 ml of petroleum ether and 20 ml of concentrated sulfuric acid and 25 g of Celite are added. The mixture is stirred, extracted with petroleum ether (150 ml and 2 x 125 ml), and the combined petroleum ether solutions are extracted with 0.25 *N* potassium hydroxide (3 x 50 ml, 15 ml ethanol is added in the first extraction). The aqueous phase is washed with petroleum ether (2 x 50 ml), acidified with 25 ml of 3 *N* hydrochloric acid, and pentachlorophenol is extracted with chloroform (3 x 70 ml). The residue after evaporation of chloroform is taken up in 5 ml of petroleum ether, treated with 2 ml of concentrated sulfuric acid, and analyzed by gas chromatography.[48]

Stark refluxed samples of fish (2 to 20 g) with 0.1 *N* potassium hydroxide (10 ml/g of fish), acidified with 500 ml of 0.5 *M* sulfuric acid, and distilled off 200 ml. The distillate was acidified with sulfuric acid and extracted with toluene.

Methyl pentachlorophenyl ether was prepared using diazomethane and determined by gas chromatography.[99]

A number of other derivatizing groups were suggested for the determination of phenols by gas chromatography with an electron-capture detector. Kawahara described pentafluorobenzyl derivatives,[61] Ehrsson et al. acylated phenols with heptafluorobutyric anhydride,[36] and Seiber et al. described the use of 2,6-dinitrophenyl derivatives.[94]

Berninger et al. separated 2- and 4-biphenylol, 2,2′- and 4,4′-biphenyldiol by thin-layer chromatography on silica with 1,2-dichloroethane-acetone 95:5 as the developing solvent. Several biphenylols were separated by thin-layer chromatography on silica gel P (Macherey-Nagel & Co) with a mixture of benzene-methanol-formic acid (90:9:1 v/v).[22] Creaven et al. separated biphenylols by thin-layer chromatography on alumina using either benzene-chloroform (19:1, v/v), *n*-hexane-methanol (49:1, v/v), or acetone-chloroform (1:1, v/v) as the solvents.[30]

Hutzinger et al. separated some chlorobiphenylol metabolites from the parent chlorobiphenyls by thin-layer chromatography on silica with either hexane-acetone (2.5:1) or benzene-ethyl acetate (12:1).[57]

### Levels of PCB's in the Environment

A large number of papers dealing with levels of PCB's in the environment has been published and several reviews of the subject have appeared.[4,6,34,116] Readers interested in an in-depth discussion of this subject should consult the original literature. The following discussion is aimed primarily at the analytical chemist to give him a general knowledge of the range of PCB concentration in various environmental samples.

While reading the literature and producing new data, the analyst should always pay attention to a few important details. First of all, the type of the PCB preparation should be noted.

The degree of chlorination gives PCB's not only different chemical properties, but also biological and toxicological properties as well. Consequently, the interpretation of the results may depend on the type of the PCB preparation. Most PCB residues in wildlife samples are quantitated as Aroclors 1254, 1260, or their equivalents. However, Aroclors 1242 and 1248 may also be encountered. In any case, the type of PCB preparation should always be clearly stated.

The results of analyses of biological samples are usually reported on wet weight or on lipid basis. This again should be clearly stated. If the percentage of lipid is given, the two types of data are easily interconvertible ($\mu$g/g lipid = 100 $\mu$g/g wet weight/% lipid). It is also important to describe the method used for the determination of lipid. Quite often authors reporting the lipid content refer in fact to hexane-extractable lipids.

Some authors prefer to report PCB concentration on lipid basis, assuming that this form of expressing the level is better related to the biological activity and enables a comparison between species. However, caution must be exercised in interpreting the results if the lipid content for example varies with the time of the year, stage of the development or the life cycle. Two populations of Atlantic herring were analyzed for PCB's using male fish of equal age, taken immediately before spawning. PCB levels on lipid basis were 12.6 $\mu$g/g in population B and 3.46 $\mu$g/g in population T. The population B consists of fish spawning in the spring after a period of low winter feeding activity and has a hexane-extractable lipid content of 1.98%. The population T spawns in the fall after an intensive summer feeding and has a lipid content of 12.7%. The average body weight of the population B is 188 g and that of the population T is 194 g. It may easily be calculated that the total body burden of PCB's is 47 $\mu$g in the population B and 85 $\mu$g in the population T. Thus, while the PCB levels on lipid basis indicate that the population B may be under a physiological stress due to the higher levels of PCB's, according to the total body burden, the population T lives in an area which is more contaminated with PCB's than that of the population B. If sampled early in spring, the population T may well have higher level of PCB's on lipid basis than the population B.[118]

The type of tissue analyzed should always be clearly identified (adipose tissue, muscle, liver, yolk, whole fish, etc.).

The weight of the analyzed specimen or organ should be given. Total PCB burden may then be calculated and provide valuable additional information or prevent erroneous conclusions. For example, in a study of the rate of excretion of Aroclor 1254 by juvenile Atlantic salmon, the whole body concentration of PCB's decreased from 32.1 $\mu$g/g wet weight at the withdrawal of the contaminated diet to 6.83 μg/g after 245 days on a control diet. However, since the fish grew in the meantime, the respective total body burdens were 221 and 255 μg, indicating that PCB's were not excreted.

The levels of PCB's in the physical environment are summarized in Table 5. Data on the concentration of PCB vapors in air are not available. The level of PCB's in particulate matter is an average value for several cities in the U.S. The data on the concentration of PCB's in rain water are quite fragmentary and more measurements are needed. A PCB fallout of 75 $\mu g/m^2$ was observed in the south of Sweden and this value is of the same order of magnitude as the estimate for North America.[4] A wide range of values is reported also for both fresh and sea water. According to some authors, the concentration of PCB's in suspended particulate matter may be much higher than that in water. Some methodological problems (discussed in the section on sampling) must be overcome before more accurate data can be obtained. It should be noted that sewage sludge is a significant source of PCB's.

The distribution of PCB levels in freshwater fish from the U.S. and from Sweden are presented in Table 6. The U.S. fish samples originated from a pesticide-monitoring network with stations covering all major river basins. It can be seen that higher PCB levels are more frequently encountered in the U.S. than in Sweden. In the U.S., higher levels are generally found in fish from the Great Lakes and from rivers in the northeast.

The distribution of PCB's in marine fish (Table 7) shows a great similarity between Sweden and Canada. The levels in marine fish are usually lower than those in freshwater fish. Some species of sharks, feeding at higher trophic levels often contain high levels of PCB's.[116] For marine mammals, levels from 1 to 250 μg/g lipid have been reported.[50] PCB levels in fish from Japan are apparently quite high.[100]

PCB levels in predatory birds and birds feeding on fish are often very high and may range from

TABLE 5

**Environmental Levels of PCB's in the Physical Environment**

| Sample | Level | Reference and remarks |
|---|---|---|
| Air | No data available | |
| particulate matter | 50 μg/g | 79 |
| rain | 20–100 ng/l | 50 |
| Freshwater | 2–20 ng/l | Areas with little industrial activity[50] |
| | 20–2,100 ng/l | industrialized areas[50] |
| Seawater | 5–20 ng/l | coastal areas[50] |
| Sewage sludge | 1–100 μg/g | 6 |

TABLE 6

**Levels of PCB's in Freshwater Fish (μg/g wet weight)**

| Range | <0.1 | 0.1–0.5 | 0.5–1 | 1–5 | 5–10 | >10 | Country |
|---|---|---|---|---|---|---|---|
| % | 18 | 21 | 17 | 37 | 4 | 3 | U.S. (Reference 47) |
| % | 79 | 15 | 3 | 3 | | | Sweden (Reference 108) |

TABLE 7

**Levels of PCB's in Marine Fish (μg/g wet weight)**

| Range | 0–0.1 | 0.2–0.5 | 0.6–1.0 | 1.1–2 | 2.1–3 | Country |
|---|---|---|---|---|---|---|
| % | 27 | 68 | 2 | 2 | 1 | Sweden (Reference 108) |
| % | 16 | 73 | 9 | 2 | 0 | Canada (Reference 116) |

about 3 to 9500 μg/g wet weight in the pectoral muscle. Similarly, the reported levels in eggs spread over more than two orders of magnitude (0.7 to 100 μg/g wet weight).[116]

The distribution of PCB's in human adipose tissue is given in Table 8. It can be seen that PCB levels in Japan and possibly also in West Germany are higher than those in the U.S. PCB concentration in human milk is 0.1, 0.06, 0.02, and 0.01 in West Germany, the U.S., Sweden, and Norway, respectively.[26,58]

Fish are the main source of PCB's in the diet (Table 9), followed by eggs and milk. Fish by-products are probably responsible for the relatively frequently encountered PCB's in the eggs since the by-products constitute a part of poultry feed. Foods packaged in cardboard and similar materials may also be contaminated by PCB's migrating from the packaging material which at times contains up to 30 μg/g of PCB's, usually of the Aroclor 1242 variety.[62]

The levels of PCB's presented in Tables 5 through 9 are due to "normal" levels of environmental contamination. Much higher levels were occasionally found as a result of industrial accidents, most frequently spills and leaks of hydraulic or heat-transfer systems.

TABLE 8

**PCB's in Human Adipose Tissue (μg/g lipid)**

| Range | 0 | <1.0 | 1–5 | >5 | Country |
|---|---|---|---|---|---|
| % | 0 | 10 | 78 | 12 | Japan (Reference 58) |
| Range | 0 | <1 | 1–2 | >2 | |
| % | 2 | 56 | 37 | 6 | U.S. (Reference 84) |
| % | 34 | 33 | 27 | 5 | U.S. (Reference 110) |
| Range | 1.7–16.6 | Average 5.7 | | | West Germany (Reference 1) |
| | 0–6.9 | 1.61 | | | Norway (Reference 26) |

TABLE 9

**PCB's in Food in the U.S.[62]**

| Food article | Fraction of samples with detectable PCB's, % | PCB's, μg/g average | PCB's, μg/g high |
|---|---|---|---|
| Fish | 54 | 1.9 | 35 |
| Cheese | 6 | 0.2 | 1 |
| Milk | 7 | 2.3 | 28 |
| Eggs | 29 | 0.5 | 4 |
| Composite diet | 3 | | 0.4 |

# REFERENCES


1. **Acker, L. and Schulte, E.,** *Dtsch. Lebensm.-Rundschau,* 66, 385, 1970.
2. **Ahling, B. and Jensen, S.,** *Anal. Chem.,* 42, 1483, 1970.
3. **Albro, P. W. and Fishbein, L.,** *J. Chromatogr.,* 69, 273, 1972.
4. **Anonymous,** Polychlorinated biphenyls – environmental impact. A review by the panel on hazardous trace substances, *Environ. Res.,* 5, 1972.
5. **Anonymous,** *Pesticide Analytical Manual,* Department of Health, Education and Welfare, Food and Drug Administration, 1972.
6. **Anonymous,** *Polychlorinated Biphenyls and the Environment,* Interdepartmental Task Force on PCB's, Washington, D.C. National Technical Information Service, COM-72-10419, Department of Commerce, Springfield, Virginia 22151.
7. **Armour, J. A. and Burke, J. A.,** *FDA Lab. Inform. Bull.,* No 918, 11, 1969.
8. **Armour, J. A. and Burke, J. A.,** *J. Assoc. Offic. Anal. Chem.,* 53, 761, 1970.
9. **Armour, J. A. and Burke, J. A.,** *J. Assoc. Offic. Anal. Chem.,* 54, 175, 1971.
10. **Asai, R. T., Gunther, F. A., Westlake, W. E., and Iwata, Y.,** *J. Agric. Food Chem.,* 19, 396, 1971.
11. **Aue, W. A. and Hastings, C. R.,** *J. Chromatogr.,* 42, 319, 1969.
12. **Aue, W. A., Kapila, S., and Hastings, C. R.,** *J. Chromatogr.,* 73, 99, 1972.
13. **Bache, C. A. and Lisk, D. J.,** *Bull. Environ. Contam. Toxicol.,* 9, 315, 1973.
14. **Bagley, G. E., Reichel, W. L., and Cromartie, E.,** *J. Assoc. Offic. Anal. Chem.,* 53, 251, 1970.
15. **Ballester, M., Molinet, C., and Castañer, J.,** *J. Am. Chem. Soc.,* 82, 4254, 1960.
16. **Barthel, W. F., Curley, A., Thrasher, C. L., Sedlak, V. A., and Armstrong, R.,** *J. Assoc. Offic. Anal. Chem.,* 52, 294, 1969.
17. **Bauer, U.,** *Vom Wasser,* 38, 49, 1971.
18. **Bauer, U.,** *GWF-Wasser/Abwasser,* 113, 58, 1972.
19. **Bellman, S. W. and Barry, T. L.,** *J. Assoc. Offic. Anal. Chem.,* 54, 499, 1971.
20. **Benthe, H. F., Knop, J., and Schmoldt, A.,** *Arch. Toxikol.,* 29, 85, 1972.
21. **Berg, O. W., Diosady, P. L., and Rees, G. A. V.,** *Bull. Environ. Contam. Toxicol.,* 7, 338, 1972.
22. **Berninger, H., Ammon, R., and Berninger, I.,** *Arzneim-Forsch.,* 18, 880, 1968.
23. **Bevenue, A. and Ogata, J. N.,** *J. Chromatogr.,* 50, 142, 1970.
24. **Bevenue, A., Kelley, T. W., and Hylin, J. W.,** *J. Chromatogr.,* 54, 71, 1971.
25. **Bevenue, A., Ogata, J. N., Kawano, Y., and Hylin, J. W.,** *J. Chromatogr.,* 60, 45, 1971.
26. **Bjerk, J. E.,** *T. norske Lageforen,* 92, 15, 1972.
27. **Carpenter, E. J., Anderson, S. J., Harvey, G. R., Miklas, H. P., and Peck, B. B.,** *Science,* 178, 749, 1972.
28. **Collins, G. B., Holmes, D. C., and Jackson, F. J.,** *J. Chromatogr.,* 71, 443, 1972.
29. **Cranmer, M. and Freal, J.,** *Life Sciences,* 9, 121, 1970.
30. **Creaven, P. J., Parke, D. V., and Williams, R. T.,** *Biochem. J.,* 96, 879, 1965.
31. **De Vos, R. H. and Peet, E. W.,** *Bull. Environ. Contam. Toxicol.,* 6, 164, 1971.
32. **Dolan, J. W., Hall, R. C., and Todd, T. M.,** *J. Assoc. Offic. Anal. Chem.,* 55, 537, 1972.
33. **Dorrenbos, H.,** U.S. Patent 3,683,029; *Chem, Abstr.,* 77, 140726, 1972.
34. **Edwards, R.,** *Chemistry and Industry,* 1340, 1971.
35. **Edwards, R.,** Personal communication, 1972.
36. **Ehrsson, H., Walle, T., and Brotel, H.,** *Acta Pharm. Suec.,* 8, 319, 1971.
37. **Elkins, H. B.,** *The Chemistry of Industrial Toxicology,* John Wiley & Sons, Inc., New York, 1959, 319.
38. **Fehringer, N. V. and Westfall, J. E.,** *J. Chromatogr.,* 57, 397, 1971.
39. **Fieser, L. F. and Fieser, M.,** *Reagents for Organic Synthesis,* John Wiley and Sons, New York, 1967, 1131.
40. **Gesser, H. D., Chow, A., Davis, F. C., Uthe, J. F., and Reinke, J.,** *Anal. Lett.,* 4, 883, 1971.
41. **Gesser, H. D., Sparling, A. B., Chow, A., and Turner, C. W.,** *J. Am. Water Works Assoc.,* 65, 220, 1973.
42. **Giam, C. S. and Wong, M. K.,** *J. Chromatogr.,* 72, 283, 1972.
43. **Gregory, N. L.,** *J. Chem. Soc.,* (B), 295, 1968.
44. **Grice, G. D., Harvey, G. R., Bowen, V. T., and Backus, R. H.,** *Bull. Environ. Contam. Toxicol.,* 7, 125, 1972.
45. **Hall, E. T.,** *J. Assoc. Offic. Anal. Chem.,* 54, 1349, 1971.
46. **Harvey, G. R. and Teal, J. M.,** *Bull. Environ. Contam. Toxicol.,* 9, 287, 1973.
47. **Henderson, C., Inglis, A., and Johnson, W. L.,** *Pestic. Monit. J.,* 5, 1, 1971.
48. **Higginbotham, G. R., Ress, J., and Rocke, A.,** *J. Assoc. Offic. Anal. Chem.,* 53, 673, 1970.
49. **Holden, A. V.,** *Nature,* 228, 1220, 1970.
50. **Holden, A. V.,** *PCB Conference II,* National Swedish Environment Protection Board, Stockholm, December 14, 1972.
51. **Holden, A. V. and Marsden, K.,** *J. Chromatogr.,* 44, 481, 1969.
52. **Holmes, D. C. and Wallen, M.,** *J. Chromatogr.,* 71, 562, 1972.

53. **Horwitz, W., Ed.,** *Official Methods of Analysis of the Association of Official Analytical Chemists,* 11th ed., Association of Official Analytical Chemists, P.O. Box 540, Benjamin Franklin Station, Washington, D.C. 20044, 1970.
54. **Huckins, J. N., Swanson, J., and Stalling, D. L.** Personal communication, 1973.
55. **Hutzinger, O., Safe, S., and Zitko, V.,** *Int. J. Environ. Anal. Chem.,* 2, 95, 1972.
56. **Hutzinger, O., Jamieson, W. D., Safe, S., and Zitko, V.,** *J. Assoc. Offic. Anal. Chem.,* 56, 982, 1973.
57. **Hutzinger, O., Nash, D. M., Safe, S., DeFreitas, A. S. W., Norstrom, R. J., Wildish, D. J., and Zitko, V.,** *Science,* 178, 312, 1972.
58. **Isono, N. and Fujiwara, K.,** *Kagaku,* 42, 396, 1972; *Fish. Res. Bd. Can.,* Translation Series No. 2587.
59. **Jacobs, M. B.,** *The Analytical Chemistry of Industrial Poisons, Hazards and Solvents,* Interscience Publishers Inc., New York, 1949, 598.
60. **Jensen, S., Renberg, L., and Olsson, M.,** *Nature,* 240, 358, 1972.
61. **Kawahara, F. K.,** *Environ. Sci. Technol.,* 5, 235, 1971.
62. **Kolbye, A. C., Jr.,** *Environ. Health Persp.,* 1, 85, 1972.
63. **Kuratsune, M. and Masuda, Y.,** *Environ. Health Persp.,* 1, 61, 1972.
64. **Lamont, T. G. and Cromartie, E.,** *J. Chromatogr.,* 39, 325, 1969.
65. **Leoni, V.,** *J. Chromatogr.,* 62, 63, 1971.
66. **Levi, I. and Nowicki, T. W.,** *Bull. Environ. Contam. Toxicol.,* 7, 193, 1972.
67. **Manri, T., Shinbori, Y., Murakawa, S., and Mochizuki, Y.,** *Radioisotopes,* 20, 371, 1971; *Chem. Abstr.,* 76, 11738, 1972.
68. **Masumoto, H. T.,** *J. Assoc. Offic. Anal. Chem.,* 55, 1092, 1972.
69. **McClure, V. E.,** *J. Chromatogr.,* 70, 168, 1972.
70. **McCully, K. A. and McKinley, W. P.,** *J. Assoc. Offic. Anal. Chem.,* 47, 652, 1964.
71. **Miles, J. R. W.,** *J. Assoc. Offic. Anal. Chem.,* 55, 1039, 1972.
72. **Mills, P. A.,** *J. Assoc. Offic. Anal. Chem.,* 51, 29, 1968.
73. **Mills, P. A., Bong, B. A., Kamps, L. R., and Burke, J. A.,** *J. Assoc. Offic. Anal. Chem.,* 55, 39, 1972.
74. **Mizutani, T. and Matsumoto, M. V.,** *Shokuhin Eiseigaku Zasshi,* 13, 398, 1972; *Chem. Abstr.,* 78, 66686, 1973.
75. **Moriarty, F.,** *Sci. Total Environ.,* 1, 267, 1972.
76. **Mulhern, B. M., Cromartie, E., Reichel, W. L., and Belisle, A. A.,** *J. Assoc. Offic. Anal. Chem.,* 54, 548, 1971.
77. **Mumma, R. O., Wheeler, W. B., Frear, D. E. H., and Hamilton, R. H.,** *Science,* 152, 530, 1966.
78. **Murphy, P. G.,** *J. Assoc. Offic. Anal. Chem.,* 55, 1360, 1972.
79. **Nisbet, I. C. T. and Sarofim, A. F.,** *Environ. Health Persp.,* 1, 21, 1972.
80. **Noren, K. and Westöö, G.,** *Acta Chim. Scand.,* 22, 2289, 1968.
81. **Pillion, E.,** *J. Gas Chromatogr.,* 3, 238, 1965.
82. **Porter, M. L. and Burke, J. A.,** *J. Assoc. Offic. Anal. Chem.,* 54, 1426, 1971.
83. **Porter, M. L. and Burke, J. A.,** *J. Assoc. Offic. Anal. Chem.,* 56, 733, 1973.
84. **Price, H. A. and Welch, R. L.,** *Environ. Health Persp.,* 1, 73, 1972.
85. **Price, N. O., Young, R. W., and Dickinson, J. K.,** *Proc. Soc. Exp. Biol. Med.,* 139, 1280, 1972.
86. **Raig, P. and Ammon, R.,** *Arzneim.-Forsch.,* 20, 1266, 1970.
87. **Raig, P. and Ammon, R.,** *Arzneim.-Forsch.,* 22, 1399, 1972.
88. **Reinke, J., Uthe, J. F., and O'Brodovich, H.,** *Environ. Lett.,* 4, 201, 1973.
89. **Reynolds, L. M.,** *Bull. Environ. Contam. Toxicol.,* 4, 128, 1969.
90. **Rote, J. W. and Murphy, P. G.,** *Bull. Environ. Contam. Toxicol.,* 6, 377, 1971.
91. **Rudling, L.,** *Water Res.,* 4, 533, 1970.
92. **Saschenbrecker, P. W. and Ecobichon, D. J.,** *J. Agric. Food Chem.,* 15, 168, 1967.
93. **Schmidt, T. T., Risebrough, R. W., and Gress, F.,** *Bull. Environ. Contam. Toxicol.,* 6, 235, 1971.
94. **Seiber, J. N., Crosby, D. G., Fouda, H., and Soderquist, C. J.,** *J. Chromatogr.,* 73, 89, 1972.
95. **Snyder, D. and Reinert, R.,** *Bull. Environ. Contam. Toxicol.,* 6, 385, 1971.
96. **Stalling, D. L.,** *Am. Chem. Soc. Div. Water, Air and Waste Chem.,* 12, 73, 1972.
97. **Stalling, D. L. and Huckins, J. N.,** *J. Assoc. Offic. Anal. Chem.,* 56, 367, 1973.
98. **Stalling, D. L., Tindle, R. C., and Johnson, J. L.,** *J. Assoc. Offic. Anal. Chem.,* 55, 32, 1972.
99. **Stark, A.,** *J. Agric. Food Chem.,* 17, 871, 1969.
100. **Tatsukawa, R. and Isono, N.,** *Shizen,* 30, 1971; *Fish. Res. Bd. Can.,* Translation Series No. 2259.
101. **Tomori, L.,** *Magy. Kem. Foly.,* 76, 437, 1970.
102. **Trout, P. E.,** *Environ. Health Persp.,* 1, 63, 1972.
103. **Uthe, J. F., Reinke, J., and Gesser, H.,** *Environ. Lett.,* 3, 117, 1972.
104. **Veith, G. D. and Lee, G. F.,** *Water Res.,* 5, 1107, 1971.
105. **Wakimoto, T., Tatsukawa, R., and Ogawa, T.,** *Kogai to Taisaku,* 7, 517, 1971; *Fish. Res. Bd. Can.,* Translation Series No. 2185.
106. **West, H. D., Lawson, J. R., Miller, I. H., and Mathura, G. R.,** *Arch. Biochem. Biophys.,* 60, 14, 1956.
107. **Westöö, G. and Noren, K.,** *Acta Chem. Scand.,* 24, 1639, 1970.

108. **Westöö, G. and Noren, K.,** *Var Foeda,* 93, 1970.
109. **Williams, D. T. and Blanchfield, B. J.,** *J. Assoc. Offic. Anal. Chem.,* 54, 1429, 1971.
110. **Yobs, A. R.,** *Environ. Health Persp.,* 1, 79, 1972.
111. **Young, S. Y. and Burke, J.,** *Bull. Environ. Contam. Toxicol.,* 7, 160, 1972.
112. **Zitko, V.,** *Bull. Environ. Contam. Toxicol.,* 5, 279, 1970.
113. **Zitko, V.,** *J. Chromatogr.,* 59, 444, 1971.
114. **Zitko, V.,** *Intern. J. Environ. Anal. Chem.,* 1, 221, 1972.
115. **Zitko, V.,** *Bull. Environ. Contam. Toxicol.,* 7, 105, 1972.
116. **Zitko, V. and Choi, P. M. K.,** PCB and other industrial halogenated hydrocarbons in the environment, *Fish. Res. Bd. Can. Tech. Rep.,* No. 272, 1971.
117. **Zitko, V., Hutzinger, O., and Safe, S.,** *Bull. Environ. Contam. Toxicol.,* 6, 160, 1971.
118. **Zitko, V., Choi, P. M. K., Wildish, D. J., Monaghan, C. F., and Lister, N. A.,** *Pestic. Monit. J.,* in press, 1974.

## Chapter 13

# RECENT DEVELOPMENTS

## A. SYNTHESIS

A number of labeled and unlabeled chlorobiphenyls have recently been described in the literature.

The free radical Gomberg-type biaryl synthesis using isoamyl nitrite was employed for the preparation of a number of tetra-, penta-, and hexachlorobiphenyls.[3,6] The melting points, UV maxima and for certain compounds the relative retention time in GC are listed in Table 1-A.

The Ullmann-synthesis of 2,2′,4,4′,5,5′-hexachlorobiphenyl from the corresponding trichloroiodobenzene was shown to yield significant quantities (3%) of the toxic 2,3,7,8-tetrachlorodibenzofuran[1] (Figure 1-A).

TABLE 1-A

**Properties of Some Tetra-, Penta-, and Hexachlorobiphenyls**

| Chlorobiphenyls | m.p. °C | $\lambda_{max}$nm (log $\epsilon$) (ethanol) | Relative retention time** |
|---|---|---|---|
| **Tetrachlorobiphenyls** | | | |
| 2,2′,3,5′-* | 49–50 | 266.5 (sh) (2.97); 275.5 (3.07); 283 (3.00) | 0.46 |
| 2,2′,4,5′-* | 65–66.5 | 274 (3.08); 281.5 (2.99) | 0.41 |
| 2,2′,5,5′-* | 85–86.5 | 268 (sh) (3.06); 275.5 (3.21); 283.5 (3.16) | 0.39 |
| 2,2′,5,6′- | 103–104.5 | 268 (sh) (2.91); 275.5 (3.00); 283.5 (2.91) | 0.33 |
| 2,3′,4′,5-* | 104–105 | 247 (4.21); 285 (sh) (3.40) | 0.63 |
| 2,3′,5,5′- | 105.5–106.5 | 245 (sh) (3.98); 283.5 (sh) (3.25) | 0.51 |
| 2,3,3′,4′- | 96–97 | 251 (4.18) | 0.74 |
| 2,2′,3,3′-* | 121–122 | 271.5 (2.88); 279.5 (sh) (2.80) | 0.54 |
| 2,3′,4,4′-* | 127–128 | 251 (4.22) | 0.65 |
| 2,2′3,4′-* | 68.5–70 | 272.5 (sh) (2.95); 280.5 (2.81) | 0.47 |
| 2,3′4′,6- | Oil | 272.5 (sh) (3.10); 281 (sh) (2.98) | 0.50 |
| 2,2′,3,6′- | 125.5–127 | 267.5 (2.89); 274.5 (2.85) | 0.38 |
| 3,3′,4,4′-* | 177–178 | 261 (4.17) | 0.99 |
| 3,3′,4,5′- | 119–120 | 258.5 (4.26) | 0.82 |
| 2,3,3′,5′- | 127.5–129 | 243.5 (4.00); 280 (sh) (3.09) | 0.58 |
| **Pentachlorobiphenyls** | | | |
| 2,2′,3,5′,6- | 98.5–100 | 270 (2.92); 277 (3.09); 284 (3.07) | – |
| 2,3,3′,4′,6- | oil | 274 (sh) (3.18); 281.5 (sh) (3.14) | – |
| 2,2′,4,5,5′-* | 76–77 | 275.5 (sh) (3.11); 281 (3.24); 289 (3.12) | 0.80 |
| 2,3′,4,4′,5-*† | 112–113 | 251.5 (4.21) | 1.27 |
| 2,2′,3′,4,5-*† | 81.5–82.5 | 272 (2.64); 281 (2.70); 289.5 (2.62) | 0.92 |
| 2,2′,3,4,5′-* | 112–114 | 276 (3.05); 284 (2.92) | – |
| 2,3,3′,4,4′-* | 117–118.5 | 251 (4.22) | – |
| 2,2′,3,3′,4- | 119–120.5 | 274 (sh) (3.08); 280 (sh) (3.05) | – |
| 2,2′,3,5,5′- | oil | 276.5 (3.07); 280 (sh) (3.08); 285 (3.20); 294.5 (3.05) | – |
| **Hexachlorobiphenyls** | | | |
| 2,2′,3,4,4′,6′- | 69.5–71 | 275 (2.85); 281.5 (2.73) | – |
| 2,3′,4,4′,5′,6- | 110–111 | 285 (sh) (2.85) | – |

*Also see Chapter 3.

**Conditions: Varian 1400 instrument fitted with an electron capture detector. Glass column (160 x 0.18 cm) containing 4% SF 96 on Chromosorb W. Oven temperature 160°, nitrogen flow 25 ml/min P,P′-DDE = 1 (24 min)

†2,2′,3′,4,5-pentachlorobiphenyl was recently reported as the 2,3′,4,4′,5-isomer and the compound is shown as such in Chapter 3. (See reference 2.)

Carbon-14 labeled 2,2′,3,3′,4,4′,5,6′-octachlorobiphenyl[4] (Figure 2-A.) and 2,2′,4,4′,5,5′-hexachlorobiphenyl[5] (Figure 3-A.) has been prepared via nitration of chlorobiphenyls of lower chlorine content and direct replacement of the nitro group by chlorine using carbon tetrachloride at elevated temperatures. 2,2′- and 2,4′- dichlorobiphenyl (C-14) have recently been prepared from ortho-chloro-aniline.

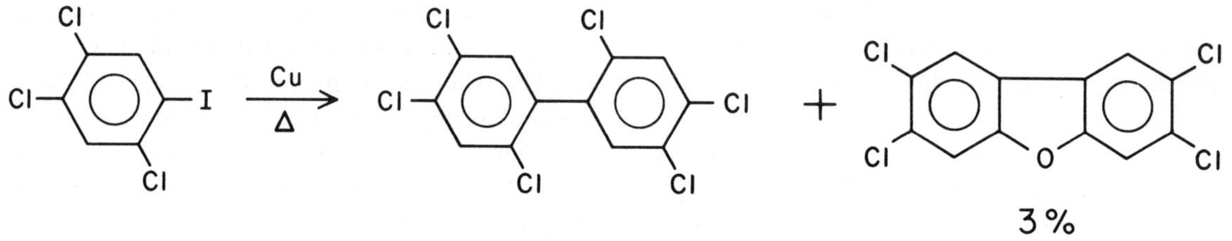


FIGURE 1-A. Formation of tetrachlorodibenzofuran during Ullmann-synthesis of 2,2′,4,4′,5,5′-hexachlorobiphenyl.

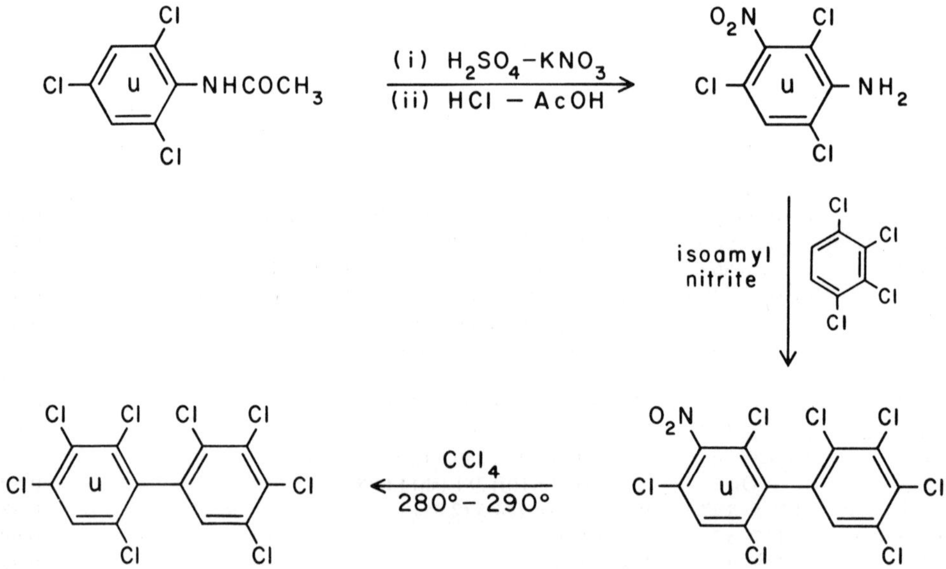


FIGURE 2-A. Synthesis of carbon-14 labeled 2,2′,3,3′,4,4′,5,6′-octachlorobiphenyl.

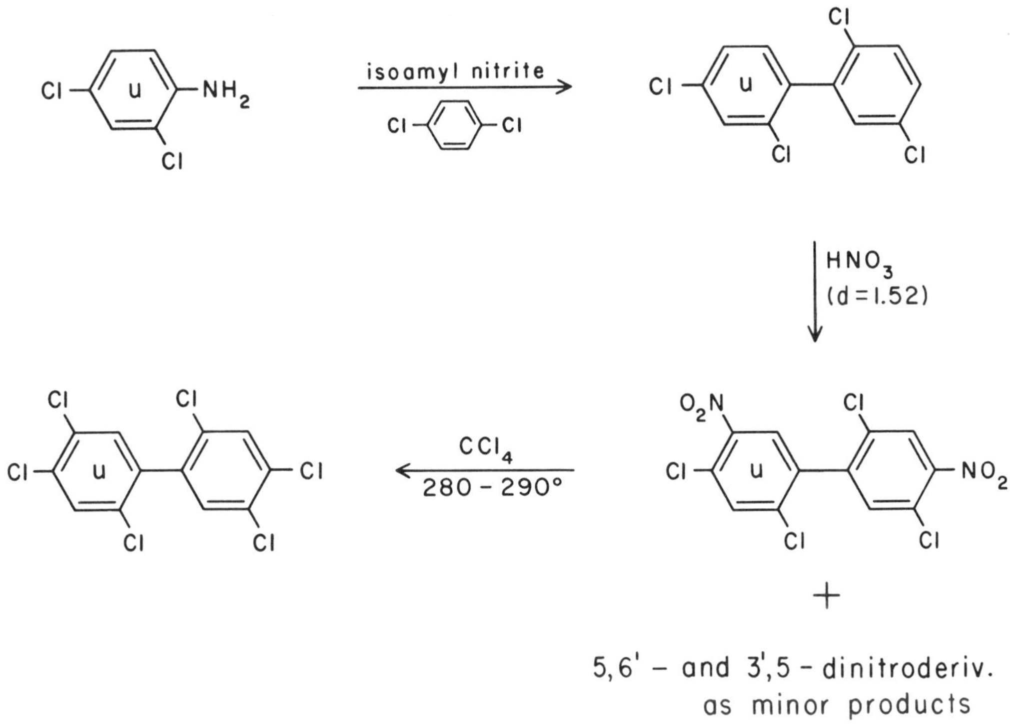


FIGURE 3-A. Synthesis of carbon-14 labeled 2,2′,4,4′,5,5′-hexachlorobiphenyl.

## REFERENCES


1. **Moron, M., Sundström, G., and Wachtmeister, C. A.,** *Acta Chem. Scand.,* 27, 3121, 1973.
2. **Saeki, S., Tsutsui, A., Oguri, K., Yoshimura, H., and Hamana, M.,** *Fukoka Igaku Zhasshi,* 62, 20, 1971; C. A. 74, 146294, 1971; National Research Council of Canada, Technical Translation, Ottawa, 1971.
3. **Sundström, G.,** *Acta Chem. Scand.,* 27, 600, 1973; private communication.
4. **Sundström, G.,** *Acta Chem. Scand.,* 27, 1109, 1973.
5. **Sundström, G.,** *Bull. Environ. Contam. Toxicol.,* 11, 39, 1974.
6. **Sundström, G., and Wachtmeister, C. A.,** *Acta Chem. Scand.,* 26, 3816, 1972.
7. **Attar, A., Ismail, R., Bieniek, D., Klein, W., and Korte, F.,** *Chemosphere,* 261, 1973.


## B. PHOTO- AND RADIOCHEMICAL REACTIONS

A number of tetrachlorobiphenyls have been irradiated in hexane and methanol at 300 nm.[7] Marked differences have been observed in the photo-reaction rates of different isomers. In hexane, dechlorination was again found to be the major reaction. In methanol, dechlorination also predominated but smaller quantities (<5%) of methoxyderivatives were formed. Dechlorination as well as substitution by methoxy occurred preferentially at the ortho- positions (Figure 1-B.).

Photo-reduction of 4-chlorobiphenyl by trimethylamine was shown to proceed via a charge transfer mechanism.[6] Biphenyl (71%) was obtained on irradiating 4-chlorobiphenyl (0.036 $M$) in acetonitrile in the presence of trimethylamine (0.1 $M$) for 70 hr with a low pressure mercury arc.

A recent thesis describes the irradiation of several chlorobiphenyls in solution in the gas phase and as thin films.[2] On irradiation in methanol 2,2′,4,4′,6,6′-hexachlorobiphenyl showed the usual dechlorination reactions. Two methoxy-derivatives (Figure 2-B.) could be isolated from the reaction mixture. Isomerization and formation of polymeric products was observed on irradiating several chlorobiphenyls in perfluorinated hydrocarbons. Similar reaction products were found

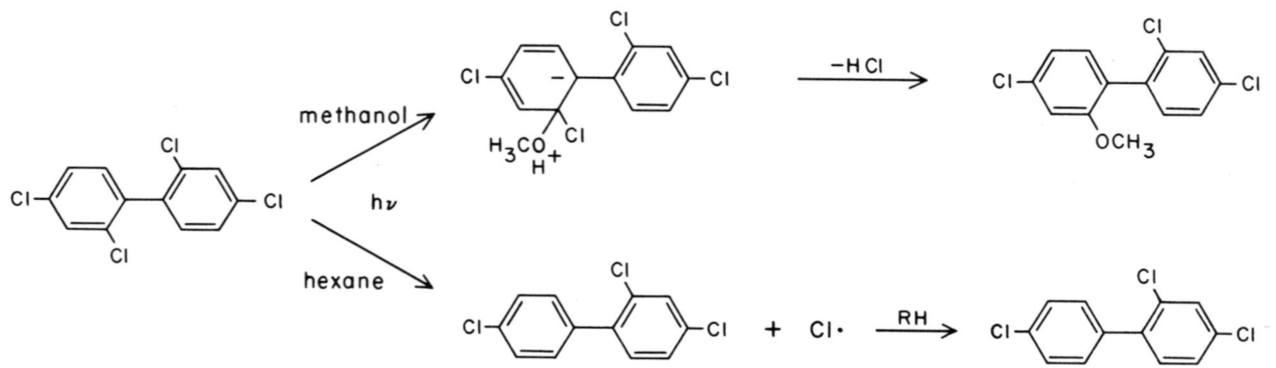


FIGURE 1-B. UV-irradiation of 2,2′,4,4′-tetrachlorobiphenyl in methanol and hexane, products formed and probable mechanism. (Courtesy of Dr. Ruzo.)

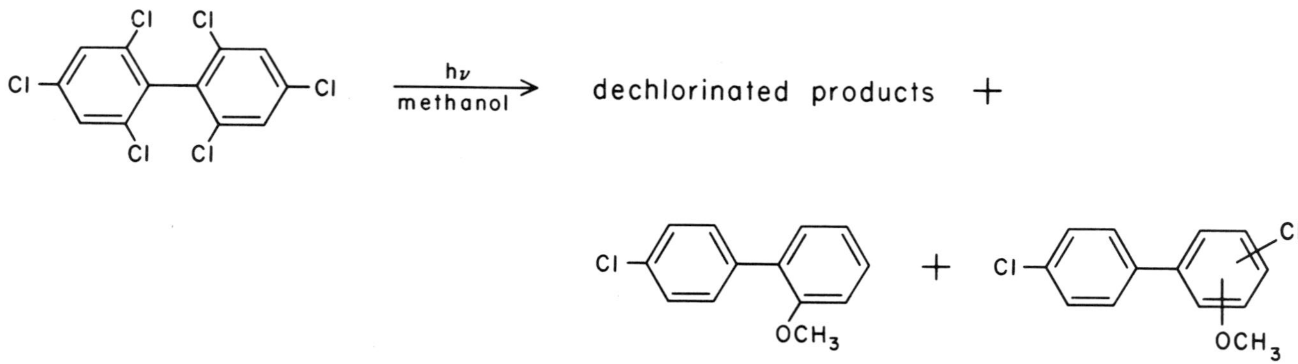


FIGURE 2-B. Oxygen containing products formed by UV-irradiation of 2,2′,4,4′,6,6′-hexachlorobiphenyl in methanol.

after irradiation of chlorobiphenyls as thin films. Under these conditions, however, chlorobiphenyls with increased chlorine content were also observed. Gas phase irradiations of 2,2′-dichlorobiphenyl and 2,2′,4,4′,6,6′-hexachlorobiphenyl gave polymeric products. In the former case phenolic derivatives were also found.

2,2′,4,4′,6,6′-hexachlorobiphenyl was also irradiated in carbon tetrachloride in the presence of oxygen atoms which were prepared photochemically from nitrogen dioxide. A number of products containing nitro or cyano groups (Figure 3-B.) were isolated or found by GC-MS in this reaction mixture.

Cobalt-60 $\gamma$-irradiation of chlorobiphenyls induces a base catalyzed chain dechlorination.[1,8,9] The products of this irradiation are biphenyl and acetone and similarities in mechanism of the irradiation with $^{60}$Co and a low pressure mercury lamp have been noted.[1]

The photochemical formation of small quantities of chlorobiphenyls from DDE in laboratory gas phase experiments was recently reported.[4,5] In the authors' opinion caution should be exercised in extrapolating these results in explaining the formation of significant quantities of chlorobiphenyls in the environment.

The photochemical lability of chlorobiphenyls was again demonstrated in an analytical method[3] (decomposition of more highly chlorinated biphenyls in samples by UV irradiation to allow analysis of chlorinated hydrocarbon pesticides) and the lability to $\gamma$-rays in a waste water treatment procedure.[10]

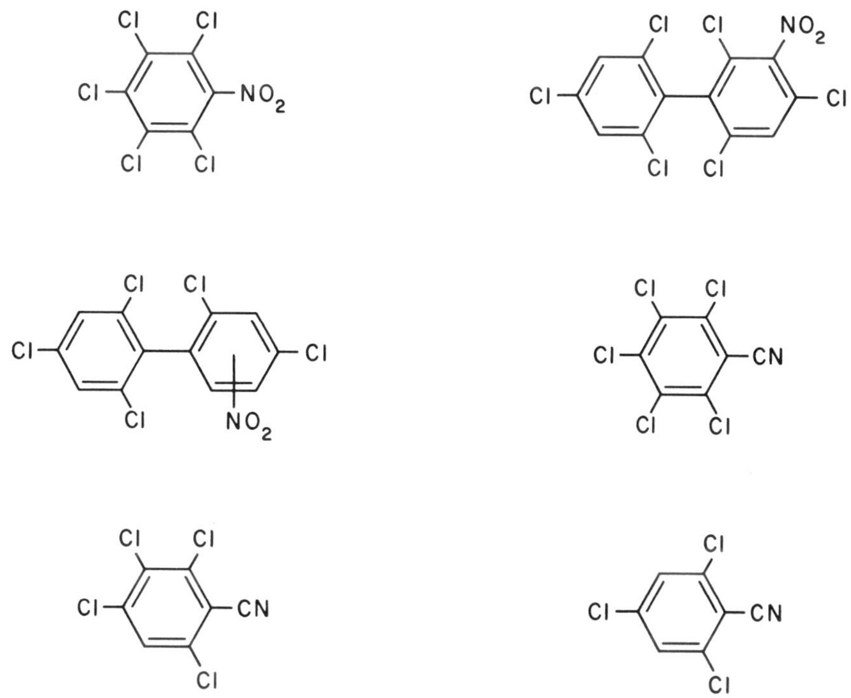


FIGURE 3-B. Products formed by UV-irradiation of 2,2′4,4′,6,6′-hexachlorobiphenyl in the presence of oxygen atoms.

## REFERENCES


1. **Arai, S., Matsui, M., Moriguchi, J., and Imamura, M.,** *Rika Gaku Kenkyusho Hokoku,* 48, 189, 1972.
2. **Hustert, K.,** Ph.D. Thesis, University of Bonn, Bonn, Germany, 1973.
3. **Leavitt, R. A., Su, G. C. C., and Zabik, M. J.,** *Anal. Chem.,* 45, 2130, 1973.
4. **Maugh, T. H.,** *Science,* 180, 578, 1973.
5. **Moilanen, K. W. and Crosby, D. G.,** 165th National Meeting of the American Chemical Society, Dallas, Texas, April 1973.

6. **Ohashi, M., Tsujimoto, K., and Seki, K.,** *J. Chem. Soc. Chem. Commun.*, 384, 1973.
7. **Ruzo, L. O., Zabik, M. J., and Schuetz, R. D.,** *J. Agric. Food Chem.*, in press.
8. **Sawai, T. and Shinozaki, Y.,** *Chem. Lett.*, 865, 1972.
9. **Sawai, T.,** *Genshiryoku Kogyo,* 18, 43, 1972; *C. A.*, 78, 78041, 1973.
10. **Sunada, T.,** *Atoms in Japan,* 7, 1972; Kinoshita, S. and Sunada, T., *Proc. 6th Intl. Conf. Adv. Water Pollut. Res.*, 607, 1973.

## C. METABOLISM

As initially observed (Chapter 7) hydroxylation proves to be the main metabolic reaction of chlorobiphenyls in a number of organisms.

In further investigations on the metabolism of biphenyl by *Pseudomonas putida,* 2,3-dihydroxybiphenyl was incubated with washed suspensions to yield 2-hydroxy-6-oxo-6-phenylhexa-2,4-dienoic acid. When incubated with crude cell-free extracts the same substrate gave benzoic acid; either 4-hydroxy-2-oxovalerate or 2-hydroxypenta-2,4-dienoate was suggested to be present in the same reaction mixture by mass spectrometry.[2]

Resting cell suspensions of *Achromobacter* PCB, a strain which has previously been shown to degrade 4-chlorobiphenyl, were found to be able to oxidize 2,3-, 2,4-, 3,4-, 3,5-, 3,3′-dichlorobiphenyl, 2′,3,4-trichlorobiphenyl, 2,2′3,3′-tetrachlorobiphenyl, and 2,3,4,5,6-pentachlorobiphenyl.[1] 2,3′,4′,5-tetrachlorobiphenyl was not oxidized under similar conditions.

Investigations with *Rhizopus japonicus* and $^{14}C$ labeled chlorobiphenyl showed that 2,2′5-trichlorobiphenyl yields three hydroxylated chlorobiphenyls, with as yet, undetermined structure, 2,2′,5,5′-tetrachlorobiphenyl gave one hydroxylated derivative, and 2,2′,4,5,5′-pentachlorobiphenyl remained unchanged.[14]

Fifteen species of lichens from rock, trees, and soil have been investigated for their ability to degrade chlorobiphenyls.[9] Pseudocyphelaria species converted 4-chlorobiphenyl [$^3H$] in relatively high yields to 4′-chloro-4-biphenylol which was further converted to 4-chloro-4′-methoxybiphenyl.

The uptake of labeled and unlabeled 4-mono-,2,4′-di-,2,2′,5-tri-,2,2′5,5′-tetra-, and 2,2′,4,5,5′-pentachlorobiphenyl was studied in tomato plants grown in soil and vermiculite (mineral salt solution). Young tomato plants contained a maximum of 10 ppb and in mature tomato plants chlorobiphenyls could not be detected (limit 0.1 to 1 ppb) in the fruit or total plant.[13]

2,2′-Dichlorobiphenyl ($^{14}C$) was shown to be converted to two isomeric monohydroxy- derivatives in higher plants (*veronica beccabunga*). Most of the radioactivity of the hydroxylated metabolites was associated with "conjugated" derivatives.[10]

It was recently suggested[7] that lack of significant metabolism of chlorobiphenyls by fish[6,8,16] may be explained by the absence of the cytochrome P-450 enzyme complex in these animals.

Several reports on the metabolism of chlorobiphenyls by mammalian species have appeared in the literature.

In the rhesus monkey 2,4′-dichlorobiphenyl ($^{14}C$) and 2,2′5-trichlorobiphenyl ($^{14}C$) was shown to be largely excreted in urine and feces.[3] The major urinary metabolites were a number of mono- and dihydroxylated derivatives.[4]

Hydroxy-derivatives of undetermined structures were also obtained from several chlorobiphenyls during in vitro studies with microsomal enzyme preparations.[5,11]

It was previously shown (Chapter 7) that 4-chlorobiphenyl in rats gave traces of a diol besides the main metabolite 4′-chloro-4-biphenylol. When this monohydroxy-derivative was fed to rats, larger quantities of a dihydroxy compound, most likely 4′-chloro-3,4-biphenyldiol, and smaller amounts of a trihydroxy-derivative were formed.[12]

The metabolic behavior of a number of dichlorobiphenyls[2a] and tetrachlorobiphenyl[15,15a] in rats has recently been reported.

Figure-1-C. shows the monohydroxy derivatives obtained from the feeding of dichlorobiphenyls.

2,3′,4,4′-Tetrachlorobiphenyl gave 3′,4,4′,6-tetrachloro-3-biphenylol as the major and 2,3′,4,4′-tetrachloro-3-biphenylol as the minor product.

The monohydroxy derivative obtained after feeding 3,3′,4,4′-tetrachlorobiphenyl was either 3,3′,4,4′-tetrachloro-2-biphenylol or 3′,4,4′,5-tetrachloro-3-biphenylol.

Gardner et al. (cf. Chapter 7) have now published the structures for the urinary metabol-

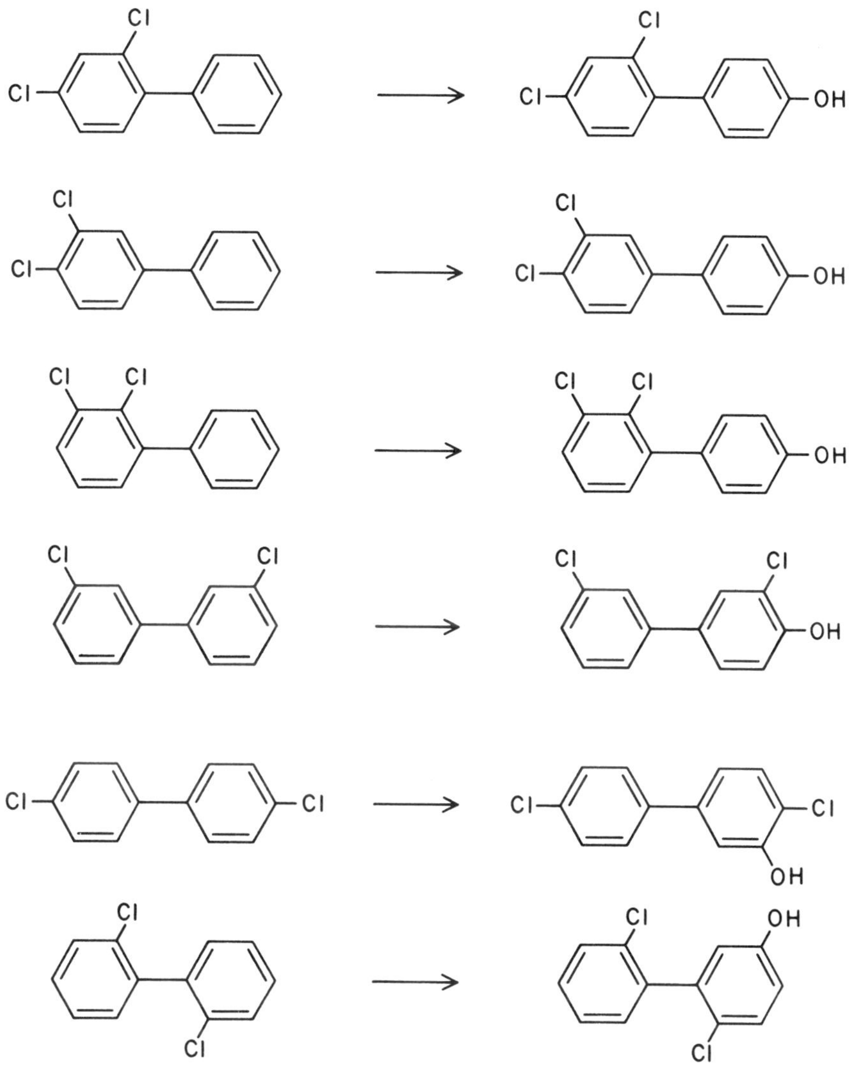

FIGURE 1-C. Monohydroxy-derivatives observed as metabolic products from dichlorobiphenyls in rats.

ites of 2,2′,5,5′-tetrachlorobiphenyl in rabbits.[1a] The products identified were *trans*-3,4-dihydro-3,4-dihydroxy-2,2′,5,5′-tetrachlorobiphenyl, 2,2′,5,5′-tetrachloro-3-biphenylol, and 2,2′,5,5′-tetrachloro-4-biphenylol. These products suggest that hydroxylation occurs via an arene oxide mechanism.

## REFERENCES


1. **Ahmed, M. and Focht, D. D.,** *Bull. Environ. Contam. Toxicol.*, 10, 70, 1973.

1a. **Gardner, A. M., Chen, J. T., Roach, J. A. G., and Ragelis, E. P.,** *Biophys. Res. Commun.*, 55, 1377, 1973.

2. **Catelani, D., Colombi, A., Sorlini, C., and Treccani, V.,** *Biochem. J.*, 134, 1063, 1973.

2a. **Goto, M., Sugiura, K., Hattori, M., Miyagawa, T., and Okamura, K.,** *New Methods in Environmental Chemistry and Toxicology*, Proc. Intl. Sym., Susono, Japan, 1973; Coulston, F., Korte, F., and Goto, M., Eds., International Academic Printing Co., 1973, Tokyo, Japan, p. 299.

3. **Greb, W., Klein, W., Coulston, F., Goldberg, L., and Korte, F.,** *Chemosphere*, 143, 1973.

4. **Greb, W., Klein, W., Coulston, F., Goldberg, L., and Korte, F.,** *Bull. Environ. Contam. Toxicol.*, submitted.

5. **Greb, W., Klein, W., Coulston, F., Goldberg, L., and Korte, F.,** *Bull. Environ. Contam. Toxicol.*, submitted.

6. **Hansen, D. J., Parrish, P. R., Lowe, J. I., Wilson, A. J., Jr., and Wilson, P. D.,** *Bull. Environ. Contam. Toxicol.*, 5, 113, 1971.
7. **Hattula, M. L. and Karlog, O.,** *Acta Pharmacol. Toxicol.*, 32, 237, 1973.
8. **Hutzinger, O., Nash, D. M., Safe, S., DeFreitas, A. S. W., Norstrom, R. J., Wildish, D. J., and Zitko, V.,** *Science*, 178, 312, 1972.
9. **Maas, W. S. G., Vey, C., and Hutzinger, O.,** 3rd International Congress of Pesticide Chemistry, Helsinki, Finland, July, 1974.
10. **Moza, P., Weisgerber, I., Klein, W., and Korte, F.,** *Chemosphere*, 2, 217, 1973.
11. **Safe, S. and Hutzinger, O.,** unpublished results.
12. **Safe, S., Ecobichon, D. J., and Hutzinger, O.,** unpublished results.
13. **Wallnöfer, P. W.,** *Nachrichtenbl. Dtsch. Pflanzenschutzdienstes* (Braunschw.), in press.
14. **Wallnöfer, P. W., Engelhard, G., Safe, S., and Hutzinger, O.,** unpublished results.
15. **Yoshimura, H. and Yamamoto, H.,** *Chem. Pharm. Bull.* (Tokyo), 21, 1168, 1973.
15a. **Yoshimura, H., Yamamoto, H., and Kinoshita, H.,** *New Methods in Environmental Chemistry and Toxicology*, Proceedings of the International Symposium, Susono, Japan, 1973; Coulston, F., Korte, F., and Goto, M., Eds., International Academic Printing Co., 1973, Tokyo, Japan, 291.
16. **Zitko, V., and Hutzinger, O.,** *PCB in the Environment*, Marcel Dekker, New York, in press.

## D. MASS SPECTROMETRY

### 1. Chlorobiphenyls (PCB's)

The development of sensitive GC-MS procedures for the identification of sub ppm levels of PCB are perhaps the most important recent results.

A Finnigan GC-MS data system was used to confirm the presence of PCB and DDE in breast milk[6] and fat and liver from polar bears,.[4] Figures 1-D. to 3-D. show the steps in identifying a heptachlorobiphenyl from polar bear fat.

A GC-MS method for the characterization of PCB's in environmental samples at the nanogram level has been reported.[8] Another GC-MS trace method makes use of chemical ionization techniques.[1]

A further example of the high resolution-photoplate method for the identification of

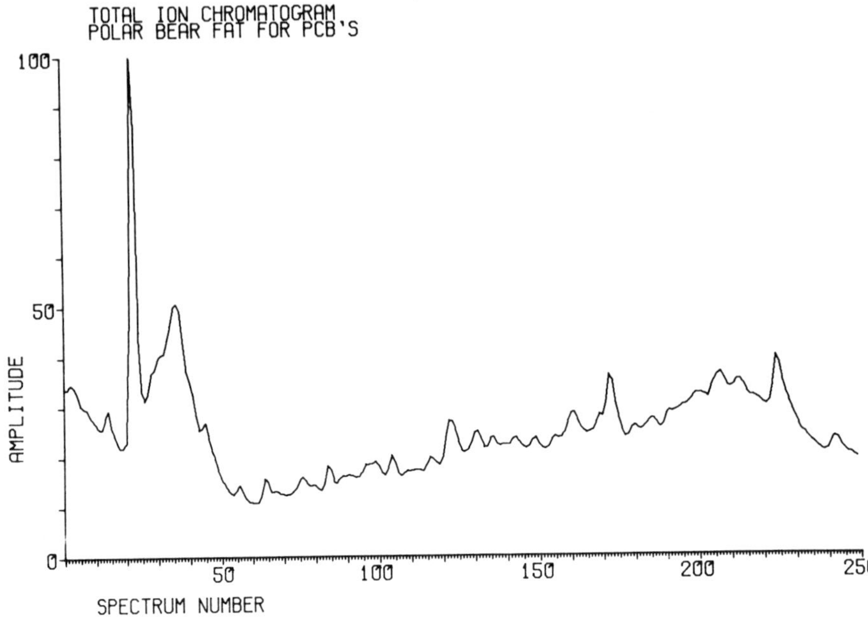


FIGURE 1-D. Total ion chromatogram of cleaned-up polar bear sample. Conditions for Figures 1-D. to 3-D.: GC, 5′ x 2 mm i.d. glass column packed with 3% OV-1 on 100/120 mesh Gas Chrom, column temperature 160 to 230°, programmed at 6°/min carrier gas helium at 20 ml/min. Interface, glass jet type separator. Mass spectrometer, Finnigan Model 3100 D, 70 eV electron energy, scan time 3 sec. Data system, Finnigan Model 6000.

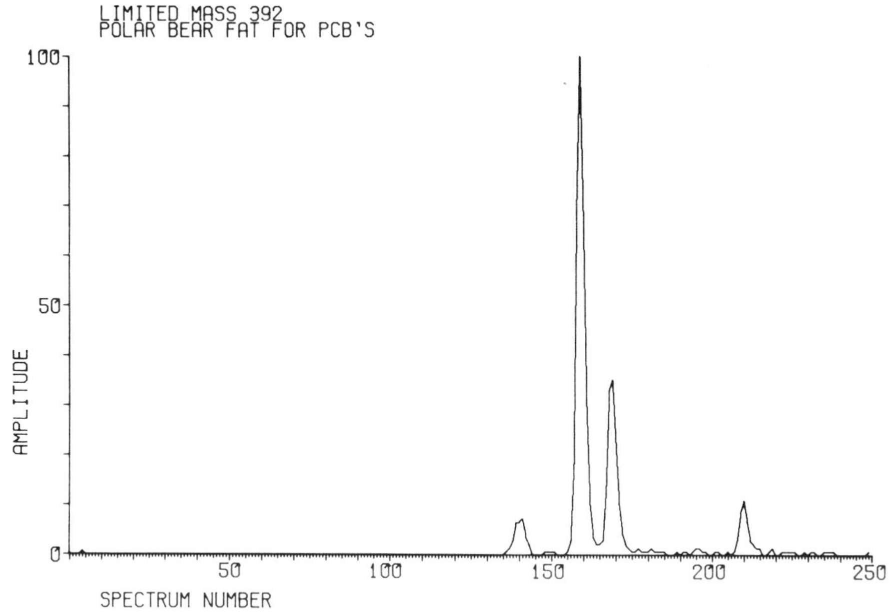


FIGURE 2-D. Limited mass scan at m/e 392 ($M^+$ for heptachlorobiphenyl) conditions of Figure 1-D.

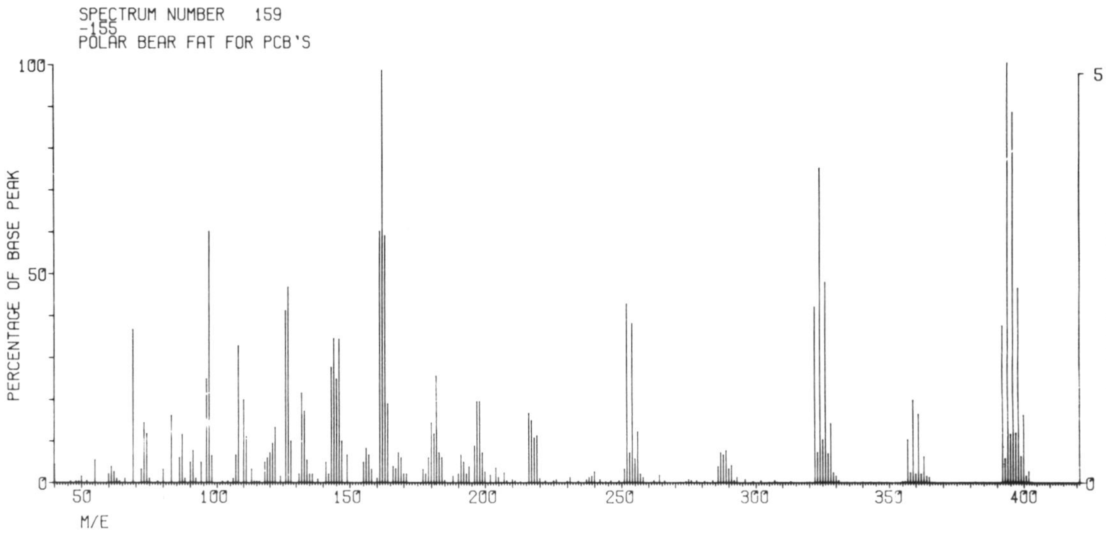


FIGURE 3-D. Recalled mass spectrum of scan 155 to 159 (see Figure 2-D.). $M^+$ m/e 392 = heptachlorobiphenyl conditions of Figure 1-D.

organochlorine compounds in crude samples was given in the study of the metabolism of chlorobiphenyls in goats.[7]

## 2. Chlorobiphenylols (Hydroxylated PCB Metabolites)

The mass spectra of a number of chlorobiphenylols have recently been examined.[3] The electron impact mass spectra of chlorobiphenylols were similar for each group of isomers and in all cases the molecular ion was the most abundant species in the mass spectrum. A summary of molecular ions for all possible chlorobiphenylmonools and diols is given in Table 1-D.

The fragmentation reactions of chlorobiphenylols is markedly dependent on both the degree of chlorine and hydroxy substitution. The influence of the position of chlorine and hydroxy

TABLE 1-D.

**Empirical Formulas and Molecular Ions of All Chlorobiphenyls with One and Two Hydroxy-Substituents**

| | Empirical formula | $M^+$ |
|---|---|---|
| Monohydroxy-Derivatives | | |
| no chlorine | $C_{12}H_{10}O$ | 170 |
| monochloro | $C_{12}H_9Cl\ O$ | 204 |
| dichloro | $C_{12}H_8Cl_2O$ | 238 |
| trichloro | $C_{12}H_7Cl_3O$ | 272 |
| tetrachloro | $C_{12}H_6Cl_4O$ | 306 |
| pentachloro | $C_{12}H_5Cl_5O$ | 340 |
| hexachloro | $C_{12}H_4Cl_6O$ | 374 |
| heptachloro | $C_{12}H_3Cl_7O$ | 408 |
| octachloro | $C_{12}H_2Cl_8O$ | 442 |
| Nonachloro | $C_{12}H\ Cl_9O$ | 476 |
| Dihydroxy-Derivatives | | |
| no chlorine | $C_{12}H_{10}O_2$ | 186 |
| monochloro | $C_{12}H_9Cl\ O_2$ | 220 |
| dichloro | $C_{12}H_8Cl_2O_2$ | 254 |
| trichloro | $C_{12}H_7Cl_3O_2$ | 288 |
| tetrachloro | $C_{12}H_6Cl_4O_2$ | 322 |
| pentachloro | $C_{12}H_5Cl_5O_2$ | 356 |
| hexachloro | $C_{12}H_4Cl_6O_2$ | 390 |
| heptachloro | $C_{12}H_3Cl_7O_2$ | 424 |
| octachloro | $C_{12}H_2Cl_8O_2$ | 458 |

group on the fragmentation reactions is presently under investigation using kinetic data. It is worth noting, however, that ions are present which can be attributed to typical phenol fragmentation reactions (i.e., loss of CO and HCO from the molecular ion or some subsequent decomposition ion) as well as ions due to loss of Cl/HCl which are typical of chloroaromatic compounds. These characteristic ions along with the abundant molecular ion species thus permit ready identification of chlorobiphenylols.

Similarly, chloromethoxybiphenyls behave like typical ethers of phenols on the one hand and chloroaromatic compounds on the other. For 2,2′5-trichloro-4′-methoxybiphenyl, for instance, the molecular ion (m/e 286) is the base peak. Important fragment ions for this compound are found at m/e 271 ($M^+-CH_3$; 30%); m/e 243 ($M^+-CH_3-CO$; 25%); m/e 173 ($M^+-CH_3-CO-Cl_2$; 22%), and weaker peaks at m/e 236 ($M^+-CH_3-Cl$);

ETHER EXTRACT OF URINE

↓ TLC

CRUDE "HYDROXY"-FRACTIONS

↓ REACT WITH DANSYL CHLORIDE

CRUDE DANSYL-CHLOROHYDROXYBIPHENYL

↓ TLC

DETECTION OF FLUORESCENT SPOTS

↓ MASS SPECTROMETRY

MOLECULAR COMPOSITION

FIGURE 4-D. Scheme for the identification of hydroxylated chlorobiphenyl metabolites in biological samples via dansylation and mass spectrometry.

m/e 207 ($M-CH_3-CO-HCl$); and m/e 201 ($M-CH_3-Cl_2$).

A fluorigenic labeling method for the analysis of hydroxylated chlorobiphenyl metabolites has been suggested.[2] This method was used to detect and determine the molecular composition of a minor metabolite (a dihydroxy-derivative) of 4,4′-dichlorobiphenyl in rats[5] using the procedures outlined in Figure 4-D. Mass spectrum of a dansyl derivative of a typical chlorobiphenyl metabolite is shown in Figure 5-D.

For the detection of hydroxylated chlorobiphenyl derivatives (chlorobiphenylols) on TLC plates, complexing spray reagents may be used as the chromogenic reagent.[2] The hydroxy group on the biphenyl nucleus renders the system electron donating and, in contrast to the starting materials (chlorobiphenyls), yellow to brownish colors due to change-transfer complexing are observed. Since the complex formation is a reversible process, mass spectra may be obtained directly from the complex (after elution from the plate) by thermal dissociation in the ion source of the mass spectrometer. An example of this is shown in Figure 6-D.

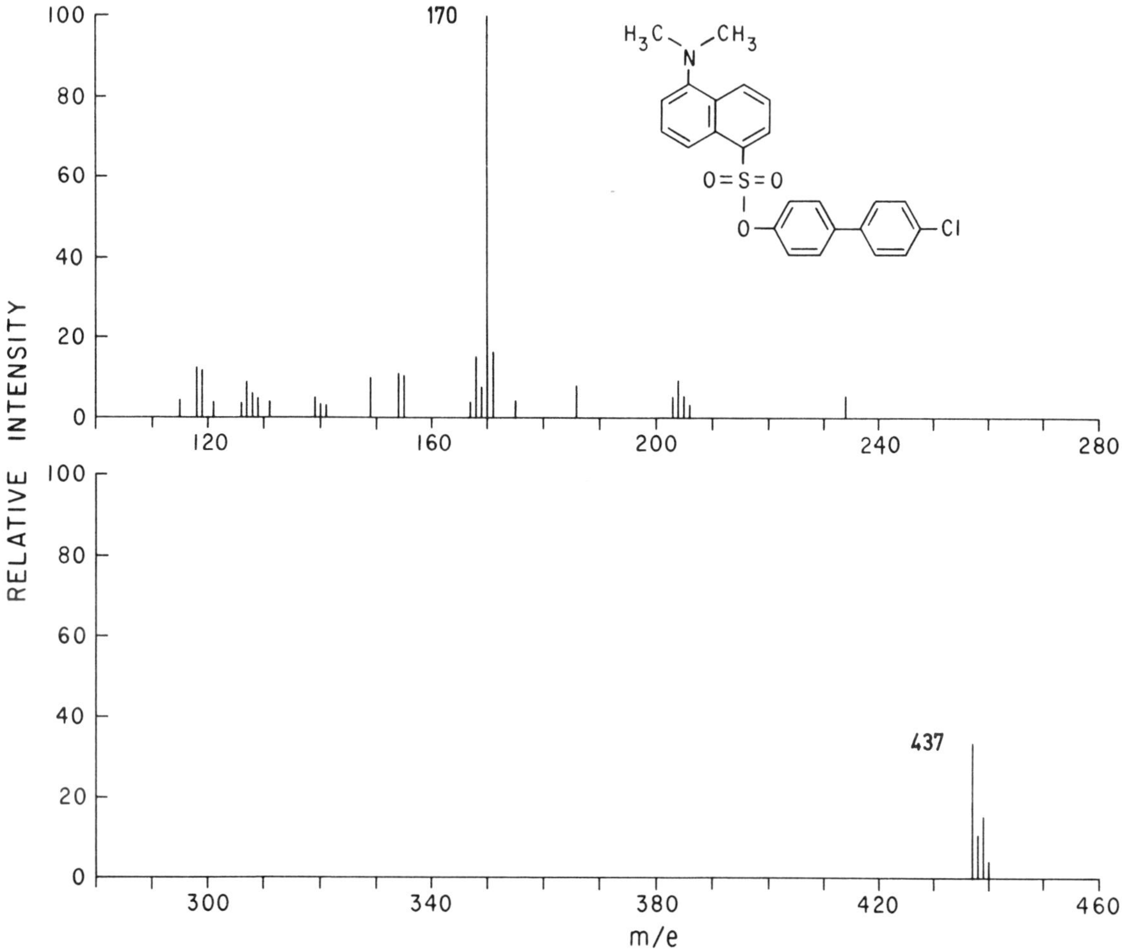


FIGURE 5-D. 70 eV mass spectrum of the dansyl-derivative of 4′-chloro-4-biphenylol.

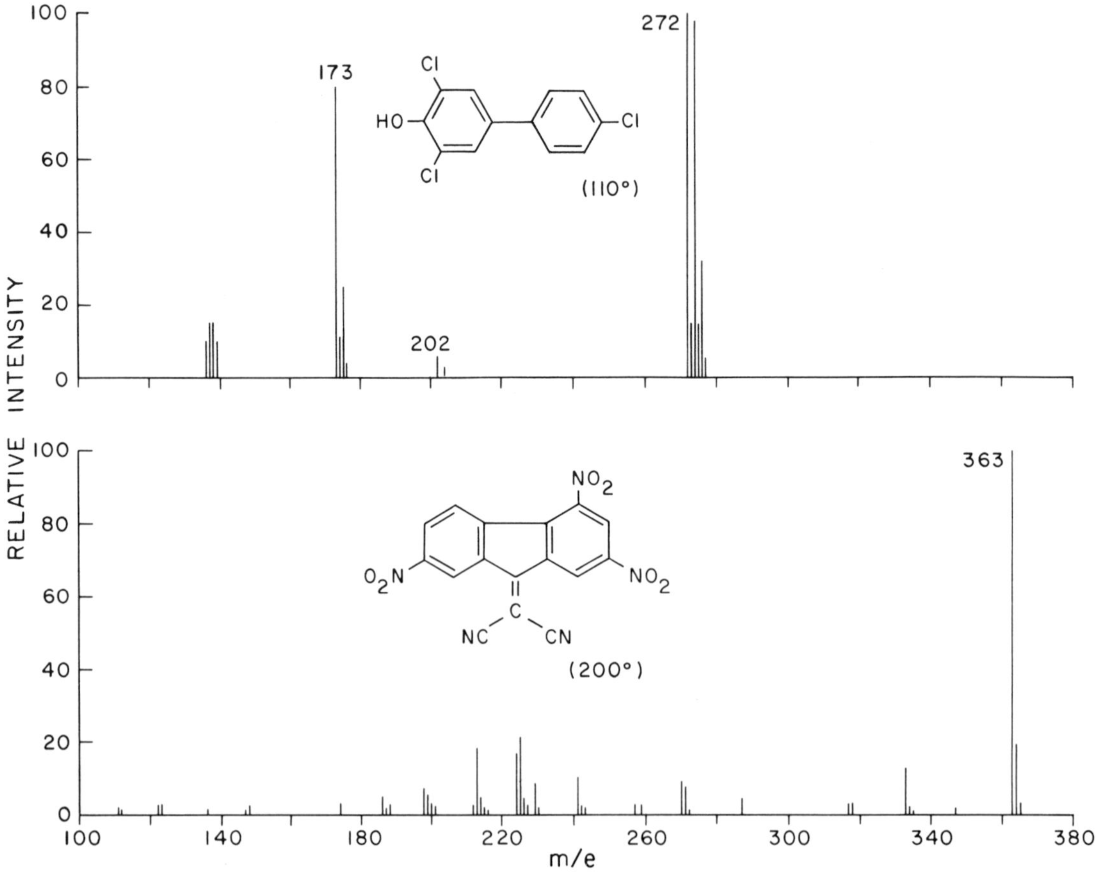


FIGURE 6-D. 70 eV mass spectra of 3,4′,5-trichloro-4-biphenylol and 9-dicyanomethylene-2,4,7-trinitrofluorene sublimed from the $\pi$ complex on the ion source of the mass spectrometer. Sample temperatures are indicated for both compounds.

## REFERENCES


1. **Dougherty, R. C., Roberts, J. D., Biros, F. J., and Ryan, J.,** 3rd International Congress of Pesticide Chemistry, Helsinki, Finland, July, 1974.
2. **Hutzinger, O., Heacock, R. A., and Safe, S.,** *J. Chromatogr.*, submitted.
3. **Hutzinger, O., Safe, S., and Zitko, V.,** *J. Assoc. Offic. Anal. Chem.*, in press.
4. **Hutzinger, O., Bowes, G. W., Safe, S., Knight, J. B., and Bonelli, E. J.,** unpublished data.
5. **Hutzinger, O., Safe, S., Ecobichon, D. J., DeFreitas, A. S. W., Norstrom, R. J., MacNeil, J. D., Wildish, D. J. and Zitko, V.,** in *PCB in the Environment,* Marcel Dekker, New York, in press.
6. **Musial, C. J., Hutzinger, O., Zitko, V., and Crocker, J.,** *Bull. Environ. Contam. Toxicol.* submitted.
7. **Safe, S., Platonow, N. S., Jamieson, W. D., and Hutzinger, O.,** 3rd International Congress of Pesticide Chemistry, Helsinki, Finland, July, 1974.
8. **Salsmans, R., Tourwé, D. and Van Binst, G.,** 3rd International Congress of Pesticide Chemistry, Helsinki, Finland, July, 1974.


## E. NUCLEAR MAGNETIC RESONANCE SPECTROSCOPY

Papers on carbon-13 and proton magnetic resonance studies of chlorinated biphenyls[4] and carbon-13 magnetic resonance of chlorinated benzenes and biphenyls[1] were presented at recent conferences.

Proton coupling constants and proton shieldings for a number of symmetric chlorinated biphenyls are shown in Tables 1-E. and 2-E. and the corresponding carbon-13 shieldings are given in Table 3-E.

Assigned carbon-13 spectra of chlorobiphenyls substituted in one ring are shown in Figures 1-E to 4-E and the observed Cmr signals for a number of tetra- to hexachlorobiphenyls are given in Table 4-E.

The conformation of 3,3′,5,5′-tetrachlorobiphenyl was recently investigated by liquid crystal nuclear magnetic resonance spectroscopy.[2]

Chlorine-35 nuclear quadruple resonance of compounds related to chlorobiphenyls have been discussed.[3]

TABLE 1-E.

**Proton Coupling Constants for Symmetric Chlorinated Biphenyls[a]**

| | J, Hz | | | | | | | | | |
|---|---|---|---|---|---|---|---|---|---|---|
| Substitution | 2,3 | 2,4 | 2,5 | 2,6 | 3,4 | 3,5 | 3,6 | 4,5 | 4,6 | 5,6 |
| 2,2′[b] | | | | | 8.07 | 1.29 | 0.42 | 7.53 | 1.68 | 7.66 |
| 3,3′[b] | | 209 | 0.46 | 1.78 | | | | 8.00 | 1.02 | 7.91 |
| 4,4′ | 8.36 | | 0.40 | 2.34 | | 2.34 | 0.40 | | | 8.36 |
| 2,2′,4,4′ | | | | | | 1.94 | 0.44 | | | 8.25 |
| 2,2′,6,6′ | | | | | 8.12 | | | 8.12 | | |
| 2,2′,4,4′,6,6′- | | | | | | | | | | |
| 2,2′,4,4′,5,5′- | | | | | | | $\simeq$0.2 | | | |
| 3,3′,4,4′ | | | 0.45 | 2.10 | | | | | | 8.35 |
| 3,3′,5,5′ | | | | | | | | | | |
| 3,3′,4,4′,5,5′ | | | | | | | | | | |

[a]At 28°C; 2.5 mol% in chloroform-*d*, except as noted.
[b]Approximately 3.5 mol% in benzene-$d_6$.

TABLE 2-E.

**Proton Shieldings for Symmetric Chlorinated Biphenyls[a]**

| | δ, ppm | | | | |
|---|---|---|---|---|---|
| Substitution | H-2 | H-3 | H-4 | H-5 | H-6 |
| 2,2′[b] | | 7.2521 | 6.8694 | 6.9071 | 7.0053 |
| 3,3′[b] | 7.2698 | | 7.0644 | 6.8456 | 6.9222 |
| 4,4′ | 7.4019 | 7.4662 | | 7.4662 | 7.4019 |
| 2,2′,4,4′ | | 7.5096 | | 7.3188 | 7.1834 |
| 2,2′,6,6′ | | 7.4482 | 7.3299 | 7.4482 | |
| 2,2′,4,4′,6,6′- | | 7.4667 | | 7.4667 | |
| 2,2′,4,4′,5,5′- | | 7.6198 | | | 7.3641 |
| 3,3′,4,4′ | 7.6184 | | | 7.5142 | 7.3581 |
| 3,3′,5,5′ | 7.5890 | | 7.5890 | | 7.5890 |
| 3,3′,4,4′,5,5′- | 7.5360 | | | | 7.5360 |

[a]At 28°C; 2.5 mol% in chloroform-*d*, except as noted.
[b]Approximately 3.5 mol% in benzene-$d_6$.

TABLE 3-E.

$^{13}$C Shieldings of Symmetric Chlorinated Biphenyls[a]

| Substitution | δ, ppm | | | | | |
|---|---|---|---|---|---|---|
| | C-1 | C-2 | C-3 | C-4 | C-5 | C-6 |
| nil | 140.6 | 126.8 | 128.4 | 126.9 | 128.4 | 126.8 |
| 2,2′ | 137.6 | 132.8 | 130.7 | 128.7 | 126.0 | 128.9 |
| 3,3′ | 140.7 | 127.4 | 134.2 | 126.8 | 129.6 | 124.8 |
| 4,4′ | 137.5 | 127.7 | 128.5 | 132.9 | 128.5 | 127.7 |
| 2,2′,4,4′ | 135.2 | 134.2 | 129.0 | 133.8 | 126.6 | 131.5 |
| 2,2′,6,6′ | 134.4 | 134.3 | 127.4 | 129.8 | 127.4 | 134.3 |
| 2,2′,4,4′,5,5′- | 135.3 | 130.7 | 131.7 | 133.9 | 131.9 | 130.6 |
| 2,2′,4,4′,6,6′- | 134.9 | 135.0 | 127.8 | 132.2 | 127.8 | 135.0 |
| 3,3′,4,4′ | 137.8 | 128.3 | 132.5 | 131.8 | 130.4 | 125.7 |
| 3,3′,5,5′ | 140.4 | 125.1 | 134.9 | 127.9 | 134.9 | 125.1 |
| 3,3′,4,4′,5,5′- | 136.6 | 126.4 | 134.2 | 133.6 | 134.2 | 126.4 |

[a]Approximately 0.6*M* in 1,2-dibromoethane-$d_4$ at 40°C.

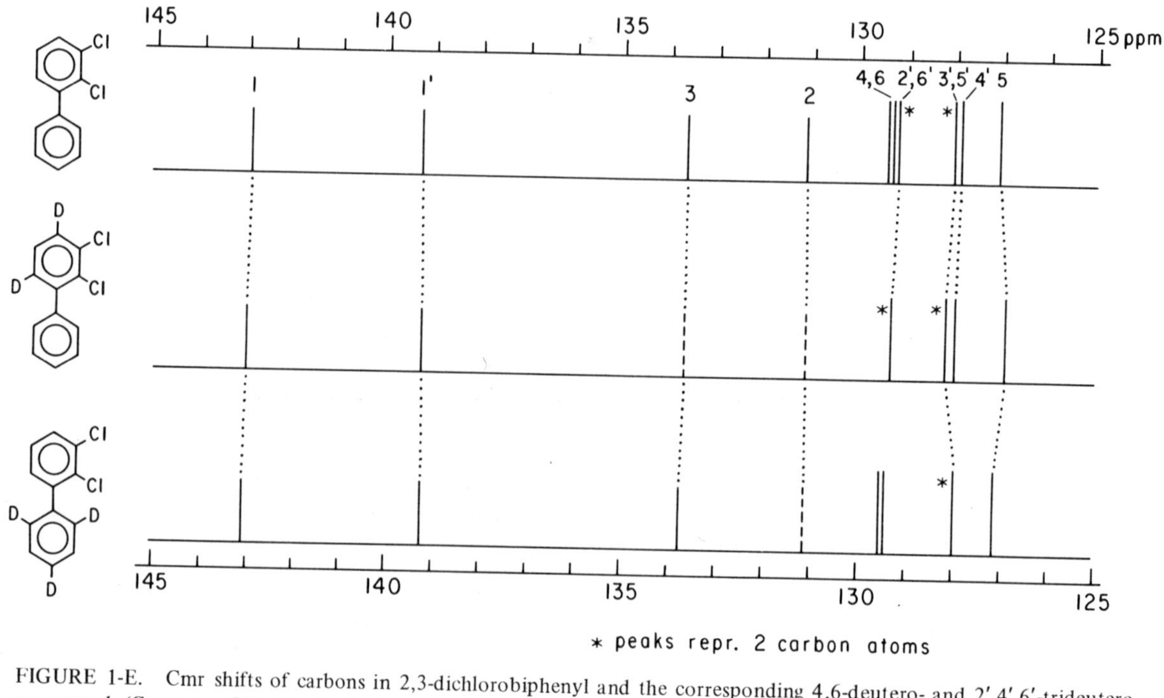


FIGURE 1-E. Cmr shifts of carbons in 2,3-dichlorobiphenyl and the corresponding 4,6-deutero- and 2′,4′,6′-trideutero compound. (Courtesy of Dr. Sundström.)

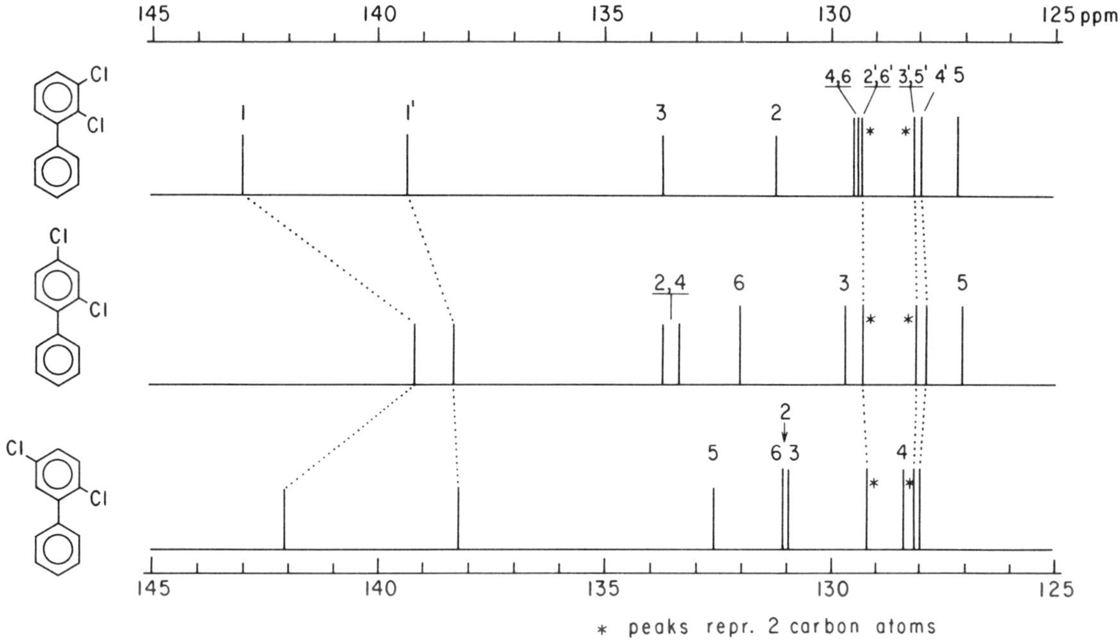


FIGURE 2-E. Cmr shifts of carbons in 2,3-, 2,4-, and 2,5-dichlorobiphenyl. (Courtesy of Dr. Sundström.)

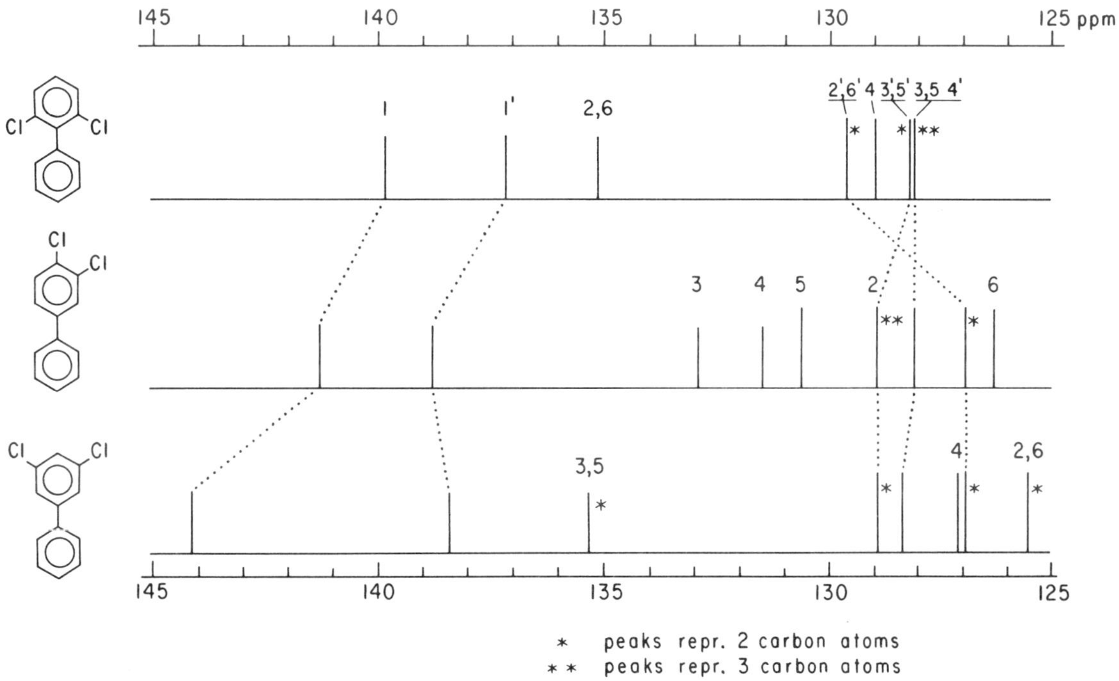


FIGURE 3-E. Cmr shifts of carbons in 2,6-, 3,4-, and 3,5-dichlorobiphenyl. (Courtesy of Dr. Sundström.)

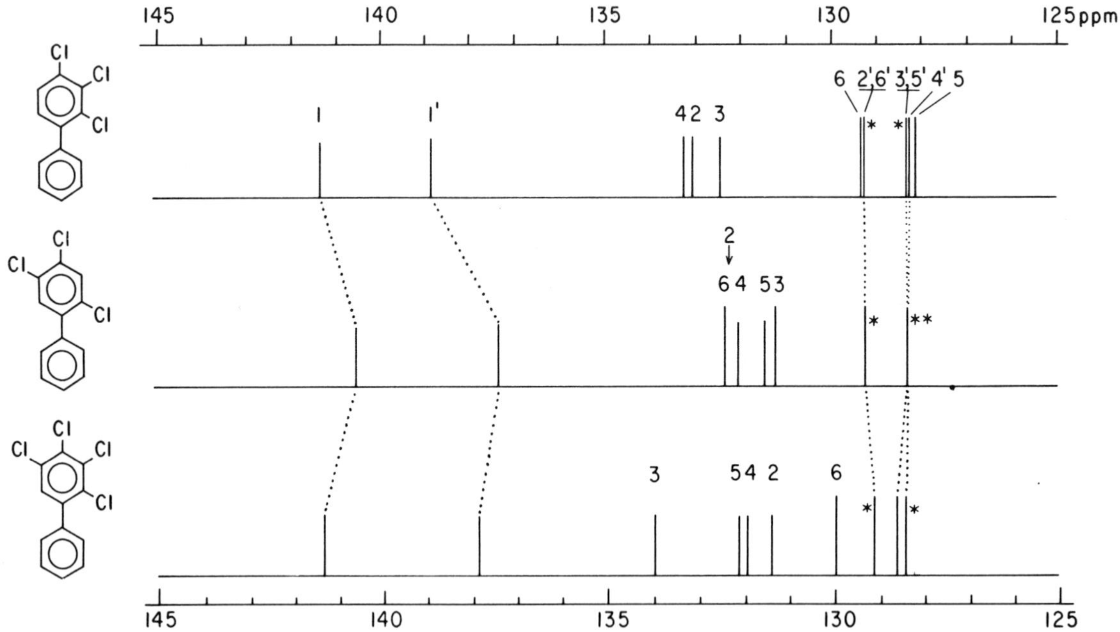


FIGURE 4-E. Cmr shifts of carbons in 2,3,4-, 2,4,5-trichlorobiphenyl, and 2,3,4,5-tetrachlorobiphenyl. (Courtesy of Dr. Sundström.)

TABLE 4-E.

**Cmr Signals of Investigated Tetra-, Penta-, and Hexachlorobiphenyls, Substituted in Both Rings**

| | | | | | | | | | | | | |
|---|---|---|---|---|---|---|---|---|---|---|---|---|
| 2,2′,3,5′- | _127.3_ | _129.3_ | _129.7_ | _130.6_ | _130.8_ | _131.0_ | 132.0 | 132.2 | 132.7 | 133.7 | 139.5 | 139.7 |
| 2,2′,4,5′- | _126.9_ | _129.4*_ | _130.5_ | _130.9_ | _131.7_ | 131.9 | 132.4 | 134.2 | 135.0 | 135.6 | 138.6 | |
| 2,2′,5,5′- | _129.5*_ | _130.5*_ | _130.7*_ | 131.7* | 132.4* | 138.4* | | | | | | |
| 2,2′,5,6′- | _127.9*_ | _129.7_ | _129.9_ | _130.5_ | _130.8_ | 132.1 | 132.6 | 135.0* | 136.1 | 137.5 | | |
| 2,3′,4′,5- | _128.7_ | _129.3_ | _130.3_ | 130.86 | _130.90_ | _131.29_ | _131.33_ | 132.7 | 132.73 | 133.0 | 138.0 | 139.7 |
| 2,3′,4,4′- | _127.8_ | _128.9_ | _130.2_ | _130.4_ | _131.5_ | _131.9_ | 132.7 | 133.5 | 135.0 | 137.0 | 138.3 | |
| 2,2′,3,3′- | _127.4*_ | _129.3*_ | _130.5*_ | 132.2* | 133.8* | 140.5* | | | | | | |
| 3,3′,5,5′- | _125.7**_ | _128.6*_ | 135.9** | 141.6* | | | | | | | | |
| 2,2′,4,5,5′- | _126.4_ | _128.3_ | _130.2_ | _131.2_ | _132.4_ | 132.7 | 132.9 | 133.7 | 137.2 | 138.0 | | |
| 2,3′,4,4′5- | _128.4_ | _130.2_ | _131.0_ | 131.1 | _131.3_ | 131.4 | _131.9_ | 132.6 | 132.8 | 132.9 | 136.9 | 137.7 |
| 2,3,3′,4′,6- | _128.6_ | _129.0_ | _130.5_ | _130.7_ | _131.6_ | 132.5 | 132.96 | 133.04 | 133.1 | 133.6 | 136.9 | 139.1 |
| 2,2′,4,4′,5,5′- | _131.2*_ | 131.5* | _132.3*_ | 132.6* | 136.2* | 139.9* | | | | | | |
| 2,2′,3,4′,5′,6- | _128.4_ | _131.1_ | _131.2_ | 131.6 | _132.1_ | 132.3 | 132.6 | 133.4 | 133.9 | 134.0 | 135.9 | 137.0 |

Italic signals: proton substituted carbons.
* signal representing two carbons
** signal representing four carbons

## REFERENCES

1. **Anderson, R., Olsson, K., and Sundström, G.,** International Symposium on Polyhalogen Compounds. Barcelona, Spain, Oct. 22 to 26, 1973 and personal communication.
2. **d'Annibale, A., Lunazzi, L., Boicelli, A. C., and Macciantelli, D.,** *J. Chem. Soc. Perkin Trans. II,* 1396, 1973.
3. **Smith, J. A. S.,** *Advances in Nuclear Quadrupole Resonance,* Vol. 1.
4. **Wilson, N. K. and Anderson, M.,** *Proc. Symp. Mass Spectrom. NMR Pestic. Chem.,* ACS Meeting, Dallas, Texas, 1973, Plenum Press, New York, in press.


## F. ULTRAVIOLET AND LUMINESCENCE SPECTROSCOPY

Absorption spectra in liquid solution as well as fluorescence and phosphorescence spectra, phosphorescence quantum yields, phosphorescence to fluorescence quantum yield ratios, and phosphorescence decay times in low temperature EPA glass were recently measured for chlorobiphenyls.[3] Chlorine substitution in positions ortho to the phenyl-phenyl bond shifts the conjugation band in the absorption spectrum and the unstructured fluorescence and phosphorescence bands to shorter wavelengths probably by preventing the biphenyl system from attaining a planar excited state. Spectra of a typical example for a hindered chlorobiphenyl (2,2′4,4′-dichlorobiphenyl) and a chlorobiphenyl without ortho-chlorine substitution (4,4′-dichlorobiphenyl) are shown in Figures 1-F. and 2-F. Chloro substitution increases the rate constant of phosphorescence in the order meta < para < ortho.

Pulsed source time resolved phosphorimetry is a useful method for the analysis of structurally similar molecules; it has been applied to halogenated biphenyls.[5] Low temperature luminescence can be used for identifying PCB's in the presence of DDT-type compounds.[2]

Increasing attention is being given to hydroxylated chlorobiphenyls because of their occurrence

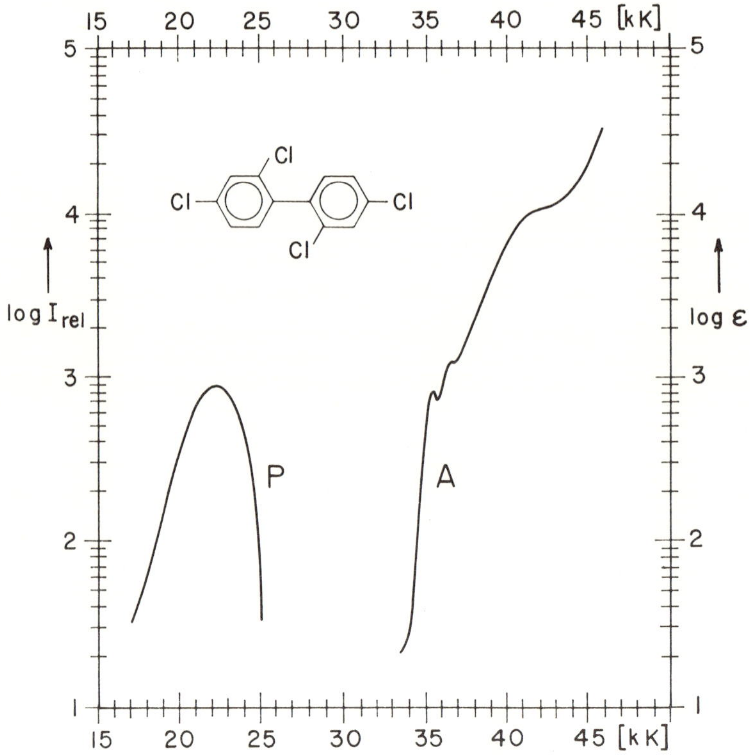


FIGURE 1-F. Absorption (A) and phosphorescence (P) spectra of 2,2′,4,4′-tetrachlorobiphenyl. (From *Z. Naturforsch.,* 27a, 756, 1972. With permission.)

as metabolites. Fluorescence properties of several hydroxybiphenyls are given in Table 1-F. (*cf.* reference 1.). The fluorescence intensity was recently shown to be decreased in the corresponding chloro-derivatives.[4]

UV absorption spectral data for mono- and dihydroxybiphenyls and a number of chlorobiphenylols[4] are given in Tables 2-F. and 3-F. Figure 3-F. shows the slight differences (small bathochromic shift at pH 7 and 9) in the spectra of 4-hydroxybiphenyl and the corresponding 3-chloro- and 3,4′-dichloro-derivative. The differences become large, however, when the position of the hydroxy group is changed.

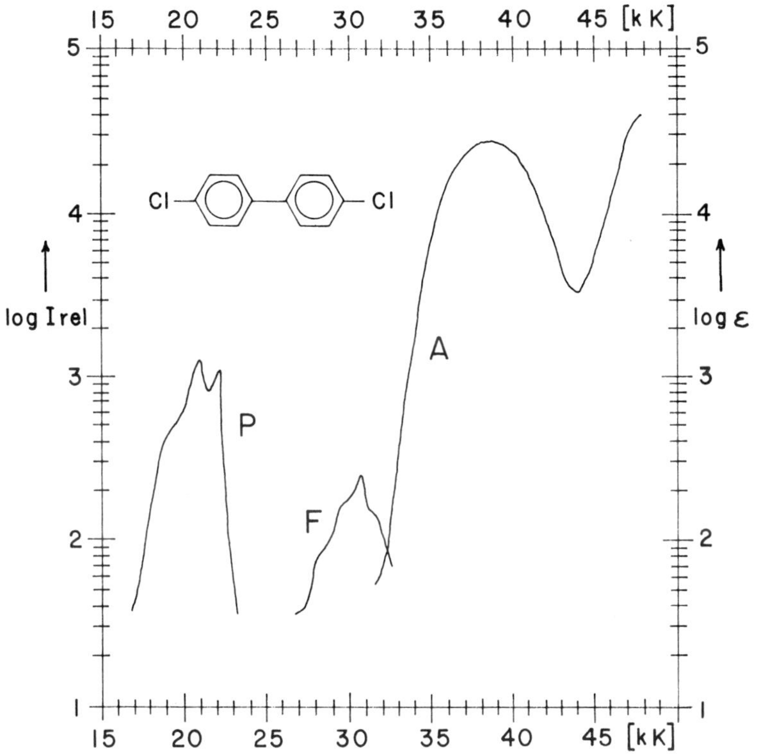


FIGURE 2-F. Absorption (A), fluorescence (F), and phosphorescence (P) spectra of 4,4′-dichlorobiphenyl. (From *Z. Naturforsch.*, 27a, 756, 1972. With permission.)

TABLE 1-F.

**Fluorescence Data for Source Hydroxybiphenyls**

| Compound* | State | Excitation $\lambda_{max}$ (mμ) | Fluorescence $\lambda_{max}$ (mμ) | Maximum** value of fluorescence intensity | pH range of maximum fluorescence |
|---|---|---|---|---|---|
| 2-Hydroxybiphenyl | Unionized | 262 | 348 | 39 | 0–1 |
| | | 295 | | 83 | |
| | Excited state ion | 262 | 415 | 67 | 2.5–9.8 |
| | | 295 | | 154 | |
| | Anion | 272 | 415 | 99 | 12–14 |
| | | 320 | | 481 | |
| 3-Hydroxybiphenyl | Unionized | 270 | 345 | 70 | 0–1 |
| | | 290 | | 78 | |
| | Excited state ion | 270 | 420 | 23 | 1.6–9.2 |
| | | 290 | | 25 | |
| | Anion | 270 | 420 | 23 | 10.1–14.2 |
| | | 320 | | 60 | |
| 4-Hydroxybiphenyl | Unionized | 288 | 340 | 184 | 1–9 |
| | Anion | 311 | 401 | 136 | 10–14 |
| 2,2′-Dihydroxybiphenyl | Unionized | 256 | 356 | 2 | -1–-0.8 |
| | | 290 | | 6 | |
| | Excited state ion | 256 | 400 | 41 | 1.0–6.6 |
| | | 290 | | 385 | |
| | Monoanion | 262 | 400 | 266 | 8.6–13 |
| | | 320 | | 1,593 | |
| 4,4′-Dihydroxybiphenyl | Unionized | 282 | 354 | 271 | 1–7 |
| | Monoanion | 300 | 438 | 19 | 10.9–11.3 |
| | Dianion | 310 | 400 | 7 | 12.1–14 |

*Ca. 10 μ molar in water containing 5% ethanol.

**Relative to biphenyl = 100

TABLE 2-F.

**Ultraviolet Spectral Data for Mono- and Dihydroxybiphenyls**

| Hydroxybiphenyl | UV absorption maxima[a] pH 7[b] | pH 9 |
|---|---|---|
| 2- | 281(5); 241(11) | 285(3.5); 315 sh |
| 3- | 280 sh; 247(14) | 305(3.5); 334(26) |
| 4- | 260(10) | 285(10); 250 sh |
| 2,2′-di- | 276(14); 240 sh | 307(19); 250 sh |
| 3,3′-di- | 285(7.5); 250(14) | 297(6) |
| 4,4′-di | 261(18) | 284(18) |
| 3,4-di | 261(17); 280 sh | 259(11); 348 sh |
| 2,5-di- | dec[c] | dec[c] |

[a]In aqueous solution at pH indicated; approximate $\epsilon$ value ($\times 10^3$) in parentheses; $\lambda$ in nm; sh = shoulder.
[b]Similar values were observed at pH 3.
[c]Decomposition in solution; values at pH 3: 248(10); 297(6.5)

TABLE 3-F.

**UV Spectra of Some Chlorobiphenylols**

| Compound | UV absorption maxima[a] | |
|---|---|---|
| | pH 7[b] | pH 9 |
| 3-chloro-4-biphenylol | 258(8) | 287(9) |
| 2-chloro-4-biphenylol | 263(18) | 286(19) |
| 4'-chloro-4-biphenylol | 262(18) | 289(20) |
| 3,5-dichloro-4-biphenylol | 287(12.5); 220(20); 209(20.5)[c] | 289(13); 221(20); 209(20) |
| 3,4'-dichloro-4-biphenylol | 262(26); 204(47) | 294(27); 211(38) |
| 2,4'-dichloro-4-biphenylol | 253(16) | 277(14) |
| 2',5'-dichloro-4-biphenylol | 255(13); 225(24.5) | 283(12); 223(22.5) |
| 4,4'-dichloro-3-biphenylol | 257(7.5); 200 sh | 315(3.5); 245(13.5) |
| 2',5'-dichloro-3-biphenylol | 280(3); 246 sh | 297(3) |
| 3,5-dichloro-2-biphenylol | 295(4); 248(9) | 322(6.5) |
| 2',5'-dichloro-2-biphenylol | 276(4) | 305 sh; 278(4) |
| 3,4',5-trichloro-4-biphenylol | 270 sh; 222 sh | 296(25.5) |
| 2,2',5'-trichloro-4-biphenylol | 254(5) | 295(13.5) |
| 3-chloro-4,4'-biphenyldiol | 262(18) | 280(18) |
| 3,3'-dichloro-4,4'-biphenyldiol | 264(23) | 290(24) |
| 2,2'-dichloro-4,4'-biphenyldiol | 276(4); 237 sh | 297 sh; 255(16.5) |
| 3,3',5-trichloro-4,4'-biphenyldiol | 278(15)[d] | 264(17.5) |
| 3,3',5,5'-tetrachloro-4,4'-biphenyldiol | 297(25)[e] | 300(27) |
| 4,4'-dichloro-3,3'-biphenyldiol | 308(8)[f] | 312(9) |
| 3,3',5,5'-tetrachloro-2,2'-biphenyldiol | 326(4.5); 260 sh[g] | 326(4.5); 260 sh |

[a]In aqueous solution at pH indicated; approximate $\epsilon$ value ($\times 10^3$) in parentheses; $\lambda$ in nm; sh = shoulder.
[b]Similar values were obtained at pH 3 except where noted.
[c]Values at pH 3: 257(11); 205(24).
[d]Values at pH 3: 264(17).
[e]Values at pH 3: 290 sh; 257(15); 218(56).
[f]Values at pH 3: 291(8); 257(13).
[g]Values at pH 3: 293(4).

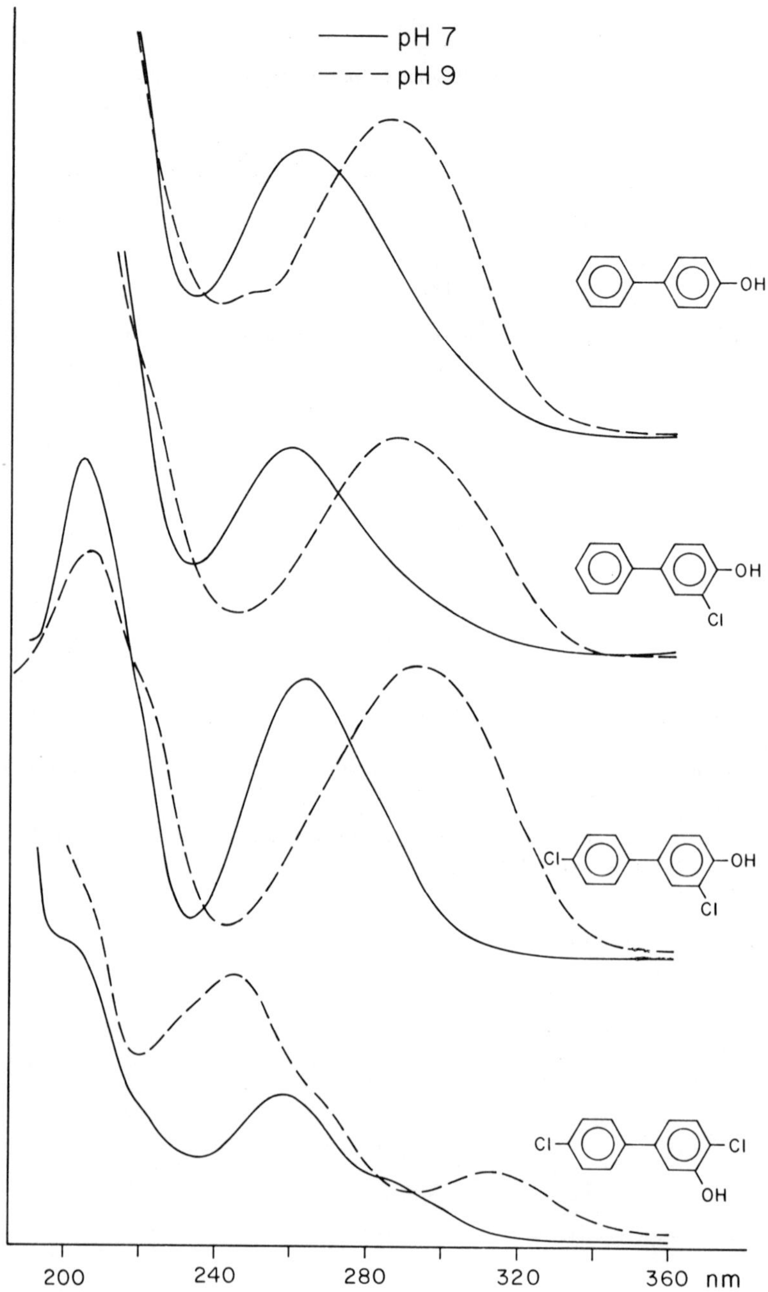


FIGURE 3-F. UV absorption spectra (in water) at different pH of several representative chlorobiphenylols.

## REFERENCES


1. **Bridges, J. W., Creaven, P. J., and Williams, R. T.,** *Biochem. J.,* 96, 872, 1965.
2. **Brownrigg, J. T., Guilfoyle, J., and Hornig, A. W.,** Paper No. 20, *Division Water Air and Waste Chemistry*, Am. Chem. Soc. Meeting, New York, August, 1972.
3. **Dreeskamp, H., Hutzinger, O., and Zander, M.,** *Z. Naturforsch.,* 27a, 756, 1972.
4. **Hutzinger, O., Safe, S., and Zitko, V.,** *J. Assoc. Offic. Anal. Chem.,* in press.
5. **O'Donnell, C. M., Harbaugh, K. F., Disher, R. P., and Winefordner, J. D.,** *Anal. Chem.,* 45, 609, 1973.

## G. ANALYSIS

Probably the most accurate quantification of PCB's short of using mixtures of appropriate chlorobiphenyls was suggested by Webb and McCall[10] and Jensen.[7] This quantitation is based on calculations of the amount of chlorobiphenyls present in individual peaks of Aroclors as these are resolved on SE-30, DC-200, OV-17, or OV-101 columns. The amount of chlorobiphenyls, *a,* is calculated from the number of chlorines, *n,* determined by mass spectrometry, and from the amount of chlorine, *m,* determined microcoulometrically.

$$a = m \frac{154 + 34.46n}{35.46n}$$

Weight percentages of chlorobiphenyls were calculated in this way by Webb and McCall for different Aroclor preparations, and values for Aroclor 1242, 1248, 1254, and 1260 are presented in Table 1-G. Using these values any laboratory may now convert peak areas to weight amounts of chlorobiphenyls. If a mixture of Aroclors occurs in environmental samples, Webb and McCall recommend using weight percentages of Aroclor 1242, 1254, and 1260 for peaks with relative retention times 11 to 70, 84 to 174, and 203 to 528, respectively.

The number of chlorine atoms in chlorobiphenyls present in individual peaks is given in Table 2-G. The reader should notice that in peaks containing chlorobiphenyls with a different number of chlorine atoms, the proportions of the more chlorinated biphenyls increase with increasing chlorine content of the parent Aroclor preparation. Thus, for example, the peak with relative retention time 70 contains 10%, 20%, 75%, and 100% of pentachlorobiphenyls in Aroclors 1242, 1248, 1254, and 1260, respectively. These differ-

TABLE 1-G.

**Mean Weight Percent of Chlorobiphenyls in Aroclor Peaks**

| Relative retention time* | 1242 | 1248 | 1254 | 1260 |
|---|---|---|---|---|
| 11 | 1.1 | | | |
| 16 | 2.9 | | | |
| 21 | 11.3 | 1.2 | | |
| 28 | 11.0 | 5.2 | | |
| 32 | 6.1 | 3.2 | | |
| 37 | 11.5 | 8.3 | | |
| 40 | 11.1 | 8.3 | | |
| 47 | 8.8 | 15.6 | 6.2 | |
| 54 | 6.8 | 9.7 | 2.9 | |
| 58 | 5.6 | 9.3 | 1.4 | |
| 70 | 10.3 | 19.0 | 13.2 | 2.7 |
| 78 | 3.6 | 6.6 | | |
| 84 | 2.7 | 4.9 | 17.3 | 4.7 |
| 98 | 1.5 | 3.2 | 7.5 } | 3.8 |
| 104 | 2.3 | 3.3 | 13.6 } | |
| 112 | | 1.2 | | |
| 117 | | | | 3.3 |
| 125 | 1.6 | 2.6 | 15.0 | 12.3 |
| 146 | 1.0 | 1.5 | 10.4 | 14.1 |
| 160 | | | 1.3 | 4.9 |
| 174 | | | 8.4 | 12.4 |
| 203 | | | 1.8 | 9.3 |
| 232 | | | 1.0 } | 9.8 |
| 244 | | | } | |
| 280 | | | | 11.0 |
| 332 | | | | 4.2 |
| 372 | | | | 4.0 |
| 448 | | | | 0.6 |
| 528 | | | | 1.5 |

*Retention time of p,p′-DDE = 100

TABLE 2-G.

**Number of Chlorines in Chlorobiphenyls in Aroclor Peaks**

| Relative retention time* | 1242 | 1248 | 1254 | 1260 |
|---|---|---|---|---|
| 11 | 1 | | | |
| 16 | 2 | | | |
| 21 | 2 | 2 | | |
| 28 | 2(25%) 3(75%) | 3 | | |
| 32 | 3 | 3 | | |
| 37 | 3 | 3 | | |
| 40 | 3 | 3(85%) 4(15%) | | |
| 47 | 4 | 4 | 4 | |
| 54 | 3(33%) 4(67%) | 3(10%) 4(90%) | 4 | |
| 58 | 4 | 4 | 4 | |
| 70 | 4(90%) 5(10%) | 4(80%) 5(20%) | 4(25%) 5(75%) | 5 |
| 78 | 4 | 4 | | |
| 84 | 5 | 5 | 5 | 5 |
| 98 | 5 | 5 | 5 | 5(60%) 6(40%) |
| 104 | 5 | 4(10%) 5(90%) | 5 | |
| 112 | | 5 | | |
| 117 | | | | 6 |
| 125 | 5(85%) 6(15%) | 5(90%) 6(10%) | 5(70%) 6(30%) | 5(15%) 6(85%) |
| 146 | 5(75%) 6(25%) | 5(85%) 6(15%) | 5(30%) 6(70%) | 6 |
| 160 | | | 6 | 6(50%) 7(50%) |
| 174 | | | 6 | 6 |
| 203 | | | 6 | 6(10%) 7(90%) |
| 232 | | | 7 | 6(10%) 7(90%) |
| 244 | | | | |
| 280 | | | | 7 |
| 332 | | | | 7 |
| 372 | | | | 8 |
| 448 | | | | 8 |
| 528 | | | | 8 |

*Retention time of p,p′-DDE = 100

ences and the chlorobiphenyl percentages, varying from preparation to preparation, still leave much room for ambiguity in this PCB quantitation. It may be worthwhile to determine the number of chlorine atoms and the amount of chlorine in peaks of typical environmental samples. Unfortunately, many laboratories do not have the necessary equipment. In addition, larger amounts of cleaned-up extracts required for this determination may make it impractical or even impossible in many instances.

Methods for the use of mixed PCB standards have recently been proposed.[2,8]

Use is made of Beroza's p-values in a gas chromatographic method for the quantitation of DDT in the presence of interfering PCB's.[12]

Armour[1] published an improved perchlorination procedure giving better than 90% recoveries of decachlorobiphenyl even from Aroclor 1242 on 1 to 20 μg scale. Vacuum hydrolysis tubes (150 × 10 mm o.d. with No. 4 Teflon Valve, Kontes Glass Co., or equivalent) were used. The charring of solvent residues by antimony pentachloride was avoided by using chloroform as the solvent. The sample was charged to the hydrolysis tube in hexane and hexane was replaced by chloroform by repeated evaporation on a steam bath (the evaporations were never carried out to less than 0.1 ml). Antimony pentachloride (0.2 ml) was added and the perchlorination was carried out at 165 to 175°C overnight. The excess of antimony pentachloride was decomposed by 6N HCl and decachlorobiphenyl was extracted with hexane.

Two procedures for the extraction of PCB's from paper have been published. Villeneuve et al.[9] extracted paper with hexane in a Soxhlet extractor for 2 hr. Young et al.[11] recommend either alkali or acid pretreatment of paper samples. In the former method, paper (10 g, cut in pieces no larger than ¼″ X ¼″) is refluxed with ethanolic potassium hydroxide (2%, 60 ml) in a 125 ml Erlenmeyer flask for 30 min. The extract is diluted with water (60 ml) and PCB's are extracted with petroleum ether. In the acid method the sample is saturated with water in a 500 ml Erlenmeyer flask (approximately 3 ml/g). The flask is cooled in tap water and concentrated sulfuric acid (5/3 of the volume of water) is added through a West condenser. After 5 min cooling, water (75 ml) is added, followed after additional cooling by ethanol (75 ml). The reaction mixture is filtered and the filtrate is extracted by hexane. The extracts from both methods may be cleaned up by column chromatography on Florisil. The methods were tested on paperboard samples spiked with Aroclor 1242 during manufacture. The alkaline extraction recovered approximately 96% of PCB's extracted by the acid procedure.

Farwell et al.[4] studied the voltammetric reduction of $5 \times 10^{-4}$ *M* solutions of chlorobiphenyls in 0.1 *M* tetraethyl ammonium bromide in dimethyl sulfoxide. The method could differentiate chlorobiphenyls according to number of chlorines and substitution patterns, and also chlorobiphenyls and chloronaphthalenes. Voltammetric reduction of chlorinated hydrocarbons is a relatively new field which certainly deserves more attention to establish its application potential.

Harvey et al.[5] reported the average concentration of PCB's in the northern North Atlantic water. The PCB's were concentrated by passing sea water through an Amberlite XAD-2 column, extracted with boiling acetonitrile, and partitioned between aqueous acetonitrile and hexane.

A quantitative TLC method for the analysis of PCB's has been reported.[3] The development of spots is carried out with silver nitrate and UV irradiation. The analytical procedure, which has a sensitivity of 0.5 ppm (in biological samples) and a precision of ±0.05 ppm, was carried out using a linear thin-layer chromatographic scanner.

The thin-layer chromatographic behavior of 27 chlorobiphenylols and hydroxybiphenyls has been examined.[6] Since TLC is a useful separation method for the more polar metabolites such as these phenolic compounds, some of the results are presented here.

The $R_f$ values for the three isomeric monohydroxybiphenyls and the corresponding 2′,5′-dichloro-derivatives are shown in Table 3-G. Generally, the $R_f$ value increases in the order 4-OH < 3-OH < 2-OH for both sets of compound in most solvents. The 2′,5′-dichloro substituted compounds generally show an increase in $R_f$ value over their parent hydroxybiphenyl. This increase is more pronounced for the 4-hydroxybiphenyl in the order 4-OH > 3-OH > 2-OH.

In Table 4-G. the $R_f$ values for a number of mono-, di-, and trichlorobiphenylols are presented. Increasing chlorine substitution usually increases the $R_f$ value of a 4-hydroxybiphenyl. For the monochloro compounds available the order is 3-Cl > 2-Cl > 4′-Cl. A similar effect can be seen in the di- and trichloro-derivatives.

The influence of *o*-chloro substitution on the $R_f$ values is shown in Table 5-G. for a number of chloro-derivatives of 4-hydroxy and 4,4′-dihydroxy biphenyl. One chlorine atom ortho to the hydroxy group usually results in a considerable increase of $R_f$ values. The second *o*-chlorine usually further increases the $R_f$ value although not so significantly.

For any particular 4-hydroxy (or 4,4′-dihydroxy) biphenyl possessing ortho chlorine substitution the $R_f$ can be decreased by the addition of ammonia to the solvent system (solvent D vs. C and solvent K vs. L). This effect is noticeable with one *o*-chlorine, it becomes significant with two *o*-chlorine, one each ortho to a different hydroxyl group (e.g., 3,3′-dichloro-4,4′-biphenyloliol $R_f$ (solvent), 0.68 (C), 0.46 (D) and 0.89 (J), 0.45 (K), and it is dramatic when two chlorine atoms ortho to the same hydroxy group are involved.

The $R_f$ values for 3,3′,5,5′-tetrachloro-4,4′-biphenyloliol decreases from 0.90 to 0.0 when ammonia is added (solvent K) to solvent system J.

TABLE 3-G.

**$R_f$ Values of the Hydroxybiphenyls and Their 2′,5′-Dichloro-Derivatives**

| Compound | $R_f$ value (× 100) in solvent* | | | | | | | | | | |
|---|---|---|---|---|---|---|---|---|---|---|---|
| | A | B | C | D | E | F | G | H | I | J | L |
| 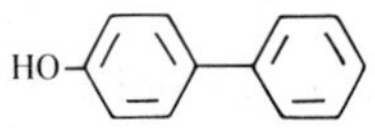 | 59 | 66 | 92 | 80 | 29 | 50 | 38 | 32 | 38 | 86 | 67 |
| 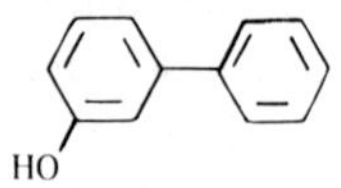 | 63 | 70 | 95 | 92 | 38 | 55 | 58 | 35 | 58 | 91 | 75 |
| 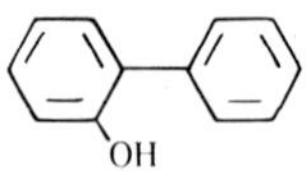 | 78 | 77 | 97 | 95 | 63 | 66 | 74 | 64 | 86 | 91 | 86 |
| 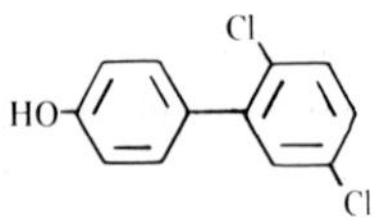 | 76 | 75 | 88 | 90 | 44 | 64 | 54 | 45 | 54 | 92 | 87 |
| 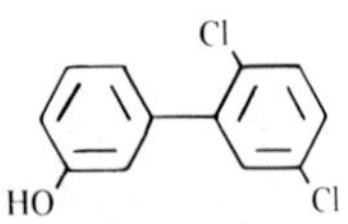 | 84 | 75 | 93 | 93 | 47 | 62 | 61 | 52 | 53 | 92 | 79 |
| 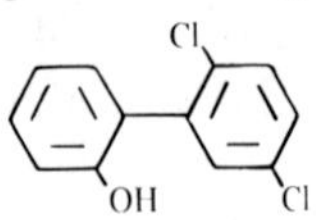 | 80 | 78 | 89 | 95 | 61 | 86 | 64 | 70 | 83 | 88 | 90 |

*Solvent systems used:

| | |
|---|---|
| A benzene-methanol-formic acid | 90:9:1 |
| B benzene-methanol | 90:9 |
| C dichloromethane-methanol | 95:5 |
| D dichloromethane-methanol-conc. ammonia | 95:5:1 |
| E carbon tetrachloride-acetone | 20:1 |
| F carbon tetrachloride-acetone-acetic acid | 40:2:1 |
| G benzene-ethyl acetate | 25:1 |
| H benzene-chloroform | 1:1 |
| I chloroform | |
| J benzene-2-propanol | 70:30 |
| K benzene-2-propanol-conc. ammonia | 70:30:4 |
| L chloroform-hexane-methanol | 45:30:5 |

TABLE 4-G.

**$R_f$ Values of Some Mono-, Di-, and Trichloro-4-Biphenylols**

| Compound | $R_f$ (× 100) in solvent* | | | | | | | | | | |
|---|---|---|---|---|---|---|---|---|---|---|---|
| | A | B | C | D | E | F | G | H | I | J | L |
| 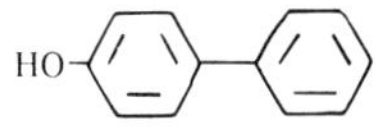 | 59 | 66 | 92 | 80 | 29 | 50 | 38 | 32 | 38 | 86 | 67 |
| 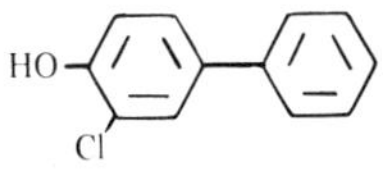 | 87 | 79 | 95 | 85 | 55 | 76 | 75 | 69 | 75 | 89 | 85 |
| 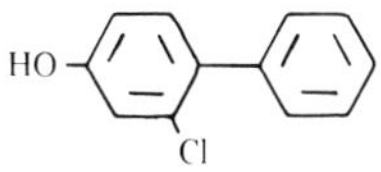 | 79 | 75 | 95 | 84 | 48 | 70 | 70 | 63 | 81 | 93 | 82 |
| 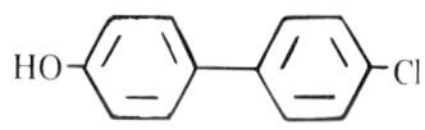 | 72 | 65 | 93 | 87 | 31 | 54 | 50 | 35 | 47 | 93 | 74 |
| 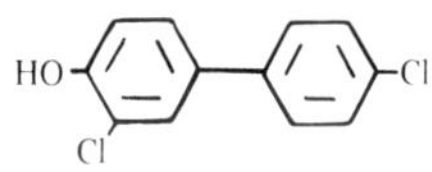 | 80 | 75 | 95 | 80 | 44 | 62 | 65 | 65 | 76 | 93 | 87 |
| 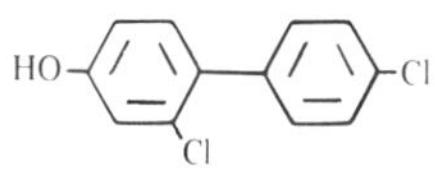 | 75 | 75 | 93 | 84 | 41 | 62 | 60 | 39 | 47 | 93 | 86 |
| 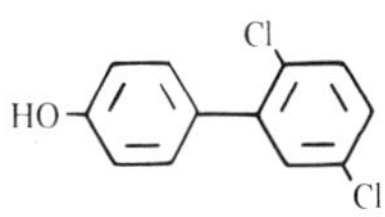 | 76 | 75 | 88 | 90 | 44 | 64 | 54 | 45 | 54 | 92 | 87 |
| 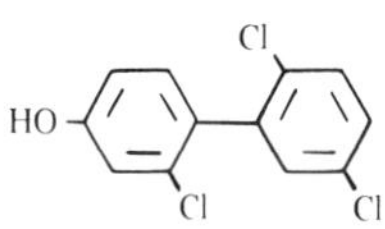 | 83 | 81 | 91 | 84 | 51 | 84 | 67 | 72 | 81 | 91 | 90 |
| 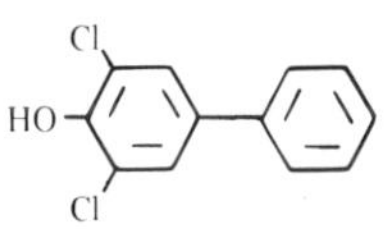 | 88 | 80 | 98 | 28 | 59 | 85 | 79 | 73 | 83 | 93 | 88 |
| 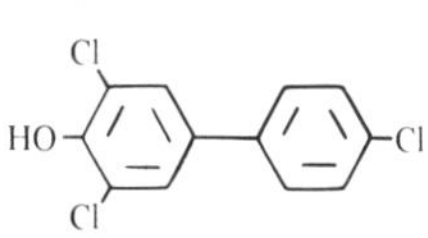 | 89 | 82 | 90 | 15 | 51 | 88 | 69 | 74 | 85 | 90 | 90 |

*The solvent systems used are given in Table 3-G.

TABLE 5-G.

**$R_f$ Values for Ortho-Chloro Substituted 4- and 4,4′-Di-Hydroxybiphenyl**

| Compound | $R_f$ value (× 100) in solvent* | | | | | | | | | | | |
|---|---|---|---|---|---|---|---|---|---|---|---|---|
| | A | B | C | D | E | F | G | H | I | J | K | L |
| 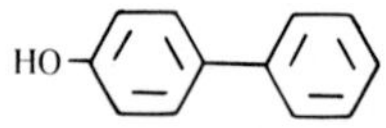 | 54 | 66 | 92 | 80 | 29 | 50 | 38 | 32 | 38 | 86 | 90 | 67 |
| 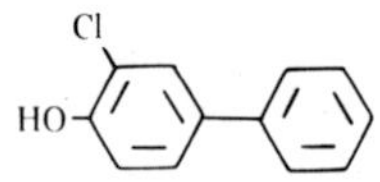 | 87 | 79 | 95 | 85 | 55 | 76 | 75 | 69 | 75 | 89 | 83 | 85 |
| 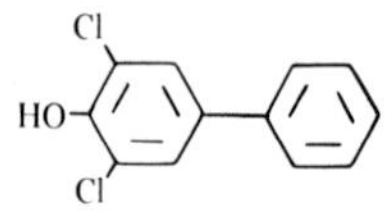 | 88 | 80 | 98 | 28 | 59 | 85 | 79 | 73 | 83 | 93 | 48 | 88 |
| 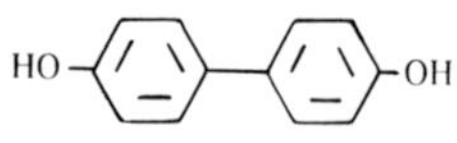 | 29 | 28 | 46 | 50 | 00 | 05 | 05 | 02 | 05 | 87 | 80 | 13 |
| 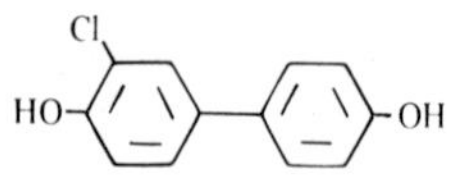 | 57 | 37 | 57 | 47 | 11 | 25 | 12 | 11 | 20 | 89 | 66 | 39 |
| 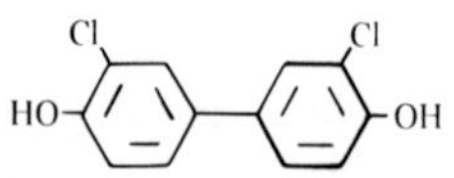 | 42 | 42 | 68 | 46 | 11 | 32 | 20 | 22 | 35 | 89 | 45 | 46 |
| 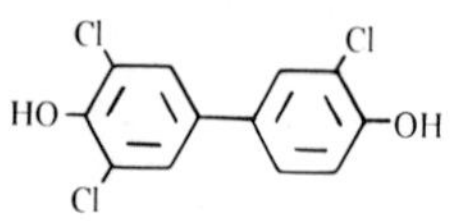 | 58 | 48 | 73 | 07 | 20 | 50 | 33 | 28 | 47 | 87 | 13 | 59 |
| 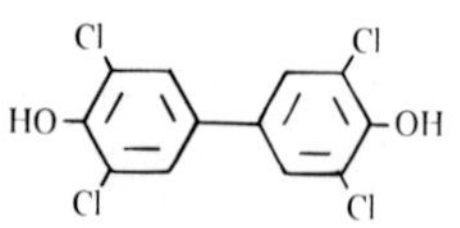 | 60 | 47 | 75 | 00 | 18 | 54 | 64 | 25 | 45 | 90 | 00 | 45 |

*The solvent systems used are given in the Table 3-G.

## REFERENCES


1. **Armour, J. A.,** *J. Assoc. Off. Anal. Chem.,* 56, 987, 1973.
2. **Beezhold, F. and Stout, V. F.,** *Bull. Environ. Contam. Toxicol.,* 10, 10, 1973.
3. **Bush, B. and Lo, F.-C.,** *J. Chromatogr.,* 77, 377, 1973.
4. **Farwell, S. O., Beland, F. A., and Geer, R. D.,** *Bull. Environ. Contam. Toxicol.,* 10, 157, 1973.
5. **Harvey, G. R., Steinhauer, W. G., and Teal, J. M.,** *Science,* 180, 643, 1973.
6. **Hutzinger, O., Heacock, R. A., and Safe, S.,** *J. Chromatogr.,* submitted.
7. **Jensen, S.,** personal communication, 1973.
8. **Ugawa, M., Nakamura, A., and Kashimoto, T.,** *New Methods in Environmental Chemistry and Toxicology,* Proceedings of the International Symposium, Susono, Japan, 1973; Coulston, F., Korte, F., and Goto, M., Eds., International Academic Printing Co. Tokyo, Japan, 1973, 253.
9. **Villeneuve, D. C., Reynolds, L. M., Thomas, G. H., and Phillips, W. E. J.,** *J. Assoc. Off. Anal. Chem.,* 56, 999, 1973.
10. **Webb, R. G. and McCall, A. C.,** *J. Chromatogr. Sci.,* 11, 366, 1973.
11. **Young, S. J. V., Finsterwalder, C., and Burke, J. A.,** *J. Assoc. Off. Anal. Chem.,* 56, 957, 1973.
12. **Zelenski, S. G., Tashiro, J., Worthen, L. R., and Olney, C. E.,** *J. Chromatogr.,* 84, 67, 1973.

## H. OCCURRENCE, LEVELS, AND ACCUMULATION IN THE ENVIRONMENT

Harvey et al.[19a] reported the average concentration of PCB in the northern North Atlantic water. The concentration of PCB was found to be 35 ppt in surface waters and 10 ppt at 200 m depth. Unfortunately the authors do not mention the type of PCB found in the samples. Data are also available on concentrations of PCB in the Rhine River and Lake Constance,[11,19] Canadian streams,[23] German[30b] and Japanese waters,[26] and the Firth of Clyde.[37a]

Investigations have been carried out on PCB levels in plankton,[17,38] fish,[8,11,12,20,31a,41] and other marine animals and general wildlife.[1,3,8,12,14,15,18,19b,19c,27a,41] A very carefully undertaken study[12] on the PCB levels of various marine animals in the English Channel and the middle North Sea gave PCB concentrations and PCB/ΣDDT values shown in Table 1-H.

Considerable attention has been given to the PCB content of paper, paperboard, copying paper, food packaging materials, and similar products[1a,6,7,10,21,22,31,34,35,37,39] as well as food and feed products.[1a,1b,4,13,21,26,29,30,31a,34,40] Polychlorinated biphenyls in solid waste and solid waste related materials have been investigated.[4a]

Levels of PCB in human fatty tissue and breast milk (on a lipid basis) continue to be reported mainly in the low ppm range.[5,8,9,24,25,27,28,36]

The uptake accumulation and distribution of PCB's by algae,[2] tetrahymena pyriformis,[4b] lake trout,[26a] carp,[38a] and ephemera danica[32] has been studied and Sodergren investigated the behavior of Clophen A50 in a model ecosystem consisting of an alga, a cyprinid, and two species of fish.[33] A partial elimination of peaks with lower chlorine content was noticed at higher trophic levels.

Very recently the structures of chlorobiphenyls present in human and animal tissues[30a] and human adipose tissue[20a] have been reported.

Finally it should be mentioned that an increased interest exists in the presence of chlorinated terphenyls in the environment, foods and packing materials.[15a,16,35,37]

TABLE 1-H.

**PCB Concentrations and PCB/ΣDDT Values of Various Species from the English Channel and the North Sea**

| Species | Number of animals | (Average) weight | Organ | PCB concentration (ppm) Wet weight | PCB concentration (ppm) Lipid | PCB/ΣDDT | Area and date |
|---|---|---|---|---|---|---|---|
| Chlamys opercularis, | 10 | 7,5* | Adductor | 0,048 | 29 | 7,4 | English Channel |
| Bivalvia | | | Digestive gland | 0,11 | 1,2 | 1,3 | 50°21'N 00°02'E |
| (queen scallop) | | | | | | | 26 Nov. 1971 |
| | 6 | 10,4* | Adductor | 0,027 | 17 | 8,1 | |
| | | | Digestive gland | 0,20 | 2,2 | 2,6 | |
| Loligo forbesi, | 4 | 260 | Mantle muscle | 0,18 | 15 | 3,1 | |
| Cephalopoda | 4 | 365 | Mantle muscle | 0,17 | 17 | 6,0 | |
| (squid) | | | | | | | |
| Trigla lucerna, | 1 | 202 | Skeleton muscle | 0,040 | 9,3 | 5,0 | English Channel |
| Pisces | | | Liver | 0,30 | 2,1 | 3,3 | 50°26'N 00°03'E |
| (sea robin) | | | Intestinal fat | 2,3 | 2,9 | 2,2 | 26 Nov. 1971 |
| | 1 | 320 | Skeleton muscle | 0,048 | 11 | 3,9 | |
| | | | Liver | 1,7 | 6,9 | 10 | |
| | | | Intestinal fat | 4,9 | 6,4 | 4,1 | |
| | 1 | 571 | Skeleton muscle | 0,046 | 9,3 | 5,6 | |
| | | | Liver | 0,78 | 2,2 | 3,6 | |
| | | | Intestinal fat | 4,8 | 5,6 | 3,8 | |
| Gadus morhua, | 4 | 917 | Skeleton muscle | 0,044 | 19 | 3,9 | Middle North Sea |
| Pisces | | | Liver | 6,4 | 19 | 4,9 | 55°30'N 04°50'E and |
| (cod) | | | | | | | 55°30'N 06°03'E |
| | 4 | 2630 | Skeleton muscle | 0,044 | 23 | 4,4 | 23 June 1972 |
| | | | Liver | 7,8 | 21 | 4,3 | |

*Soft tissue

# REFERENCES


1. **Addison, R. F., Kerr, S. R., Dale, J., and Sergeant, D. E.,** *J. Fish. Res. Board Can.,* 30, 595, 1973.
1a. **Anon.,** *Fed. Reg.,* 38, 18096, 1973.
1b. **Baluja, M. G.,** *Medicamenta (Madr.),* 61, 125, 1973.
2. **Bauer, U.,** *Schr. Ver. Wasser-, Boden-, Lufthyg., Berlin-Dahl,* 37, 211, 1972.
3. **Bowes, G. W. and Jonkel, C.,** in *PCB in the Environment,* Marcel Dekker Inc., New York, in press.
4. **Bradley, R. L., Jr.,** *J. Milk Food Technol.,* 36, 155, 1973.
4a. **Carnes, R. A., Doerger, J. U., Sparks, H. L.,** *Arch. Environ. Contam. Toxicol.,* 1, 27, 1973.
4b. **Cooley, N. R., Keltner, J. M., Jr., and Forester, J. J.,** *Protozool.,* 20, 443, 1973.
5. **Curley, A., Burse, V. W., and Jennings, R. W.,** *Nature,* 242, 338, 1973.
6. **Derra, R.,** *Wochenbl. Papierfabr.,* 101, 315, 1973.
7. **Derra, R.,** *Papier* (Darmstadt), 27, 237, 1973.
8. **Doguchi, M.,** *New Methods in Environmental Chemistry and Toxicology,* Proceedings of the International Symposium, Susono, Japan, 1973; Coulston, F., Korte, F., and Goto, M., Eds., International Academic Printing Co., Tokyo, Japan, 1973, 269.
9. **Doguchi, M., Fukano, S., and Ushio, F.,** *Bull. Environ. Contam. Toxicol.,* in press.
10. **Easty, D. B.,** *Tappi,* 56, 131, 1973.
11. **Eichner, M.,** *Z. Lebensm.-Unters. Forsch.,* 151, 376, 1973.
12. **Ernst, W., Goerke, H., Eder, G., and Schaefer, R. G.,** Submitted to *Bull. Environ. Contam. Toxicol.,* and private communication.
13. **Flanagan, V. P. and Ferretti, A.,** *J. Lipid Res.,* 14, 306, 1973.
14. **Fog, M. and Kraul, I.,** *Acta Vet. Scand.,* 14, 350, 1973.
15. **Frank, R., Ronald, K., and Braun, H. E.,** *J. Fish. Res. Board Can.,* 30, 1053, 1973.
15a. **Freudenthal, J. and Greve, P. A.,** *Bull. Environ. Contam. Toxicol.,* 10, 108, 1973.
16. **Fries, G. F. and Marrow, G. S.,** *J. Assoc. Off. Anal. Chem.,* 56, 1002, 1973.
17. **Giam, C. S., Wong, M. K., Hanks, A. R., and Sackett, W. M.,** *Bull. Environ. Contam. Toxicol.,* 9, 376, 1973.
18. **Greichus, Y. A., Greichus, A., and Emerick, R. J.,** *Bull. Environ. Contam. Toxicol.,* 9, 321, 1973.
19. **Greve, P. A.,** *Sci. Total Environ.,* 1, 173, 1972.
19a. **Harvey, G. R., Steinhauer, W. G., and Teal, J. M.,** *Science,* 180, 643, 1973.
19b. **Heppleston, P. B.,** *Mar. Pollut. Bull.,* 4, 44, 1973.
19c. **Hidaka, K., Ohe, T., and Fujiwara, K.,** *Shokuhin Eiseigaku Zasshi,* 13, 523, 1972; C. A., 79, 028100, 1973.
20. **Huschenbeth, E.,** *Schr. Ver. Wasser-, Boden-, Lufthyg., Berlin-Dahl,* 37, 103, 1972.
20a. **Jensen, S. and Sundström, G.,** *Ambio,* in press.
21. **Leoni, V., D'Alessandro de Luca, E., and Simeone, A. M.,** *Rass. Chim.,* 25, 99, 1973.
22. **Liukkonen, H., Rajama, J., and Blomberg, K.,** *Lab. Pract.,* 22, 350, 1973.
23. **Miles, J. R. W. and Harris, C. R.,** *Pestic. Monit. J.,* 6, 363, 1973.
24. **Minagawa, K., Sakai, Y., and Takizawa, U.,** *Jap. J. Hyg.,* in press.
25. **Musial, C. J., Hutzinger, O., Zitko, V., and Crocker, J.,** *Bull. Environ. Contam. Toxicol.,* submitted.
26. **Nose, K.,** *Noyaku Seisan Gijutsu,* 30, 29, 1973.
26a. **Parejko, R. and Johnston, R.,** *U.S. Natl. Tech. Inform. Serv.* PB report No. 214768/4, 1973.
27. **Pesendorfer, H., Eichler, I., and Glofke, E.,** *Wien. Klin. Wochenschr.,* 85, 218, 1973.
27a. **Saschenbrecker, P. W.,** *Can. J. Comp. Med.,* 37, 203, 1973.
28. **Savage, E. P., Tessari, J. D., Malberg, J. W., and Wheeler, H. W.,** *Bull. Environ. Contam. Toxicol.,* 9, 222, 1973.
29. **Savage, E. P., Tessari, J. D., and Malberg, J. W.,** *Bull. Environ. Contam. Toxicol.,* 10, 97, 1973.
30. **Sawyer, L. D.,** *J. Assoc. Off. Anal. Chem.,* 56, 1015, 1973.
30a. **Schulte, E. and Acker, L.,** *Naturwissenschaften,* 61, 79, 1974.
30b. **Selenka, F.,** *Schr. Ver. Wasser-, Boden-, Lufthyg., Berlin-Dahl,* 37, 113, 1972.
31. **Shahied, S. I., Stanovick, R. P., and McInturff, D. E.,** *Bull. Environ. Contam. Toxicol.,* 10, 80, 1973.
31a. **Smith, W. E., Funk, K., and Zabik, M. E.,** *J. Fish. Res. Board Can.,* 30, 702, 1973.
32. **Sodergren, A. and Svensson, B. J.,** *Bull. Environ. Contam. Toxicol.,* 9, 345, 1973.
33. **Sodergren, A.,** *Oikos,* 24, 30, 1973.
34. **Stanovick, R. P., Shahied, S. I., and Missaghi, E.,** *Bull. Environ. Contam. Toxicol.,* 10, 101, 1973.
35. **Thomas, G. H. and Reynolds, L. M.,** *Bull. Environ. Contam. Toxicol.,* 10, 37, 1973.
36. **Ueta, M.,** *Igakuno Ayumi,* in press.
37. **Villeneuve, D. C., Reynolds, L. M., and Thomas, G. H.,** *J. Assoc. Off. Anal. Chem.,* 56, 999, 1973.
37a. **Waddington, J. I., Best, G. A., Dawson, J. P., and Lithgow, T.,** *Mar. Pollut. Bull.,* 4, 26, 1973.
38. **Williams, R. and Holden, A. V.,** *Mar. Pollut. Bull.,* 4, 109, 1973.
38a. **Yoshida, T., Takashima, F., and Watanabe, T.,** *Ambio,* 2, 111, 1973.
39. **Young, S. J. V., Finsterwalder, C., and Burke, J. A.,** *J. Assoc. Off. Anal. Chem.,* 56, 957, 1973.
40. **Zimmerli, B., Merek, B., and Sulser, H.,** *Mitt. Geb. Lebensmittelunters. Hyg.,* 64, 70, 1973.
41. **Zitko, V. and Hutzinger, O.,** *Schr. Ver. Wasser-, Boden-, Lufthyg., Berlin-Dahl,* 37, 121, 1972.

## I. BIOLOGICAL EFFECTS

A review article on recent advances in the toxicology of pesticides includes a section on polychlorinated biphenyls;[10] an article in Japanese covers selected aspects of biological effects of PCB.[11]

Chlorinated biphenyls were shown to induce microsomal enzyme activity in the house fly[14] and cytochrome P448 in the liver.[2] Hepatic function of rats, to which pure chlorobiphenyls had been fed, was assessed by phenobarbital sleeping times and a number of in vitro assays of enzyme activities.[9] Toxicity of chlorobiphenyls increases with simultaneous administration of alkylbenzene sulfonic acid as assayed by liver enzymes, liver weight, and cholesterol levels.[8]

Evidence was presented for the transplacental passage of chlorobiphenyls in the rat.[6] The effect of 2,2′-dichlorobiphenyl on embryonic development[15] and the embryotoxic effect of three chlorobiphenyls in the chicken were studied.[4] Infant rhesus monkeys are less susceptible to PCB poisoning than the adult animal.[1]

Further studies with chlorinated biphenyls include their effect on serum proteins,[16] on the estrous cycle in the mouse,[12] on Japanese quail[3] tetrahymena pyriformis[5] and on their possible role in liver tumorigenesis in mice[7] and mortality of gannets.[13]

The acute toxicity of 3′,4,4′,6-tetrachloro-3-biphenylol (a metabolite of 2,3′,4,4′-tetrachlorobiphenyl in rats) was shown to be higher than that of the parent 2,3′,4,4′-tetrachlorobiphenyl by intraperitoneal injection. The $LD_{50}$ of the phenolic metabolite was 0.43 g/kg whereas the tetrachlorobiphenyl had an $LD_{50}$ of 2.15 g/kg.[17]

## REFERENCES


1. **Abrahamson, L. J. and Allen, J. R.,** *Environ. Health Perspect.,* 1, 81, (1973).
2. **Alvares, A. P., Bickers, D. R., and Kappas, A.,** *Proc. Natl. Acad. Sci. U.S.A.* 70, 1321, 1973.
3. **Call, D. J.,** Ph.D. thesis, University of South Dakota, Vermillion, S. Dakota, 1972. *C.A.,* 79, 49627, 1973.
4. **Carlson, R. W. and Duby, R. T.,** *Bull. Environ. Contam. Toxicol.,* 9, 261, 1973.
5. **Cooley, N. R., Keltner, J. M., Jr., and Forester, J.,** *J. Protozool.,* 20, 443, 1973.
6. **Curley, A., Buse, V. W., and Grim, M. E.,** *Food Cosmet. Toxicol.,* 11, 471, 1973.
7. **Ito, N., Nagasaki, H., and Arai, M.,** *New Methods in Environmental Chemistry and Toxicology,* Proceedings of the International Symposium, Susono, Japan, 1973; Coulston, F., Korte, F., and Goto, M., Eds., International Academic Printing Co., Tokyo, Japan, 1973, 141.
8. **Itokawa, Y., Kamchara, K., and Fujiwara, V.,** Experientia, 29, 822, 1973.
9. **Johnstone, G. J., Ecobichon, D. J., and Hutzinger, O.,** *Toxicol. Appl. Pharmacol.,* in press.
10. **Kay, K.,** *Environ. Res.,* 6, 202, 1973.
11. **Ohi, G.,** *Shokuhin Eiseigaku Zasshi,* 13, 499, 1972; *C. A.,* 78, 155189, 1973.
12. **Orberg, J. and Kihlstrom, J. E.,** *Environ. Res.,* 6, 176, 1973.
13. **Parslow, J. L. F.,** *Mar. Pollut. Bull.,* 4, 41, 1973.
14. **Rhee, K. S. and Plapp, F. W., Jr.,** *Arch. Environ. Contam. Toxicol.,* 1, 182, 1973.
15. **Török, P.,** *Chemosphere,* 2, 173, 1973.
16. **Wassermann, M., Wassermann, D., Kedar, E., and Djavaherian, M.,** *Bull. Environ. Contam. Toxicol.,* 10, 42, 1973.
17. **Yoshimura, H., Yamamoto, H., and Kinoshita, H.,** *New Methods in Environmental Chemistry and Toxicology,* Proceedings of the International Symposium, Susono, Japan, 1973, Coulston, F., Korte, F., and Goto, M., Eds., International Academic Printing Co., Tokyo, Japan, 1973, 291.

# INDEX

## B



## C

## D

## E



## F

# I

## K



## L



## M



## N



## O

## P

## T

## U



## V



## W